SpringerWienNewYork

Yearbook on Space Policy

Edited by the
European Space Policy Institute
Director: Kai-Uwe Schrogl

European Space Policy Institute
Kai-Uwe Schrogl, Wolfgang Rathgeber,
Blandina Baranes, Christophe Venet (eds.)

Yearbook on Space Policy 2008/2009

Setting New Trends

SpringerWienNewYork

European Space Policy Institute, Vienna, Austria
Kai-Uwe Schrogl
Wolfgang Rathgeber
Blandina Baranes
Christophe Venet

SpringerWienNewYork is a part of
Springer Science + Business Media
springer.at

Cover illustration: "Satellite of the SAR-Lupe satellite system, completed by the launch of the fifth satellite on 22 July 2008 (Grafik: OHB-System AG)"

Typesetting: Thomson Press (India) Ltd., Chennai
Printing: Holzhausen Druck GmbH, 1140 Wien

Printed on acid-free and chlorine-free bleached paper

SPIN: 80016141

With 31 Figures and 7 Tables

CIP data applied for

ISSN 1866-8305
ISBN 978-3-7091-0317-3 SpringerWienNewYork

Preface

The current global financial and economic crisis does not leave the space sector unaffected. The concrete effects, however, are different from what might have been expected. First of all, no sudden decline in the space market or in programme development took place. This is mainly due to the long-term planning and the long duration of project implementation. Secondly, and this is even more striking, space emerged in the high-level political debates to counter the crises as one promising area to focus on. The European Council of December 2008 stressed this and promoted that space as a "lead market" should be an important element in the European Economic Recovery Plan and the European Plan for Innovation. The 5th European Space Council already in September 2008 had emphasised the contribution of space to implement the Lisbon Strategy and the ESA Council at Ministerial Level in November 2008 had initiated substantive new programmes. Through this, space emerged from the crisis with an even stronger political standing and reputation highlighting its economic potential as never before. The coming months, way into 2011, will show whether this promise can be kept and whether governments actually maintain their high engagement in space programmes and as contractors for satellite systems facing a huge wave of replacements during the next years.

This incidence-driven trend of highlighting the economic potential of space is currently complemented by an already longer-lasting but now thriving tendency. It is the use of space assets for Europe's security, where the quest for European approaches, concepts and policies gained considerable momentum during the past year: from the growing use of satellite data for European military missions to the establishment of a "structured dialogue" between the European Commission, the European Space Agency (ESA) and the European Defence Agency (EDA) up to the identification of new areas in need for European concepts like space for internal security. In order to stress and signify this accelerating movement the cover of this edition pinpoints at the potential of space assets for Europe's security.

These are the two most significant new trends in space policy which this new edition of the "Yearbook on Space Policy" analyses in a thorough way. Many more smaller trends and developments have shown up in the reporting period between July 2008 and June 2009. They are all reflected in this Yearbook, which is consequently subtitled "New Trends in Space Policy". Based on its European focus, the Yearbook also identifies the most notable developments around the world comprising the leading as well as the emerging space powers. Now, with this third edition, the "Yearbook on Space Policy" can present to the reader a multiyear development of space activities, the governmental and commercial sectors as well as policies, programmes and technologies. This continuous inspection is laid out in the first and the third part of the Yearbook, with the first providing the description and analysis and the latter data and chronology. These two parts are prepared in-house by ESPI.

The second part of the Yearbook again brings in the views of ten distinguished analysts in the field of space policy touching topics or events which stirred the space sector between mid-2008 and mid-2009. Of course, the two mega-trends outlined above are covered by specific contributions, but new types of programmes (like the integrated applications) or regulatory issues of particular timeliness and relevance (like national space legislation) are discussed as well. This Yearbook also contains in its second part more contributions from abroad Europe than ever before. Scholars from the U.S., Canada and India have accepted to share their analyses of particularly important issues like the expectations in the new U.S. administration's space policy making. This multitude of issues and views has by now also become a trademark of the Yearbook, enhancing an intellectual dialogue between Europe and abroad. This dialogue was not a virtual one, since an authors' conference in the framework of ESPI's Autumn Conference of its European Space Policy Research and Academic Network (ESPRAN) in September 2009 provided for an extensive in-depth exchange of ideas and views on the draft manuscripts. This served as an efficient instrument for attuning the contributions to this edition.

The ten contributors of the articles for part 2 are also the persons ESPI would like to thank in a particular way for their engagement in this project. In addition to that, the editorial team is indebted to the members of the Advisory Council of the Institute, chaired by Herbert Allgeier, which in this composition acts as the Editorial Advisory Board to the Yearbook series. Thanks are also due to Wendelin Pohl for his zestful support.

Kai-Uwe Schrogl, Wolfgang Rathgeber, Blandina Baranes, Christophe Venet
ESPI editorial team

Table of contents

PART 1
The Year in Space 2008/2009

Chapter 1. European space activities in the global context. *Wolfgang Rathgeber and Christophe Venet*

Chapter 2. Developments in space policies, programmes and technologies throughout the world and in Europe.

Wolfgang Rathgeber and Christophe Venet

PART 2
Views and Insights

PART 3
Facts and Figures
Christophe Venet

List of acronyms

3D: 3 Dimensions
3DTV: 3 Dimensions Television

A

AATSR: Advanced Along Track Scanning Radiometer
ACI: Airports Council International
ADF: Australian Defence Force
ADM: Atmospheric Dynamics Mission
AEHS: Advanced Extremely High Frequency Satellite
AFRL: Air Force Research Laboratory
AIS: Automatic Identification System
ALC: African Leadership Conference on Space Science and Technology for Sustainable Development
ALOS: Advanced Land Observing Satellite
APRSAF: Asia-Pacific Regional Space Agency Forum
APSCO: Asia-Pacific Space Cooperation Organisation
ARATS: Association for Relations Across the Taiwan Straits
ARMC: African Resource Management and Environmental Constellation
ARTA: Ariane 5 Research and Technology Accompaniment Programme
ARTES: Advanced Research in Telecommunications Systems
AR5: 5th Assessment Report
ARV: Advanced Re-entry Vehicle
ASAL: Agence Spatiale Algérienne (Algerian Space Agency)
ASAT: Anti-Satellite
ASI: Agenzia Spaziale Italiana (Italian Space Agency)
ATM: Air Traffic Management
ATV: Automated Transfer Vehicle
AU: African Union
AVIC: Aviation Industries of China

B

BAE: British Aerospace
BGAN: Broadband Global Area Network
BLS: Boeing Launch Services

BNSC: British National Space Centre
BRIC: Brazil Russia India China

C

C4ISR: Computerised Command, Control, Communications, Intelligence, Surveillance, Reconnaissance
CALT: China Academy of Launch Vehicle Technology
CASA: Construcciones Aeronáuticas Sociedad Anónima
CASC: China Aerospace Corporation
CASTC: China Aerospace Science and Technology Corporation
CBERS: China Brazil Earth Resources Satellites
CD: Conference on Disarmament
CDR: Critical Design Review
CEA: Commissariat à l'Energie Atomique (French Atomic Energy Commissariat)
CEO: Chief Executive Officer
CEOS: Committee on Earth Observation Satellites
CFE: Commercial and Foreign Entities
CFSP: Common Foreign and Security Policy
CGEA: Community General Export Authorization
CGWIC: China Great Wall Industry Corporation
CHAMP: Challenging Mini-Satellite Payload
CIP: Competitiveness and Innovation Framework Programme
CMA: China Meteorological Administration
CMSEO: China Manned Space Engineering Office
CNES: Centre National d'Etudes Spatiales (French Space Agency)
CNNC: China National Nuclear Corporation
CNSA: China National Space Administration
CoC: Code of Conduct
COF: Columbus Orbital Facility
COFUR: Cost Of Fulfilling User Requests
COPUOS: Committee on the Peaceful Uses of Outer Space
CoReH$_2$O: Cold Regions Hydrology High-resolution Observatory
CORONAS: Complex ORbital Observations Near-Earth of Activity of the Sun
COSMO-Skymed: Constellation of small Satellites for the Mediterranean basin Observation
COSTIND: Commission for Science, Technology and Industry
COTS: Commercial Orbital Transportation Services
CSA: Canadian Space Agency

CSIS: Center for Strategic and International Studies
CSSC: China State Shipbuilding Corporation

D

DARPA: Defense Advanced Research Projects Agency
DBS: Direct Broadcast Services
DGA: Direction Générale de l'Armement (French Military Procurement Agency)
DHS: Department for Homeland Security
DLR: Deutsches Zentrum für Luft- und Raumfahrt (German Space Agency)
DMO: Defence Material Organisation
DMSP: Defense Meteorological Satellite Program
DOC: Department of Commerce
DoD: Department of Defense
DOR: Differential One-way Range
DRC: Democratic Republic of Congo
DSP: Defense Support Program
DSTO: Defence Science and Technology Organisation
DTH: Direct-to-Home

E

EADS: European Aeronautic Defence and Space Company
EarthCARE: Earth Clouds, Aerosol and Radiation Explorer
EC: European Commission
ECA: Evolution Cryotechnique Type A
ECB: European Central Bank
e-CORCE: e-Constellation of Observation by Recurrent Cellular Environment
EDA: European Defence Agency
EDEM: European Defence Equipment Market
EDRS: European Data Relay Satellite
EEA: European Environment Agency
EELV: Evolved Expandable Launch Vehicle
EERP: European Economic Recovery Plan
EGB: EUROBOT Ground Prototype
EGNOS: European Geostationary Navigation Overlay Service
EISC: European Interparliamentary Space Conference
EJSM: Europa Jupiter System Mission
ELINT: Electronic signals Intelligence

eLORAN: enhanced LOng RAnge Navigation
ELV: Expandable Launch Vehicle
EMS: Electromagnetic Sciences
EnMAP: Environmental Mapping and Analysis Programme
EO: Earth Observation
EPS: EUMETSAT Polar System
ERA: European Research Area
ERC: European Research Council
ERS: European Remote Sensing Satellite
ESA: European Space Agency
ESDP: European Security and Defence Policy
ESP: European Space Policy
ESPI: European Space Policy Institute
ESTRACK: ESA Tracking Network
EU: European Union
EUFOR: European Union Force
EULEX: European Union Rule of Law Mission in Kosovo
EUMETSAT: European Organisation for the Exploitation of Meteorological Satellites
EU NAVFOR: European Union Naval Force Somalia
EUSC: European Union Satellite Centre
EVA: Extravehicular Activity

F

FAA: Federal Aviation Administration
FAO: Food and Agricultural Organisation
FBI: Federal Bureau of Investigations
FCC: Federal Communications Commission
FLPP: Future Launcher Preparatory Programme
FOC: Full Operational Capability
FP7: Framework Programme for research and technological development 7
FSS: Fixed Satellite Services
FY: Fiscal Year
FY: Feng Yung

G

GAC: GMES Advisory Council
GAD: General Armaments Department
GAGAN: GPS-Aided Geosynchronous Augmented Navigation System
GAO: Government Accountability Office

GBAORD: Government Budget Appropriations or Outlay on R&D
GCM: GMES Contributing Missions
GDP: Gross Domestic Product
GEMS: Gravity and Extreme Magnetism Small Explorer
GEO: Geostationary Orbit
GEO: Group on Earth Observations
GEOSS: Global Earth Observation System of Systems
GERD: Gross Domestic Expenditure on R&D
GES: Global Exploration Strategy
GES: Growth Environment Score
GIANUS: Global Integrated Architecture for iNnovative Utilisation of space for Security
GIO: GMES Initial Operations
GIOVE: Galileo In-Orbit Validation Element
GIP: Galileo Interinstitutional Panel
GIS: Geographic Information System
GJU: Galileo Joint Undertaking
GLONASS: Global Navigation Satellite System
GMES: Global Monitoring for Environment and Security
G-MOSAIC: GMES services for Management of Operations, Situation Awareness and Intelligence for regional Crises
GMSK: Gaussian Minimum Shift Keying
GOCE: Gravity field and steady-state Ocean Circulation Explorer
GOES: Geostationary Operational Environmental Satellite
GOSAT: Greenhouse Gases Observing Satellite
GPS: Global Positioning System
GSA: GNSS Supervisory Authority
GSC: GMES Space Component
GSC: Guyana Space Centre
GSLV: Geosynchronous Satellite Launch Vehicle
GTO: Geostationary Transfer Orbit
G8: Group of Eight
G20: Group of Twenty

H

HDTV: High Definition Television
HR: High Resolution
HSPG: High-Level Space Policy Group
HTV: H-2 Transfer Vehicle

I

IAASS: International Association for the Advancement of Space Safety
IADC: Inter-Agency Space Debris Coordination Committee
IAEA: International Atomic Energy Agency
IBEX: Interstellar Boundary Explorer
ICAO: International Civil Aviation Organization
ICBM: Intercontinental Ballistic Missile
ICG: International Committee on Global Navigation Satellite Systems
ICT: Information and Communication Technologies
IEA: International Energy Agency
IFAD: International Fund for Agricultural Development
IGN: Institut Géographique National (French National Geographic Institute)
IGS: Integrated Geo Systems
IGT: Innovation Growth Team for Space
IGY: International Geophysical Year
IHY: International Heliophysical Year
ILS: International Launch Services
IMF: International Monetary Fund
IMINT: Imagery Intelligence
IMO: International Maritime Organisation
INMARSAT: International Maritime Satellite Organisation
INSPIRE: Infrastructure for Spatial Information in Europe
IOV: In-Orbit Validation
IP: Internet Protocol
IPCC: Intergovernmental Panel on Climate Change
IRIS: Interface Region Imaging Spectrograph
ISA: Israeli Space Agency
ISAF: International Security Assistance Force
ISC: International Space Company
ISECG: International Space Exploration Coordination Group
ISRO: Indian Space Research Organisation
ISS: International Space Station
ITAR: International Traffic in Arms Regulations
ITU: International Telecommunication Union
IXO: International X-Ray Observatory

J

JAPCC: Joint Air Power Competence Center
JAXA: Japan Aerospace Exploration Agency

JEM: Japanese Experiment Module
JMA: Japan Meteorological Agency

K

KEW: Kinetic Energy Weapon
KMA: Korea Meteorological Administration
KSLV: Korea Space Launch Vehicle
KT: Kaitouzhe

L

LaRa: Lander Radio-Science
LCD: Liquid Crystal Display
LCROSS: Lunar CRater Observing and Sensing Satellite
LEO: Low-Earth Orbit
LEOP: Launch and Early Orbit Phase
LIMES: Land and See Monitoring for Environment and Security
LM: Long March
LMCLS: Lockheed Martin Commercial Launch Services
LRO: Lunar Reconnaissance Orbiter
LWS: Living With a Star

M

MACC: Monitoring Atmospheric Composition and Climate
MAI: Moscow Aviation Institute
MAVEN: Mars Atmosphere and Volatile Evolution
MDA: Missile Defense Agency
MDG: Millennium Development Goals
MEJI: Mars Exploration Joint Initiative
MELiSSA: Micro-Ecological Life Support System Alternative
MEO: Medium Earth Orbit
MERIS: Medium Resolution Imaging Spectrometer
MHI: Mitsubishi Heavy Industries
MoD: Ministry of Defence
MONUC: Mission de l'Organisation des Nations Unies en République démocratique du Congo (UN Mission in the Democratic Republic of the Congo)
MoonLITE: Moon Lightweight Interior and Telecom Experiment
MoU: Memorandum of Understanding
MPA: Maritime Patrol Aircraft
MPLM: Multipurpose Laboratory Module
MR: Medium Resolution

MRM: Mini Research Module
MRO: Mars Reconnaissance Orbiter
MSAS: MTSAT Satellite-Based Augmentation System
MSC: Meteorological Service of Canada
MSG: Meteosat Second Generation
MSI: Multi-Spectral Imager
MSL: Mars Science Laboratory
MSS: Mobile Satellite Services
MSV: Mobile Satellite Ventures
MTCR: Missile Technology Control Regime
MTG: Meteosat Third Generation
MTI: Moving Target Indicator
MUOS: Mobile User Objective System
MUSIS: Multinational Satellite-based Imagery System

N

NAO: National Applications Office
NASA: National Aeronautics and Space Administration
NASDA: National Development Space Agency of Japan
NATO: North Atlantic Treaty Organisation
NBO: Network-Based Operations
NDPG: National Defence Program Guidelines
NEO: Near-Earth Objects
NERC: Natural Environment Research Council
NEREUS: Network of European Regions Using Space Technologies
NFIRE: Near Field Infrared Experiment
NGDI: National Geospatial Data Infrastructure
NGO: Non-governmental Organisation
NGST: New Generation Space Telescope
NOAA: National Oceanic and Atmospheric Administration
NORAD: North American Aerospace Defense Command
NPOESS: National Polar-orbiting Operational Environmental Satellite System
NRO: National Reconnaissance Office
NSSA: National Security Space Authority

O

OECD: Organisation for Economic Co-operation and Development
OHB: Orbitale Hochtechnologie Bremen
OLCI: Ocean Land Colour Instrument

OPEC: Organisation of Petroleum Exporting Countries
ORFEO: Optical and Radar Federated Earth Observation
ORS: Operationally Responsive Space
OSTM: Ocean Surface Topography Mission

P

PBEO: Programme Board for Earth Observation
PCT: Patent Cooperation Treaty
PDR: Preliminary Design Review
PFI: Public Financing Initiative
PLA: People's Liberation Army
PNT: Positioning, Navigation and Timing
POES: Polar Operational Environment Satellites
PPP: Public Private Partnership
PREMIER: Process Exploration through Measurement of Infrared Emitted Radiation
PRS: Public-Regulated Service
PSA: Programme on Space Applications
PSLV: Polar Satellite Launch Vehicle

Q

QDR: Quadrennial Defense Review

R

R&D: Research & Development
RCA: République Centrafricaine
RISAT: Radar Imaging Satellite
RLV-TD: Reusable Launch Vehicle Technology Demonstrator
RSCC: Russian Satellite Communications Company
RTD: Research and Technology Development

S

SA: Société Anonyme (Public Limited Company)
SAFER: Services and Applications for Emergency Responses
SALMON: Stand Alone Mission of Opportunity
SAR: Synthetic Aperture Radar
SAOCOM: Satelite de Observation y Communicacion (Observation and Communications Satellite)

SBSS: Space-Based Surveillance System
SDA: Satellite Data Association
SDCM: System of Differential Corrections and Monitoring
SDI: Strategic Defense Initiative
SDO: Solar Dynamics Observatory
SDSTB: State Defense Science and Technology Bureau
SEF: Straits Exchange Foundation
SELENE: SELenological and ENgineering Explorer
SES: Single European Sky
SES: Société Européenne des Satellites
SHF: Super High Frequency
SHSP: Strategic Headquarters for Space Policy
SIA: Satellite Industry Association
SIASGE: Sistema Italo Argentina de Satelites para la Gestion de Emergencias (Italian-Argentinian Satellite System for Emergency Management)
SICRAL: Sistema Italiano per Comunicazioni Riservate ed Allarmi (Italian Military Communications System)
SIGINT: Signals Intelligence
SIPRI: Stockholm International Peace Research Institute
SLSTR: Sea Land Surface Temperature Radiometer
SMC: Space and Missile Systems Center
SMDC: Space and Missile Defense Command
SME: Small and Medium Entreprise
SMEX: Small Explorer
SMOS: Soil Moisture and Ocean Salinity
SMP: Systèmes Midi-Pyrénées
SNC: Sierra Nevada Corporation
SOHO: Solar and Heliospheric Observatory
SPOT: Satellite pour l'Observation de la Terre (Earth Observation Satellite)
SPS PS: Standard Positioning Service Performance Specification
SS2: Space Ship 2
SSA: Space Situational Awareness
SSC: Swedish Space Corporation
SSL: Space Systems/Loral
SSN: Space Surveillance Network
SSOT: Sistema Satelital para Observacion de la Tierra (Satellite System for EO)
SSTL: Surrey Satellite Technology Ltd.
S&T: Science and Technology
STAR: Satellite Technology for the Asia-Pacific Region
START: Strategic Arms Reduction Treaty

STERO: Solar TErrestrial RElations Observatory
STFC: Science and Technology Facilities Council
STSS: Space Tracking Surveillance System
SWIR: Shortwave Infrared

T
TCBM: Transparency and Confidence Building Measures
THEO: Thai Earth Observation System
TIP: Tender Information Package
TSAT: Transformation Communications Satellite
TSB: Technology Strategy Board
TSSM: Titan Saturn System Mission
TV: Television

U
UAE: United Arab Emirates
UHF: Ultra High Frequency
UK: United Kingdom
ULA: United Launch Alliances
UN: United Nations
UNCCC: United Nations Climate Change Conference
UNEP: United Nations Environment Programme
UNESCO: United Nations Educational, Scientific and Cultural Organization
UNFCCC: United Nations Framework Convention on Climate Change
UNGA: United Nations General Assembly
UNGIWG: United Nations Geographic Information Working Group
UNIDIR: United Nations Institute for Disarmament Research
UNISPACE: United Nations Conference on the Exploration and Peaceful Uses of Outer Space
UNOOSA: United Nations Office for Outer Space Affairs
UNSC: United Nations Security Council
UNSDI: United Nations Spatial Data Infrastructure
UN-SPIDER: UN Platform for Space-based Information for Disaster Management and Emergency Response
US: United States
USAF: United States Air Force
USGS: United States Geological Survey
USN: Universal Space Network
USSTRATCOM: United States Strategic Command
UV: Ultraviolet

V

VC: Venture Capital
VERTA: Vega Research and Technology Accompaniment
VHR: Very High Resolution
VN: Vereinte Nationen (United Nations)
VNIR: Visible and Near Infrared

W

WEU: Western European Union
WFP: World Food Programme
WGS: Wideband Global Satcom
WHO: World Health Organisation
WIPO: World Intellectual Property Organisation
WK2: White Knight 2
WRS: World Radiocommunication Seminar
WTSA: World Telecommunication Standardisation Assembly

PART 1

THE YEAR IN SPACE 2008/2009

European space activities in the global context

Wolfgang Rathgeber and Christophe Venet

1. Global political and economic trends

The most striking issue in 2008/2009 was the financial and economic crisis with its worldwide consequences, affecting both advanced economies and emerging countries. The period was also marked by the confirmation of several ongoing trends, such as the political and economic rise of China, the resurgence of Russia on the international scene and the limited growth of Western economies. By contrast, the election of Barack Obama as the 44th American President unleashed new expectations about U.S. international policies. Finally, central transnational problems, such as climate change, the energy crisis or terrorism, remained at the top of policy agendas. In the following sections, global challenges and their respective global responses will be considered and analysed for their impact on space activities and space policies. A particular focus will be put on Europe.

1.1. Global economic outlook

In 2008 and in the first half of 2009, the world economy witnessed its most dangerous financial shock since the 1930s. Global growth slowed down substantially in 2008, and by the beginning of 2009, global economy entered in a severe recession. In addition, energy and commodity prices remained very high in the first part of 2008, before declining sharply in the second half of the year. The International Monetary Fund (IMF) kept revising its projections downwards throughout the period. The world output in 2008 was estimated at 3.2%, compared to 5% in 2007. The world output is projected to decline by 1.3% in 2009 and to recover only gradually starting from 2010 (+1.9%).[1] Advanced economies were hit the hardest, witnessing a 7.5% decline in real GDP in the fourth quarter of 2008, but emerging economies are badly suffering as well (−4% real GDP growth in the fourth quarter of 2008).[2] As a whole, the IMF predicts that the stabilisation of financial markets will take longer than previously envisaged. The low confidence in the markets will continue to limit access to credit in 2009 and 2010, in both developed and emerging economies.

Extraordinary measures were taken by governments to stabilise the markets and support demand, both at national and international levels. State intervention was widely used as a regulating tool, on the background of growing criticism towards the neoliberal economic model and pressure from the IMF. In October 2008, six central banks injected 180 billion U.S. dollars into the monetary markets in a concerted action.[3] An emergency G20 meeting was held in November 2008 and participants committed themselves to increase the transparency, the reliability and the effective regulation of the financial markets, to improve the confidence in and for the markets, to increase international cooperation and to reform the IMF, the World Bank and other financial institutions.[4] In February 2009, G7 finance ministers further affirmed their rejection of protectionism to fight the crisis[5] and in April 2009, a global plan for recovery and reform as well as the largest fiscal and monetary stimulus in modern times was adopted at the London G20 Summit.[6] These concerted actions were supplemented by national stimulus measures in all the important economies, but these policy responses have only had a limited effect so far.

The financial and economic crisis has a potential impact on the space sector, even if its scope is difficult to assess. One has to consider institutional spending and commercial activities separately. As a whole, the institutional demand does not seem to have been affected by the crisis and institutional space budgets for the Fiscal Year (FY) 2009 increased in almost all the space faring nations. Even more, space was considered a useful tool to fight the crisis, as it is an R&D intensive sector and therefore represents an investment in the future. This was testified by the decisions taken at the Brussels European Council of 11–12 December 2008 and at the 6th Space Council of 29 May 2009 in Europe,[7] and by the allocation of additional 1 billion U.S. dollars to NASA (National Aeronautics and Space Agency) as part of the stimulus bill in the U.S. In the commercial sector, the main actors seemed to have remained unaffected by the crisis, as shown by the robust 2008 results in the areas of satellite manufacturing and satellite services. Prolonged difficulties to access credit and investment could however hamper the activity of actors in the space sectors, especially SMEs.

1.2. Political developments

In order to exemplify the links between major global political trends and space policy, a specific focus will be put on six key issue-areas: security, environment, energy, resources, knowledge and mobility.[8] This conceptual approach emphasises the fact that space policy is at the crossroads of different overlapping political issue-areas. These six issue-areas are closely interrelated and provide a coherent and exhaustive picture of major global trends in 2008/2009.

1.2.1. Security

A narrow definition of security focusing on military aspects is adopted here, in order to avoid overlapping with the other issue-areas. Space and security are increasingly tied together. Beyond the classical space-based security applications of Earth observation, navigation and communication, cross-cutting space applications and added-value services are also more and more used for security purposes. In 2008 and 2009, the fight against terrorism, nuclear proliferation and intrastate wars remained the main security concerns. In addition, the ongoing conflicts in Iraq and Afghanistan were increasingly marked by asymmetric warfare, while new conflicts erupted in the unstable regions of Caucasus and the Middle East.

In Iraq, the war entered in its 7th year in March 2009, and the total number of military deaths passed 4000. Despite continued bombings, the country witnessed an overall amelioration of the security situation from 2008 on. On 27 November 2008, the U.S. and Iraq signed a security agreement, scheduling the complete withdrawal of U.S. troops for 2010. In parallel, the new U.S. administration confirmed the strategic shift from Iraq to Afghanistan. As the security situation continued to worsen in 2008 and 2009, U.S. President Obama unveiled its new Afghanistan strategy in March 2009. Four thousand additional U.S. troops will be sent to the country, while other participating countries in International Security Assistance Force (ISAF) also decided to increase their troop contingents.[9]

In the Middle East, a series of positive steps such as the definitive end of the Lebanon war in August 2008 and the normalisation of relations between Syria and Lebanon in October 2008 were followed by resumed violence at the end of the year. Israel launched a massive military operation in the Gaza stripe in December 2008 after Hamas broke a six-month truce by firing rockets on the Hebrew State. U.S. President Obama indicated his willingness to contribute to a conflict settlement in the region, as he called for the creation of a Palestinian State and opposed further Israeli colonisation activities, but peace prospects remain darkened by the continuing internal conflict between Hamas and Fatah.

Other conflict areas in 2008 and 2009 included the Democratic Republic of Congo (DRC), where violence between the government and the rebels in the Kivu province resumed in August 2008 and went on throughout 2009, and the Horn of Africa, where the civil wars in Somalia and Darfur continued to escalate.

Additionally, the Gulf of Aden witnessed an exponential increase of piracy acts in the last months, which prompted the EU to send a military naval force under UN mandate in the region in December 2008. Pakistan also remained unstable, as the country had to cope with the growing tension with its Indian neighbour following the Mumbai bombings in November 2008, as well as with the Taliban insurrection at the border with Afghanistan.

Russia pursued its attempt to play a major role on the international scene, among others by trying to enhance its influence in the "near abroad".[10] The Georgia crisis, which erupted in the summer of 2008, has to be viewed in this frame. Georgia launched a military offensive against South Ossetia on 7–8 August 2008, Russia intervened militarily in support of Ossetia and the conflict spread to Abkhazia. A peace plan was negotiated after the conflict, but Russia violated the terms of the agreement by keeping troops in South Ossetia and Abkhazia. NATO exercises in Georgia in May 2009 as well as the Russian veto to the prolongation of the UN Observation Mission in Georgia in June 2009 didn't contribute to ease the tensed situation.

China continued to raise its international profile in 2008 and 2009. Internally, it had to deal with the increasing issue of "separatism", particularly in Xinjiang and Tibet. It postponed the 11th EU-China Summit scheduled for December 2008 to protest against the Dalai-Lama's visit to several EU countries, but the Summit finally took place in Prague in May 2009. Externally, the tension eased between mainland China and Taiwan in the second half of 2008. After two meetings in June and November, direct passenger airline services were established between the island and the continent, Taiwan opened itself to mainland tourists and cooperation in food safety issues was initiated.[11]

At the level of international organisations, three new European Security and Defense Policy (ESDP) missions were launched under the aegis of the EU: the EU Monitoring Mission to Georgia in October 2008, the EU Rule of Law mission EULEX to Kosovo in December 2008 and the naval mission EU NAVFOR to Somalia in December 2008. Additionally, a new impetus was given to ESDP during the European Council of 11–12 December 2008.[12] On its side, NATO organised a Summit in Strasbourg on the occasion of its 60th anniversary on 3–4 April 2009. This event was marked by the integration of two new Member States, Croatia and Albania, and by the return of France in the NATO integrated structures. A declaration on Alliance security was also adopted, reaffirming the principle of security indivisibility, the commitment to transatlantic solidarity and laying down the fundaments of NATO's transformation in order to be able to meet new security challenges.[13]

Finally, nuclear proliferation remained a central security threat in 2008. Tensions resurfaced between North Korea and the U.S. at the end of 2008, after disagreements over the six parties' talks on the disarmament process and after International Atomic Energy Agency (IAEA) inspectors were expelled from the country. The situation worsened when North Korea conducted a rocket launch in the beginning of April, officially to orbit a satellite. South Korean and U.S. officials however, believed this to be a disguised missile test for a powerful ICBM, which could be capable of reaching the U.S. West Coast. The Iranian nuclear crisis is still facing a stalemate as Iran refuses to give up uranium enrichment, despite several diplomatic offers by the 5 + 1 group[14] and new UN resolutions. In the frame of its National Missile Defense programme, the U.S. found an agreement with Poland to deploy the missile defence system on the Baltic coast. On a more positive note, President Obama decided to give a new impetus to U.S.-Russia disarmament talks, and both countries began renegotiating the START (Strategic Arms Reduction Treaty) in May 2009.

1.2.2. Environment

Space has a crucial role to play in the implementation of sustainable policies for the environment. It can contribute to accurate global data measurements and collection, as well as to monitoring greenhouse gases or the melting of the polar ice-cap, thus giving valuable information to fight against climate change and to respond to natural disasters.

Climate change was a main issue of concern in 2008, as the effects of global warming became more and more visible. For the first time, the northeast and northwest passages in the Arctic Ocean were free of ice in the summer of 2008 (monitored by satellites). As the global consciousness about the emergency of the issue rose, the process of international cooperation to fight global warming accelerated in 2008 and 2009, with the organisation of several international conferences. 2009 will be a negotiation year towards a new climate treaty that will replace the Kyoto Protocol, which expires in 2012.

The Intergovernmental Panel on Climate Change (IPCC) remained a key actor in disseminating knowledge on climate change and therefore supporting the decision-making process in this field. During its 29th session in Geneva on 31 August–4 September 2008, which marked its 20th anniversary, it proposed the drafting of a special report on managing the risk of extreme events to advance climate change adaptation. Its 30th session in April 2009 was mainly dedicated to the preparation of the 5th Assessment Report (AR5),[15] to be completed by 2014.[16]

The UN Conference on the Framework Convention on Climate Change (UNFCCC) will take place in Copenhagen at the end of 2009. Its purpose is the adoption of an agreement on an effective international response to climate change – if political will reaches this goal, otherwise more modest results like non-binding understandings might be the results. Preparatory steps towards this major event included the UN Climate Change Conference (UNCCC) which took place in Poznan, Poland, on 1–12 December 2008 and the Bonn UN climate change talks in April and June 2009. At this occasion, participants from government, industry, business, research and environmental organisations met to negotiate a global agreement, focusing on the key issues of emission reductions by industrialised countries and the improvement of emission trading.[17] During the Copenhagen conference, the contribution of satellite technologies to the fight against climate change will be on the agenda. In particular, satellites can be useful to verify the implementation of international treaties by monitoring greenhouse gas emissions.[18]

In addition to the UN, the G20 and G8 were active players in the fight against climate change. In preparation to the G8 Summit in Hokkaido, Japan, the G8 environment ministers met in Kobe in May 2008. They called for decisive actions to preserve biodiversity, agreed on the long-term goal of halving greenhouse gas emissions by 2050 and adopted the Kobe 3 R action plan (reduce-reuse-recycle). The G20 organised a workshop on climate change in February 2009, with the theme "Financing for climate change". At the London Summit in April 2009, the G20 members further reaffirmed their commitment to fight climate change and to reach an agreement at the UNFCCC.[19] However, despite these declarations, no quantitative goal for emission reductions was set up and emerging countries are still reluctant to halve their emissions by 2050.

The EU also continued to put forward its policy agenda on climate change. The "climate-energy" legislative package was discussed at the European Council on 11–12 December 2008, before being adopted by the European Parliament on 17 December 2008 and by the Council on 6 April 2009. The package will help transforming Europe into a low-carbon economy and increase its energy security. The EU committed itself to reduce its overall emissions to at least 20% below the 1990 levels by 2020, to increase the share of renewable energy to 20% by 2020, and to reach a 20% cut in overall energy use.[20] After laying down EU's position on the negotiations for a post-Kyoto treaty in a communication of 28 January 2009,[21] the European Commission (EC) also set up a strategic roadmap to fight climate change in a White Paper entitled "Adapting to climate change: Towards a European framework for action".[22] Finally, the strong link between climate change and international security was highlighted in a paper issued by the EC and the High Representative, recommending the development of an EU preventive security policy with regards to climate change.[23]

1.2.3. Energy

Space can provide tools for strategic decision-making in the energy sector. It can support the planning and monitoring of electrical power grids, pipelines and of the operational needs of the energy sector, including weather forecast, or seismic activity. Remote sensing activities are also useful to assess and monitor exploration surveys.

The International Energy Agency (IEA) identifies three priorities a balanced energy policy making should focus on: energy security, economic development and environmental protection.[24] In 2008 and 2009, the issue area of energy policy was marked by two main events: the global energy crisis in the second half of 2008 and the gas conflict between Russia and the Ukraine in 2009. These two developments underlined the need for new impetus in energy policies, along the three lines mentioned above.

Rising oil prices reached historical levels in the first half of 2008 and triggered the energy crisis. In the second half of 2008, the negative impact of the financial crisis on the international oil demand lead to a strong decrease of oil prices, which lost 70% of its value within five months.[25] The Organisation of Petroleum Exporting Countries (OPEC) agreed on production reductions three times in four months to try to hold up the collapse of oil prices.

Tensions between Russia and Ukraine rose when Russia cut its gas supply to Kiev on 1 January 2009. As gas supplies to the West are transiting through Ukrainian territory, Europe was affected by this crisis as well. After long diplomatic negotiations concluded by an EU brokered agreement, the gas supply to Europe was fully re-established on 20 January. Tensions between Russia and Ukraine resumed in the spring of 2009, shedding light once again on the crucial issue of energy security for Europe. In June 2009, the EC issued a draft report, calling for upgraded coordination measures in case of a renewed gas crisis.[26]

More generally, the EU continued to pursue an ambitious energy policy based on three pillars: security of supply, competitiveness and sustainability. A directive on renewable energies was discussed in the course of 2008, with the establishment of national action plans for each Member State. In December, the integrated energy/climate package was adopted, and the third liberalisation package of the European Commission, aiming at liberalising the energy sector in the EU, is under discussion. The security and geopolitical dimension of energy policy has also been acknowledged by the EU, as steps have been undertaken to integrate energy aspects into the relations with third countries.

Given the transnational character of energy-related policy issues, it is necessary to develop global answers to tackle them. Energy was on the agenda of the G8 Summit in Hokkaido, Japan, as the IEA submitted a report to the G8 members presenting strategies to enhance energy security, to speed up the development of cleaner energy and to promote energy efficiency.[27] The IEA further gave inputs to the G8 Environment Ministers' meeting in April 2009 and to the G8 Energy Ministers' meeting in May 2009. The UNCCC also tackled the energy issue, by calling for a clean energy deal, encompassing the development of renewable energies and the amelioration of efficiency for traditional energy sources. Finally, no global energy policy can be implemented without the U.S. In this respect, the new U.S. administration is likely to bring a change in U.S. energy policy, as President Obama proposed an alternative energy plan during his campaign, putting emphasis on low-carbon energy sources.

Whereas Western economies struggle to develop sustainable energy policies, emerging economies, in particular China, have to face growing energy needs. In order to satisfy the energy demand of its heavy industry, China had to turn to international markets. Considering Russia, Central Asia and the Middle East as unstable providers, China tried to further increase the diversification of its energy supplies by augmenting its imports from Africa and Latin America.

1.2.4. Resources

Space assets can provide valuable contributions for issue areas such as crop modelling, agriculture monitoring, humanitarian response to food problems or the development of water master plans. Earth observation systems, as well as space-based communication and information technologies can also enhance the effectiveness of humanitarian responses to food crisis.

The growing scarcity of natural resources – mainly water and food – highlights the need for transnational solutions in the management of resources. The issue-area is closely related to environmental concerns, as global warming is threatening food and water sustainability, and has links to energy issues, the best example being the impact of biofuels on crop prices. 2008 and 2009 were marked by the food crisis, and by the preoccupying state of water resources.

The global food crisis pushed 130–150 million people to extreme poverty in the last two years, with 44 million more malnourished. According to the World Bank, food prices rose by 83% in the last three years, the Food and Agricultural Organisation (FAO) cites a 45% increase of its world food price index between January and August 2008. The immediate factors explaining the crisis lied in the high oil prices, the diversion of 5% of the world's cereals to agrofuels or the doubling of per-capita meat consumption in some developing countries. Even if food prices decreased sharply in the second half of 2008 because of the financial crisis, the long-term and deeper factors fuelling the crisis, namely the conflicts over trade agreements between the North and the South, remain unsolved. The failure of the negotiations in the frame of the Doha development rounds in July 2008 illustrated this situation.[28] In their Agricultural outlook 2009–2018, the OECD (Organisation for Economic Co-operation and Development) and the FAO indicate that the rise in food commodity prices will continue to ease in 2009. They also identify four critical factors that influence food protection: land availability, productivity gains, water usage and climate change.[29]

As a consequence, food security was an important topic on the international agenda in 2008 and 2009. Two UN High-Level Conferences tackled this issue, in Rome in June 2008 and in Madrid in January 2009, where the UN called for a global food security plan. The three UN food agencies[30] further called for a UN Food Summit to take place in 2009, as the food crisis is still threatening millions of lives. The issue of food security was raised as well at the G8 Hokkaido Summit, where a statement on food security was adopted, stressing the need to prevent future crisis.[31] The first ever G8 Agricultural Ministers Summit took place on 18–20 April 2009 in Italy. It stressed the importance of increasing private and public investment in sustainable agriculture, while noting that only well functioning markets can improve food security.[32] As for concrete actions, the World Food Programme (WFP) had to raise its 2009 budget by 2 billion U.S. dollars, up to 5.2 billion U.S. dollars, to account for the increase of its beneficiaries, while the EU granted 1 billion euros over two years as a rapid response to soaring prices of food in developing countries in December 2008.

The growing scarcity of water resources represented a second important issue on the global agenda in 2008 and 2009. The increased water demand, fuelled by population growth and economic development, is putting pressure on water prices around the globe. According to the UN Environment Programme (UNEP), the total usable freshwater supply for ecosystem and humans is about 200,000 km^3 – less than 1% of all freshwater resources, freshwater itself representing only 2.5% of the global water resources. FAO statistics indicate that by 2025, 1.8 billion of people will be living in countries or regions with absolute water scarcity, and two-thirds of the world population could be under stress conditions. Finally, the World Health Organisation (WHO) indicates that one in six people worldwide – 894 million – don't have access to safe freshwater.[33] In the light of this figures, there is a growing pressure to recognise access to safe drinking water and sanitation as basic human rights.

Global answers have been provided through various events in 2008 and 2009. The year 2008 was declared the International Year of Sanitation by the UN Secretary General. The World Water Week took place in Stockholm, Sweden, in August 2008, as well as the International Expo "Water and sustainable development" in Zaragoza, Spain, between 14 June and 14 September 2008. A Charter was issued at the end of the expo, giving recommendations for a comprehensive vision for water.[34] Other important events included the 5th Water Forum which took place in Istanbul, Turkey in March 2009 and the release by UNESCO (United Nations Educational, Scientific and Cultural Organization) of the UN World Water Development Report in March 2009, giving a comprehensive picture of the state of the world's freshwater resources. Finally, the EU signed a key document with businesses, Non-governmental Organisations (NGOs) and national ministries in June 2008. The paper, entitled "Water for a sustainable Europe – our vision for 2030" contains 10 points, calling for a true "water democracy" and using transparent and efficient pricing as a tool for a sustainable water use.[35]

1.2.5. Knowledge

In knowledge-based societies, fundamental science and applied science are crucial to stimulate research, and therefore investment and growth. Space can contribute to this research endeavour by better understanding Earth and its place in the universe. Applied research in space is also beneficial for Earth-based application, and communication satellites can help bridging the "digital divide", therefore contributing to the sharing of knowledge. The financial and economic crisis provided a strong incentive for political initiative in the field of innovation. In particular, space as an intensive R&D sector was increasingly seen as an investment in the future. In general, the need to shift towards a knowledge-driven economy was made necessary by the new challenges of globalisation, and therefore represents the overarching political goal of all industrialised economies.

In Europe, the Brussels European Council on 11–12 December 2008 set guidelines for a European innovation policy by approving the European Economic Recovery Plan (EERP), and calling for the launch of a European Plan for Innovation and for the development of the European Research Area. Space technologies and services were identified as one of the main areas for future investments.[36] The 6th Space Council on 29 May 2009 focused on the concrete contribution of space to economic recovery, competitiveness and innovation.[37] Furthermore, the EC published a report entitled "A more research-intensive and integrated European Research Area"[38] on 22 January 2009. Its stated goal was to make Europe more innovative, in order to become the world largest knowledge-based economy. It identified several shortcomings in the European research policy, pointing out that Europe still lags far behind the U.S. and Japan in terms of R&D. EU's chief instrument for funding research over the period 2007–2013 is the 7th Framework Programme (FP7), which is centred on innovation and knowledge for growth, and intends to build the internal market of knowledge. In line with these political goals, 2009 was declared European Year of Creativity and Innovation by the European Parliament and the Council.

The specific contribution of the space sector to innovation and growth was also increasingly acknowledged in the U.S. A total amount of 1 billion U.S. dollars was allocated to NASA in the framework of the stimulus bill. The aim is to create jobs in the sector by stimulating innovation, in particular in the private sector.

More specifically and space-related, the International Heliophysical Year (IHY) took place in 2007 and 2008 to coincide with the fiftieth anniversary of the International Geophysical Year (IGY) in 1957–1958.[39] In the same line, 2009 was declared the International Year of Astronomy by the 62nd General Assembly of the UN, to coincide with the 400th anniversary of the first recorded astronomical observations with a telescope by Galileo. 2008 and 2009 were marked by a series of scientific achievements in the field of space. In October, the Hubble telescope was rebooted in space, and another repair mission was successfully conducted in May 2009. The International Space Station (ISS) further celebrated its 10th anniversary. The first component of the ISS, Zarya, was launched on 20 November 1998, and after 29 additional construction flights, the station is now the largest space object ever built. The international cooperation between NASA, the Russian Federal Space Agency (Roscosmos), the Japan Aerospace Exploration Agency (JAXA) and 11 Member States of ESA lead to the installation of 19 research facilities onboard the ISS and opened the way for space exploration throughout the solar system. Important scientific missions were launched, including NASA's IBEX (Interstellar Boundary Explorer) mission, which will fly to the outer solar system and collect new data on solar wind, NASA's Kepler mission, which will explore the structure and diversity of planetary systems and ESA's Herschel and Planck spacecrafts, which will investigate the question of the beginning of the universe and its evolution. On Earth, an important event was the start of the large Haldron collider, the world's largest and highest-energy particle accelerator. It was built to test various predictions of high-energy physics, but had to be stopped after 10 days due to a defect of the cooling system.

1.2.6. Mobility

Satellite-based navigation systems increase traffic safety and security. The European Galileo satellite navigation system in particular will greatly improve all modes of transport. Telecommunication satellites combined with space-based navigation systems also play a role in ensuring a sustainable mobility. Taking into account environmental concerns is the major challenge of sustainable mobility policies, which emphasises the interdependence of mobility policies with other issue areas such as energy, environment or resources.

Mobility could be understood in a broad sense, including not only land, air and sea traffic, but also migration and refugee streams. In the following section, mobility will be limited to transportation issues, and focus on shipping, air transportation and land transportation.

Ninety percent of trade worldwide is done by shipping, which underlines the vital importance of the shipping industry and of maritime services for the global economy. One of the main issues of concern for shipping in 2008 was piracy, as the number of piracy acts increased exponentially, particularly along the Somali coast. Two UN Security Council (UNSC) resolutions were adopted in 2008 to fight against piracy and to protect the commercial ships.[40] Another important trend in shipping is the growing concern about sustainability and environment. Although shipping is the most environmental-friendly transportation means, sea pollution is a serious issue. The International Maritime Organisation (IMO) is tackling these issues by issuing a series of legal instruments. In September 2008, the latest IMO Convention entered in force, the International Convention on the Control of Harmful Anti-fouling Systems on Ships.[41]

In September 2008, Airports Council International (ACI) released its global traffic forecast 2008–2027.[42] After several years of strong growth, global passenger traffic is expected to increase at a slower pace in 2008 and 2009, following the financial crisis. The 5 billion passengers mark will be surpassed by 2009, and by 2027, 11 billion passengers are expected to travel by air. The potential for domestic traffic growth remains very high in Eastern Asia emerging economies, whereas the biggest market, North America, witnessed a negative growth in 2008. Concerns about future energy policies, fuelled by the rise of oil prices in the first half of 2008, lead to increased interest in biofuels for airplanes. In February 2008, an Airbus A380 was the first plane to fly with a biofuel powered engine, and since then several airlines tested biofuels.

The car industry violently suffered from the financial crisis. An emergency plan was launched in the U.S. to save the country's automotive industry in December 2008, and several major car producers, like Toyota, had record losses in 2008. The car sales between September and December 2008 decreased by 30% and the American, European and Japanese markets are expected to stay weak for a longer period. In this sector too, the growing trend of "green investments" continued in the course of 2008 and 2009, with more and more major companies investing in sustainable concepts such as the hybrid technology.

The European Union continued to work towards the establishment of a coherent transportation policy in 2008 and 2009. The EC issued a communication on sustainable transport in June 2009, laying down the vision of an integrated, technology-lead and user-friendly sustainable transport system after 2010.[43] More specifically, the second package of the Single European Sky (SES II) Act was adopted by the European Parliament in March 2009. The SES I package came into force in 2004, with a focus on congestion and safety in air transportation and in June 2008, the EC released a communication on the SES II package centred on environment and cost-efficiency issues.[44]

1.3. Main science and technology indicators relevant for space activities

In order to assess recent trends in the space sector, it is useful to consider a limited number of key science and technology (S&T) indicators, giving information on the economic environment in which the space sector is embedded. Since the 1960s, the input-output framework is the more commonly used tool to measure science activities in social reports. Inputs are defined as "investments in the resources necessary to conduct scientific activities, like money and scientific and technical personnel", while "outputs are what comes out of these activities: knowledge and inventions".[45] In Europe, developing a more effective research system through the implementation of the European Research Area remains a central objective, both to stimulate investment in research and to facilitate structural change towards a more knowledge-based economy. Research indeed is a key competitive asset in the global world of science and technology, which is increasingly marked by the emergence of new players in S&T, most notably in Asia.[46]

1.3.1. Science and technology inputs

Two easily measurable indicators of S&T inputs will be analysed in the following section, with a particular focus on Europe: R&D spending and the number of researchers. As for the global political framework, the Lisbon strategy, adopted in 2000 as EU's response to globalisation, sets a clear objective: transform Europe into a competitive knowledge-based economy. The objective of 3% R&D intensity[47] was set up, but was not met yet, as overall R&D intensity amounted 1.85% in 2007, the exact same figure as in 2000. Only two countries, Sweden and Finland, reached the target of 3% R&D intensity.[48]

The increasingly competitive global environment for S&T is reflected in key figures indicating that EU's position in the international comparison is deteriorating. In absolute terms, the U.S. spent 2.97% of their GDP for R&D activities in 2006 and Japan 3.4% of its GDP, compared to EU's 1.85%.[49] These differences are mainly due to the lower level of business R&D intensity in the EU.[50] Indeed, when considering the data on Government Budget Appropriations or Outlay on R&D (GBAORD) measuring public funding for R&D, figures for the three regions are similar: 0.67% of the GDP for the EU in 2007, 1.03% of the GDP for the U.S. and 0.68% of the GDP for Japan.[51] In relative terms, EU's world share in Gross Domestic Expenditure on R&D (GERD) diminished by 7.6% between 2000 and 2006, and it represented 24.4% of global GERD in 2006.[52] The shift in the overall GERD shares between 2000 and 2006 also testifies the increasing importance of Asian developed economies and emerging economies in the S&T area (Figure 1).

As for the share of researchers in the labour force within the European Research Area (ERA), it grew by 1.9% a year in average since 2000. This represents a similar figure as in Japan, and twice the growth rate of the U.S. However, the EU has proportionally less researchers than those two countries. In 2006, researchers represented 5.6‰ of the labour force in the EU 9.3‰, in the U.S. and 10.7‰ in Japan. These differences can mainly be traced back to the far lower share of researchers in the business sector in the EU.[53]

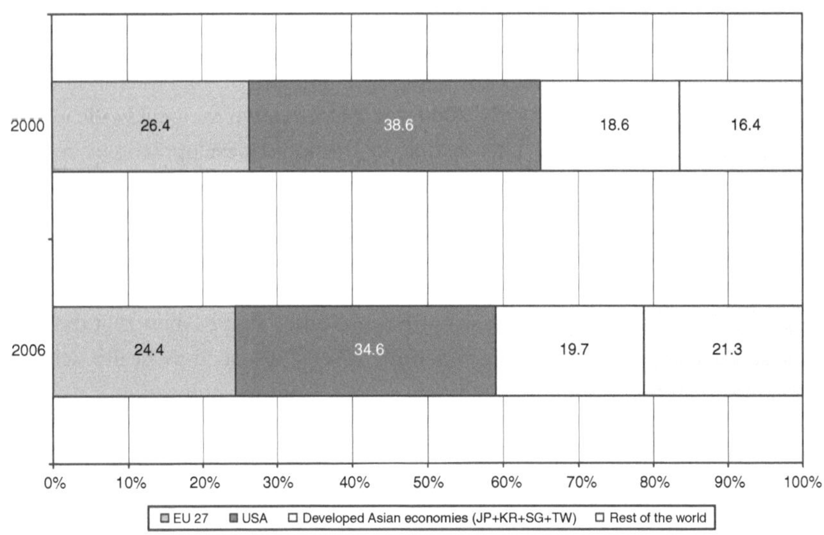

Fig. 1: *Global shares in GERD (source: European Commission).*[54]

Despite these figures, some positive evolutions took place in the EU in the last years. In real terms, R&D expenditure in EU-27 has grown by 14.8% between 2000 and 2006, compared to 10.1% in the U.S. and 21.9% in Japan. R&D expenditure grew in real terms in every EU Member State during the same period, in particular in the most recent Member States. The total real growth of R&D expenditures between 2000 and 2006 exceeds 100% in the Baltic countries and in Cyprus, and is greater than 60% in Hungary, Romania, the Czech Republic, Spain and Ireland.[55]

As a whole, and despite significant progress in R&D spending, the overall R&D intensity of the EU remained unchanged since 2000. This highlights the need to fully implement the ERA strategy to further support and promote research-intensive sectors. In this respect, the ERA Green Paper issued by the EC in 2007 identifies six priorities: realising a single labour market for researchers, developing world class infrastructures, strengthening research infrastructures, sharing knowledge, optimising research programmes and priorities and opening to the world through international cooperation in S&T.[56]

1.3.2. Science and technology outputs

Patent applications and scientific publications are commonly used indicators to assess S&T outputs. In these areas as well, figures suggest that a global shift is ongoing in the S&T area, in favour of emerging Asian economies.

In 2006, the EU remained the largest producer of scientific publications in the world, with a share of 37.6% of the global peer-reviewed scientific articles, compared to 31.5% in the U.S. Despite a growth of the total number of publications of around 18% since 2000, the EU's share in the worldwide number of publications decreased. This is due to the rapid development of research capacities in emerging countries, in particular in China, where the total number of scientific publications grew by 178% between 2000 and 2006. EU's share in high-impact publications is lower that the U.S. share and regarding this indicator, China made significant progresses as well, catching up with Japan and South Korea between 2000 and 2003. Finally, scientific publication figures show that the EU is not specialised in faster-growing scientific disciplines, such as health sciences, materials sciences, environmental sciences or geological engineering.[57]

Patent applications provide a measure of inventiveness. Between 2000 and 2005, the number of patent applications with EU inventors filed under the Patent Cooperation Treaty (PCT)[58] has increased by 13%, compared to 9.6% for patent applications with U.S. inventors. In the same period, the number of PCT patent applications from Asian countries has increased dramatically: 100% in

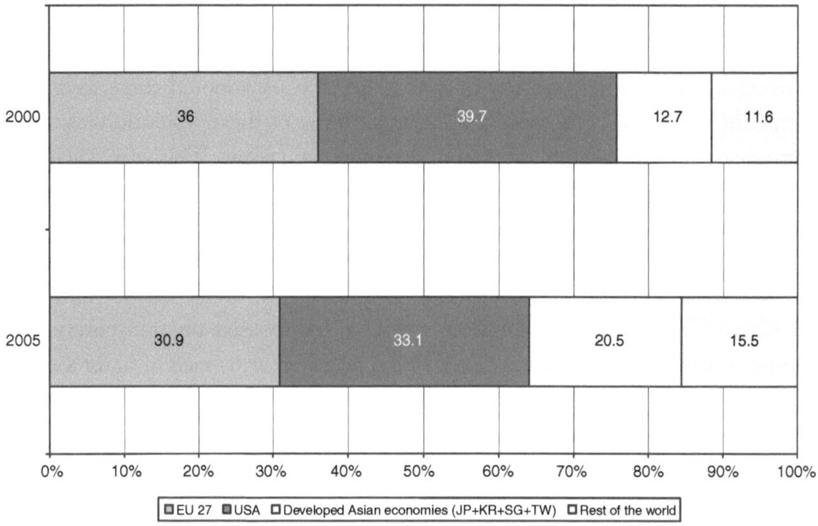

Fig. 2: *Global Shares in patent applications (source: European Commission).*[59]

Japan, 161% in South Korea, 137% in China, 241% in India. As a consequence, the world share of the EU and the U.S. in patent applications has decreased (Figure 2).

2. Worldwide space policies and strategies

2008/2009 was marked by the flourishing of cooperation endeavours, in particular those involving emerging space faring nations. At the national level, 2008/2009 was a transition period for the U.S., Japan and Russia, while India and China continued to work on the implementation of their ambitious space programmes – as did Europe with regard to the European Space Policy.

2.1. The United Nations system

The United Nations system has three priorities regarding space: the promotion of international cooperation in space with a particular focus on developing countries, the use of satellite applications within the UN system and the development of a legal framework for space-related activities.[60] In 2008 and 2009, three bodies or group of bodies within the UN continued to work towards these goals: the UN General Assembly (UNGA), the UNGA Committees and a series of other UN entities and organs.

2.1.1. United Nations General Assembly (UNGA)

During its 63rd plenary session in 2008, the UNGA adopted three resolutions pertaining to space. The topic and basic patterns of these resolutions is similar every year, but special emphasis can be put on certain issues. This year, the issue of security was again a high priority, mirroring growing concerns about a possible arms race in outer space.

The resolution on the prevention of an arms race in outer space was adopted on 2 December 2008 by 177 countries, while the U.S. voted against and Israel abstained. In this year's resolution, the UNGA focused on transparency and confidence-building measures (TCBM) to avoid an arms race in outer space. It recalled that the existing legal framework is not a guarantee against an arms race in space, and therefore stressed the need for more transparency and better information. While recognising the virtues of bilateral efforts, the resolution called space faring nations, especially the most important ones, to continue negotiating in the framework of the Conference on Disarmament (CD), which was considered the sole multilateral disarmament forum.[61]

The second resolution adopted concerned transparency and confidence-building measures in outer space activities. It was adopted on 2 December 2008 by 180 countries, the U.S. voting again against and Israel abstaining. Emphasising the danger to international peace and security by an arms race in outer space, the resolution invited the Member States to continue submitting proposals on TCBM in outer space to the Secretary General. The European draft Code of Conduct and the Chinese-Russian draft treaties were specifically mentioned as constructive contributions within the CD.[62]

The last resolution on international cooperation in the peaceful use of outer space was adopted by consensus without a vote. It recalled all the central aspects and challenges of the peaceful use of outer space, most notably space debris and a potential arms race in space and noted the central importance of international cooperation to tackle these issues. Finally, it reviewed all the important steps which have been undertaken in 2008 in the field of international cooperation for outer space issues, including conferences and forums, sessions of the Committee on the Peaceful Uses of Outer Space (COPUOS) and progresses in programme implementation.[63]

2.1.2. UNGA Committees

Two of the three resolutions adopted in plenary session were introduced in the Disarmament and International Security Committee (First Committee), while the last resolution was introduced in the Special Political and Decolonisation Committee. The 2008/2009 activities of COPUOS were marked by the sessions of its two subcommittees and by its plenary session.

The prevention of an arms race in outer space was again a topic on the top of the agenda of the First Committee in 2008. As a consequence, it discussed two resolutions on that issue: the resolution on the prevention of an arms race in outer space and the resolution on TCBM in outer space activities. The debates were marked by enduring divergences, mainly between the U.S. on one side and China and Russia on the other side. The two latter countries emphasised the importance of preventing the placement of weapons in space to guarantee international security. This was also the rationale behind the introduction of their draft treaty in the CD. The U.S., in contrary, refused to negotiate on the basis of a treaty, arguing that a definition of "space-based weapons" is impossible and that the implementation of reliable verification mechanisms will be problematic. Instead, the U.S. advocated voluntary TCBM.[64]

The draft resolution on international cooperation in the peaceful use of outer space was introduced by Colombia in the Special Political and Decolonisation Committee, and had the same format as in previous years.[65]

The Scientific and Technical Subcommittee of COPUOS held its 46th session on 9–20 February 2009. Key topics on the agenda included possible dangers from near-Earth objects (NEO), the safety framework for nuclear power sources in outer space and space debris mitigation. Besides, other issues were treated as well, such as a review of the implementation of UNISPACE III (United Nations Conference on the Exploration and Peaceful Uses of Outer Space), space-based disaster management support, recent developments in Global Navigation Satellite Systems (GNSS) or the International Heliophysical Year.[66]

The Legal Subcommittee took place from 23 March to 3 April 2009. Issues tackled included the status and application of the five UN treaties on outer space, with a specific focus on activities being carried on the Moon and other celestial bodies, a review and possible revision of the principles relevant to the use of nuclear power sources in space, the definition and delimitation of outer space and capacity building in space law. Main item for discussion was national space legislation for which a working group was established, which will meet for three years.[67]

COPUOS finally concluded the year session period with its 52nd plenary meeting on 3–12 June 2009. It focused on the implementation of the recommendations of UNISPACE III, the spin-off benefits from space technology, space and society, space and water, space and climate change, the use of space technology in the UN system and international cooperation for the use of space-derived geospatial data for sustainable development.[68]

2.1.3. Other UN bodies and organs monitoring outer space activities

The importance of space in the work of the UN system is testified by the number of space-related initiatives running under the aegis of the United Nations Office for Outer Space Affairs (UNOOSA) and by the number of UN bodies relying on space for their activities.

UNOOSA serves as the secretariat for COPUOS, and carries out various specific activities. It first manages the UN Platform for Space-based Information for Disaster Management and Emergency Response (UN-SPIDER), which was established by the GA in 2006. In 2008 and 2009, UN-SPIDER continued to be implemented as an open network of providers of space-based solutions. Several workshops and regional meetings were organised in this framework, while a 4th advisory mission was launched in Togo on 2 July 2009.[69] UN-SPIDER also continued to expand, with new regional support offices set up in Iran, Romania, Algeria and Nigeria.[70]

Through the United Nations Programme on Space Applications (PSA), UNOOSA then conducts international workshops, training courses and pilot projects on various topics. In 2008 and 2009, issues tackled included integrated space technology applications to water resource management, the use of space tools and solutions for monitoring the atmosphere and land cover, satellite-aided search and rescue, the use of space technology for tele-epidemiology or integrated applications of GNSS.[71]

UNOOSA also acts as the secretariat of the International Committee on Global Navigation Satellite Systems (ICG), which was established on a voluntary basis to promote cooperation in matters of satellite navigation. It held its third meeting in Pasadena, USA, on 8–12 December 2008, while the next one will take place in Saint Petersburg, Russia, in September 2009. In 2008, substantial progress was made on the Workplan through four working groups on compatibility and interoperability, enhancement of performances of GNSS services, information dissemination and capacity building. The Workplan was finally adopted by the Providers Forum, together with its Terms of References.[72]

Finally, UNOOSA is the secretariat and coordinator of the Inter-Agency Meeting (IAM) on outer space activities, which aims at increasing coordination among the UN specialised agencies and the UN PSA. The 29th session of the IAM took place in Vienna on 4–6 March 2009. Like every year, the IAM reviewed and finalised the report of the Secretary General on the coordination of space-related activities within the UN system. The document identified four priorities for the period 2009–2010: the further strengthening of the IAM, the reinforcement of the contributions to the UN Spatial Data Infrastructure (UNSDI), the enhancement of the use of space-based assets in support of disaster management and the reinforcement of the contributions made by the UN entities to the Global Earth Observation System of Systems (GEOSS).[73]

A number of UN bodies are relying on space applications to fulfil their missions. One of those is UNESCO which continued to use remote sensing systems to monitor its world heritage sites under the "Open Initiative". On 2 December 2008, JAXA joined the initiative, which will allow UNESCO to use the Daichi Earth observation (EO) satellite. The International Year of Astronomy was also launched at UNESCO headquarters in Paris on 15–16 January 2009, and a series of events, research colloquiums and workshops will be organised by UNESCO in this frame.[74]

A further UN entity relying on space applications is the United Nations Geographic Information Working Group (UNGIWG), which brings together the Geographic Information Systems (GIS) specialists and cartographers from all UN agencies, funds and programmes. At its 9th annual meeting in Vienna on 5–7 November 2008, it decided to implement the UNSDI on a project basis. UNSDI is understood as a comprehensive and decentralised geospatial information network aiming at facilitating decision-making.[75]

The International Telecommunication Union (ITU) on its side continued its regulatory work on space-related telecommunication issues in 2008 and 2009. Its policy priorities for this period included standardisation work, climate change, cybersecurity and the bridging of the digital divide. On the background of the economic crisis, ITU also put forward the vital role Information and Communication Technologies (ICT) can play in the process of economic recovery. ITU's various bodies tackled these issues. The World Telecommunication Standardisation Assembly (WTSA) took place in Johannesburg on 21–30 October 2008 and adopted a roadmap for future developments of communication standards. The 2008 session of the ITU Council took place in Geneva on 12–21 November 2008 and endorsed the organisation's operational plan for the years ahead. More issue-specific events included the World Radiocommunication Seminar (WRS) in December 2008, which dealt with international regulations on the use of radio-frequency spectrums and satellite orbits, and the World Summit on Information Society in May 2009, focusing on the potential of ICT for the implementation of the Millennium Development Goals (MDG) and on their contribution to fight the economic crisis.[76]

Finally, space security issues were at the core of the agenda of two UN bodies. The United Nations Institute for Disarmament Research (UNIDIR) continued to analyse developments in the CD and to propose options to break the deadlock. In 2008, it introduced an options paper in the CD, reviewing past proposals and proposing new options. Besides this, UNIDIR organised two events centred on the question of space security: a conference in New York on 20 October 2008 entitled "Prospects for preserving a cooperative security framework in outer space" and another one in Geneva on 15–16 June 2009 on the theme "Moving toward a safer space environment".[77]

The CD was in session from 28 July to 12 September 2008, from 19 January to 27 March 2009 and from 19 May to 2 July 2009. Its work in 2008 in the field of space security ended without significant process, as no agreement could be found on the draft treaty banning space weapons introduced by Russia and China. During the 2009 session, the prevention of an arms race in outer space was again a central topic on the agenda, and discussions focused among other things on the Russian-Chinese draft treaty, on TCBM and on the EU draft code of conduct. Even though little progress was achieved on the content, the CD adopted a working programme for the first time in ten years on 19 May 2009.[78]

2.2. The Group on Earth Observation

The Group on Earth Observations (GEO) is a voluntary partnership of governments and international organisations whose task is to coordinate efforts to build a GEOSS. In 2008 and 2009, the GEO continued to organise workshops and symposia to advance the construction of a GEOSS. In addition, general orientations to the GEO work plan were provided during the 2008 plenary session and during the meetings of its Executive Committee and Technical Committees.

The 5th plenary session of GEO (GEO-V) took place on 19–20 November 2008 in Bucharest, Romania. At this occasion, four new members were welcomed: Turkey, Estonia, the Bahamas and Peru. The GEOSS implementation progress was reviewed, a Data Sharing Principles Task Force was established and the work plan for 2009–2010 was adopted as a living document. Key aspects of the new work plan include establishing the GEO web-portal, promoting free access to data, advancing the GEO Biodiversity Observation Network and building a system for monitoring carbon.

An important political support was given to the GEOSS in the G8 Declaration on Environment and Climate Change adopted at the G8 Summit in Hokkaido in July 2008. It stated that G8 members will accelerate their efforts within the GEOSS given the growing demand for EO data.[79]

2.3. Regional cooperation in space activities

Regional cooperation activities outside Europe continued to develop in 2008/2009, mainly in Asia and Africa. This is due to the fact that developing countries, realising the potential of space applications to reach their development objectives, tend to pool their resources in international organisations. In this respect, it is possible to distinguish regional cooperation activities in space under the umbrella of the UN, rather concentrated on capacity-building and information dissemination, and cooperation through ad hoc institutions, aiming at the concrete use of space applications.

Within the UN system, the regional centres for space science and technology education, affiliated to the UN, constitute regional focal points to disseminate knowledge on space-based applications. In 2008 and 2009, the five centres in India, Morocco, Nigeria, Brazil and Mexico continued to organise workshops and courses on remote sensing, space science, meteorology and GIS.[80]

In Asia, two space-related organisations coexist, respectively sponsored by China and Japan: the Asia-Pacific Space Cooperation Organisation (APSCO) on the one hand and the Asia-Pacific Regional Space Agency Forum (APRSAF) on the other hand. APSCO formally started its operations in December 2008, after the idea of such an organisation has been put forward as early as 1992 by China, Thailand and Pakistan. The APSCO Convention was signed in 2005 by nine countries,[81] but the main initiator and supporter of the organisation is China. Its aim is to promote multilateral cooperation in space science and technology.[82]

APRSAF was founded in 1993 under Japanese leadership to enhance the development of the members' space programmes and to exchange views on future cooperation in space activities. The flagship programme of APRSAF is the STAR (Satellite Technology for the Asia-Pacific Region) programme, which was introduced by Japan in 2007. In 2009, an international project team began a 3-years study for a STAR EO system in order to meet the future needs of the Asia-Pacific region in terms of EO. The 15th session of APRSAF took place in Vietnam in December 2008 under the theme "Space for sustainable development".[83]

African countries are increasingly interested in using space-based applications for their development goals. In this perspective, the African Leadership Conference on Space Science and Technology for Sustainable Development (ALC) was launched. The first two conferences took place in Nigeria in 2005 and in South Africa in 2007, while the third one is scheduled in November 2009 in Algeria. These three countries agreed to launch the ALC as a forum to enhance high-level cooperation among African countries in space affairs. The 3rd ALC will mainly focus on space science and technology in Africa's development, IGOs and NGOs active in the space field in Africa, regional space conferences and the African Resource Management and Environmental Constellation (ARMC) project.[84]

The ARMC will be a constellation of Low-Earth Orbit (LEO) EO satellites designed to address core remote-sensing needs of African decision-makers and the data generated will be shared among the participants. A declaration of intent signed by South Africa, Nigeria and Algeria in June 2008 launched the project. Since then, the user requirements of the ARMC have evolved to include medium resolution imagery and very high resolution imagery, in addition to the already identified needs for Synthetic Aperture Radar (SAR) and thermal infrared image data. The technologies needed will be designed, developed and manufactured by the African technology base. A Memorandum of Understanding (MoU) is under preparation and will be signed at the 3rd ALC.[85]

2.4. Europe

In 2008 and 2009, the three main European space actors, namely the EU, ESA and the European Organisation for the Exploitation of Meteorological Satellites (EUMETSAT), continued to focus on the implementation of the European Space Policy (ESP). In particular, the two flagships of ESP, Galileo and GMES (Global Monitoring for Environment and Security), were at the core of the agenda. In this respect, the growing role of the EU in space policy issues continued to be a striking feature throughout 2008 and 2009.

2.4.1. European Space Agency

The main event for ESA in 2008 was the Council meeting at Ministerial level that took place in The Hague on 25–26 November 2008. By dealing with the implementation of the ESP, giving impetus to future programmes and taking decisions on the next phases of ongoing projects, the Council set the course of the institution's space policy orientations for the years ahead.

Despite the financial and economic turmoil, the Council was only little affected by this. In fact, the 5th and 6th Space Council meetings had given full support to the ESP and had established clear objectives and priorities, ESA confirmed its role as a global player in 2008 with the launch of the Columbus module and the Automated Transfer Vehicle (ATV) to the ISS, national programmes are increasingly planned in a complementary way with ESA programmes and the EU is becoming a key player in strengthening Europe as a space power.[86]

ESA asked for 10.41 billion euros to finance its programmes (a 20% increase compared to the last Council Meeting in 2005), and the stakeholders granted 9.94 billion euros. Four resolutions were adopted, on the objectives for the next ten years and the orientations of the next Council Meeting in 2011, on the resource level and basic activities, on the financing of the Guyana Space Center (GSC) and on the evolution of ESA.[87]

Thirty programmes were funded at the Council, mainly in the area of space applications. These programmes were the translation of the five political priorities identified within the ESP: space applications serving Europe's public policies, enterprises and citizens, meeting Europe's security needs, competitive and innovative industries, contributing to the knowledge-based society and securing access to technologies, systems and capabilities for independence and cooperation.[88]

Another important issue on the agenda of ESA in 2008/2009 was the future of its launcher programmes. The Council funded several programmes enabling an independent European access to space: the Ariane 5 post-ECA programme, the Ariane 5 Research and Technology Accompaniment Programme (ARTA) extension for 2011–2013, the Vega Research and Technology Accompaniment Programme (VERTA), the Future Launcher Preparatory Programme (FLPP) and the funding of the GSC. These priorities reflected the current developments in this field, which included the preparation of the GSC for Soyuz launches, the progresses on the Vega small launcher programme and the reflection on the future European launcher policies. In this regard, the publication of a French report on the future of ESP in the area of launchers in May 2009 represented an important milestone.[89]

In terms of scientific activities, ESA had a very successful year 2008/2009. With the attachment of the Columbus module to the ISS, ESA officially became a co-owner of the ISS. The successful ending of the ATV Jules Verne mission in September 2008, the 6 month ISS mission of ESA astronaut Frank de Winne which started in May 2009 and the selection of a new class of six astronauts in May 2009 further testified the leading role of ESA in manned space activities in general and in the use of the ISS in particular.

2009 was also marked by the launch of the ambitious scientific missions Herschel and Planck in May, destined to explore the origins of the universe, and by the launch of the first mission of ESA's Earth Explorers, GOCE (Gravity field and steady-state Ocean Circulation Explorer) in March. The ExoMars mission on its side faced serious budgetary constraints, and ESA explored various options to keep the project on track, including cooperation with Russia or with NASA. In parallel, ESA continued to implement the GMES and Galileo programmes.

ESA's membership also evolved during the period 2008/2009, as the Czech Republic became the organisation's 18th Member State on 8 July 2008. Regarding ESA's infrastructure, a new ESA research centre will be established at the Harwell Science and Innovation Campus in Oxfordshire in the U.K. Furthermore, it was decided at the 2008 ESA Council that a third deep space station will be built. The 35 m antenna will be part of the ESA Tracking (ESTRACK) Network and will be located in Argentina.[90]

2.4.2. European Union

The increasing awareness of the central importance of space for the economic and social growth as well as for the political profile of the EU on the international scene was the motor of the growing EU involvement in space affairs. In 2008/2009, the EU continued to pursue its priorities in space: defining an adequate institutional structure for the ESP, making Europe a global actor in space, implementing the Galileo and GMES programmes, establishing a regulatory framework for electronic communications networks and Mobile Satellite Services (MSS), and strengthening an industrial policy for space. In addition, the 5th Space Council on 26 September 2008 identified four new priorities within the ESP: space and climate change, the contribution of space to the Lisbon strategy, space and security and space exploration.[91] Finally, two other factors had an impact on the decisions and orientation of the ESP in 2008, the Irish rejection of the Lisbon Treaty on 13 June 2008 on the one hand, and the rapid spread of the financial and economic crisis on the other hand.

Article 189 of the Lisbon Treaty gives the ESP a constitutional basis and allows the EU not only to adopt and define a European space programme, but also to implement it.[92] With the Irish rejection of the Lisbon Treaty, the only available framework in the meantime until the entering into force of the Lisbon Treaty was the formalised relationship between ESA and EC. In this regard, governance issues were of crucial importance to define a political guidance for the ESP. At an informal meeting of European ministers responsible for space in Kourou on 20–22 July 2008, a division of labour between the EU and ESA was acknowledged. While ESA has a recognised expertise in scientific and technological matters, the EU should take the lead for the political guidance of the ESP, considering the challenges it faces in this field.[93] The ESP Progress Report released by the EC on 11 September 2008 tackled the governance issue as well. It first stated that the Framework Agreement between ESA and the EC which entered into force in 2004 was extended in 2008 until 2012. It also announced that the European space programme, which represents the practical implementation of the ESP, will be further developed in 2009 through the High Level Space Policy Group (HSPG).[94] Finally, the 5th Space Council recalled the overarching aim of ameliorating the institutional governance and coordination in Europe.

Besides the institutional setup, the EU aimed at giving a political impetus to the ESP, with the stated goal of making Europe a global actor in space. The French Presidency of the Council was particularly active in this regard, and space was identified as one of its top three priorities.[95] The global ambitions of Europe in the space sector were affirmed by the EU Member States at the informal meeting in Kourou, by the Council in the 5th European Space Council resolution and by the

EC, which developed elements for a European Strategy for International Relations in Space in an annex to the ESP progress report.[96] The development and enhancement of cooperation relations with other space faring nations is an element of this global strategy. Europe is also increasingly seeking for strategic independence and initiated a series of concrete steps towards this goal in 2008. At the Ministerial Council, ESA first decided to launch a European Space Situational Awareness (SSA) programme. A second initiative was the creation of an ESA/EC/European Defense Agency (EDA) task force aiming at working towards European strategic independence in the area of critical space technologies.[97]

On the operational level, Europe continued to work towards the implementation of its two flagship programmes Galileo and GMES, despite several financial and organisational obstacles. Following the re-profiling of the Galileo programme in 2007, marked by a switch towards full EU public funding, its implementation continued in 2008 with the creation of a new Galileo Interinstitutional Panel (GIP) at the governance level and the launch of the GIOVE B satellite on the operational side. A regulation on the European satellite navigation programmes EGNOS (European Geostationary Navigation Overlay Service) and Galileo was released on 9 July 2008, tackling all the issues at hand: the different phases of Galileo, funding issues, the compatibility and interoperability of both systems, ownership questions and the public governance of both programmes.[98] The procurement for the first constellation of Galileo satellites was also launched in July 2008 and the first contracts were signed in June 2009 with EADS Astrium and OHB/SSTL. Furthermore, the European Court of Auditors examined the Galileo programme, and concluded that the management of the development and validation phase was inadequately prepared and conceived in a report issued in June 2009.[99] The EC and ESA also conducted separate audits on the cost overruns of the In-Orbit Validation (IOV) phase, concluding that these were justifiable given the contract modifications.[100]

GMES on its side was still facing problems, in particular concerning its financing and institutional setup. At the informal meeting in Kourou, the European ministers responsible for space reaffirmed the need to strengthen the GMES programme, among other by securing its financing. The 5th and 6th Space Council stressed the need for a GMES action plan, including aspects of governance, sustainable funding and data policy. Following the resolution of the 5th Space Council, the EC issued a communication on 12 November 2008, announcing a legislative proposal for a GMES programme that will be user-driven and public-driven.[101] The Competitiveness Council adopted a roadmap towards a GMES Programme in December 2008, seeking for clarification on certain matters, including the mechanisms and rules of the GMES decision process, cost estimates for all GMES components, a strategy of international cooperation and a definition of the role of the entities contributing to GMES.[102] The EC proposal for a regulation on GMES and its initial operations was released on 20 May 2009, with the objective of establishing a legal basis for the GMES programme and EC funding of the initial operations.[103] In the meantime, GMES services in pre-operational mode were launched in September 2008 and the programme is planned to move into operational phase in 2011.

On the regulatory level, two issues were tackled in 2008 and 2009. First, at a Transport, Telecommunication and Energy Council meeting on 27 November 2008, an agreement was reached on the review of the EU's regulatory framework for electronic communications networks and services. The Council adopted a common position on the EC proposal for a directive aiming at improving the effectiveness of the regulatory framework for e-communications.[104] The second issue concerned the selection and authorisation of systems providing MSS. A decision was issued on 30 June 2008, with the goal of providing satellite operators with a single authorisation and therefore to create a single European market in this area.[105] This in turn, should allow providing MSS that can extend to the most remote areas. Four companies entered the competition, and in May 2009, the EC announced that Inmarsat and Solaris Mobile[106] had been selected.

A further priority was the strengthening of the European industrial basis in the space sector and the development of the economic potential of space. In particular, the GMES and Galileo programmes can potentially generate a European downstream market, especially for SMEs. The need to stimulate these commercial developments was highlighted in the 5th Space Council Resolution, the EC ESP Progress Report and in all the EU documents specifically dealing with GMES. The economic and financial crisis acted as a strong incentive for political initiatives in this field. The Brussels European Council identified space technologies and services as one of the main technologies for the future.[107] The 6th Space Council on 29 May 2009 focused on the concrete contribution of space to innovation, competitiveness and economic recovery. It recommended space to be considered in the fund allocation for the EERP, stated that space should be included in the Lead Market Initiative and highlighted the potential of satcom for future growth. More generally, it stressed the need to improve cross-fertilisation of knowledge, innovation and ideas between the space and non-space sectors.[108]

In addition to these activities, the 5th Space Council identified four new priorities for the ESP. It first recognised that space could improve our understanding of climate change and its consequences, and therefore called for a definition of GMES services that would fulfil this task. The Council then recalled that space has been identified as one of the priorities to develop a European knowledge-based society, which is at the core of the Lisbon strategy. It further focused on space exploration, which should be understood both as a political and a global endeavour. Finally, the Council acknowledged the growing security relevance of space, which was also highlighted by a resolution from the European Parliament in July 2008[109] and by the European Council declaration on strengthening the capabilities of ESDP in December 2008.[110] As a consequence, the EU continued to develop space-based assets for CFSP/ESDP in 2008 and 2009, with an emphasis on strengthening the cooperation between civilian and military projects, defining military uses for GMES and Galileo and developing an autonomous SSA capability. In terms of diplomatic initiatives in the field of security, the finalising of an EU draft Code of Conduct (CoC) during the French Presidency was a major topic.[111]

2.4.3. Other European institutions

Three other European institutions were active in space matters in 2008/2009: the Western European Union (WEU), the European Interparliamentary Space Conference (EISC) and the Network of European Regions Using Space Technologies (NEREUS).

Since the Treaty of Amsterdam, WEU's operational capacities were transferred to the EU and the Assembly of the WEU plays the role of the European Security and Defense Assembly. In this capacity, it acts as an important discussion platform for European security-related issues. The Technological and Aerospace Committee is particularly relevant for security-related space activities. At its 55th plenary session on 2–4 December 2008, it adopted a report on the Multinational Space-based Imaging System (MUSIS), recognising the crucial need for Earth observation systems in the new geostrategic environment and calling for an effective harmonisation of future optical and radar observation systems at the European level. It further adopted a report on SSA at its 56th session on 2–4 June 2009, stressing the need for European strategic independence in this field.[112]

EISC gathers members of national Parliaments and serves as a parliamentary discussion platform on ESP-related issues. The 10th EISC took place in Prague on 13-14 October 2008 under Czech Presidency. It put emphasis on the definition of an industrial policy at EU level for space activities, recalling the central role of SMEs in the space field and the need to create opportunities for them. It further called for a European launch service procurement policy and for a strengthened EC/ESA cooperation in the field of critical technologies.[113] In 2009, the U.K. took over the EISC Presidency, under the theme of "problem-solving", to investigate which space applications could help solving concrete problems.

NEREUS aims at enhancing regional cooperation in space matters within the EU. After the preparatory phase and the signature of the political charter on

18 December 2007, the network entered in the operational phase in 2008. A permanent secretariat was set up in Brussels, and the General Assembly which took place on 7 July 2008 marked the official launch of the activities. A first international conference was organised by the Italian region of Basilicata on 30–31 October 2008 on the theme "EO and new technologies for environmental monitoring, assessment and management". Finally, the first General Assembly of NEREUS took place in Brussels on 13 March 2009. Its main objective was to set up working group on five themes: EO, telecommunications, navigation and localisation, education, formation and communication and inter-regional projects.[114]

2.4.4. EUMETSAT

In 2008/2009, EUMETSAT continued to work on the implementation of its operational programmes, mainly focusing on its contribution to GMES and on the Meteosat Third Generation (MTG). In parallel, it further developed its international cooperation ties, in particular with National Oceanic and Atmospheric Administration (NOAA). Finally, its membership evolved, with new Member States and the extension of Cooperating States Agreements.

The 64th EUMETSAT Council took place in Darmstadt on 1–2 July 2008 focusing on GMES and a Jason-2 follow-up mission. It was agreed that EUMETSAT data and products will be made available free of charge to the five core GMES services during the pre-operational period of GMES, running from 2008 to 2013. An approach was also adopted to implement the GMES Sentinel 4 and 5 instruments on the next generation missions of EUMETSAT. As for the follow-up of Jason-2, a preliminary programme proposal was made. The 65th Council took place on 9 October 2008 and dealt with the MTG programme, launching the preliminary design activities. During the 66th Council on 9–10 December 2008, the preliminary programme proposal for the Sentinel-3 Third Party Programme was adopted. At the 67th Council, held on 30 June and 1 July 2009, the Jason-3 altimetry programme and GMES were on the agenda. A draft programme proposal was adopted for Jason-3, stating that the programme should be fully financed by the end of 2009. Concerning GMES, EUMETSAT approved a framework agreement with ESA on the Sentinels 4 and 5.[115]

In 2008/2009, EUMETSAT also continued to strengthen its ties with its international partners. It extended its data exchange agreement with NOAA and signed a cooperation agreement with the China Meteorological Administration (CMA) at the 64th Council. It further endorsed the concept of cooperation with NOAA on a Joint Polar System at the 66th Council, while the cooperation agreement with ISRO (Indian Space Research Organisation) was extended. EUMETSAT is also an important actor of the ESP, due to its contribution to GMES. In this regard, the 6th Space Council invited ESA and the EC to a dialogue with EUMETSAT, in order to explore its role in space observation missions.[116]

EUMETSAT membership also evolved significantly in 2008/2009. Hungary became a full Member State in October 2008, as well as Poland and Latvia in 2009. At the same time, the Cooperating State Agreement with Romania was extended until the end of 2010 while negotiations for full membership began in early 2009. Furthermore, the ratification of the Cooperating State Agreement with Serbia is under way. Finally, the Czech Republic will join EUMETSAT in 2010 after having signed the accession agreement in June 2009.[117]

On the operational level, the Jason-2 altimetry satellite dedicated to ocean topography and launched in June 2008 started to disseminate operational data in December 2008. Similarly, the NOAA-19 satellite with EUMETSAT instruments onboard was launched in February 2009. A launch contract was signed with Arianespace in July 2008 for the launch of the last MSG (Meteosat Second Generation) in 2013. Finally, a new reception station was inaugurated in Ethiopia and will supply the African Union (AU) Commission with real-time data, images and products from EUMETSAT. It is the first station installed in the framework of the African Monitoring of the Environment for Sustainable Development (AMESD) programme.[118]

2.4.5. National governments

While most European countries allocate the main part of their space budget to ESA, the four major European space faring nations, France, Germany, Italy and the U.K. have substantial domestic space activities. In 2008/2009, they continued to actively implement their national space programmes, mainly through their respective space agencies, while pursuing active policies of international cooperation.

2.4.5.1. France

While the French space policy is increasingly tied to the European context, as exemplified during the French EU Council Presidency, France showed new ambitions to be back on the scene as the European leader in space. Besides the development of its projects through Centre National d'Etudes Spatiales (CNES), France worked towards the implementation of the French Space Operation Act in 2008/2009.

One of the most direct translations of France's renewed ambitions in space was the announcement of future increases in the space budget. At the ESA Ministerial Council, France agreed to provide 2.3 billion euros, a 34% increase above its commitments at the last ESA Council. After having been stalled at 585 million euros a year since 2003, France's contribution to ESA will climb again to reach 770 million euros by 2011. At the national level, a higher financial commitment is already written down into CNES's 2009–2011 spending plan.[119] In 2008, France had the largest European space budget, with 1.93 billion euros, including 1.49 billion euros for civilian purposes and 429 million euros for military space.

France's priorities at the ESA Council gave indications on the focuses of its space activities. Launchers remain a core activity for France, as it agreed to spend 506 million euros on the Ariane 5-related programmes and on the GSC. France will also spend almost 400 million euros for the exploitation phase 3 of the manned flight programme, centred on the ATV. It further dedicated 450 million euros for EO, and it will spend 54 million euros to perform Soyuz launches from the GSC.[120]

The French Space Operation Act was adopted on 3 June 2008 and set out the legal framework for French space activities.[121] Its application was subjected to the promulgation of ten decrees by the Council of State, which were passed on 9 June 2009. Another central topic was the status of the GSC. Two acts were passed by the French Senate (in December 2008) and by the French National Assembly (in April 2009) on the GSC, both being ratifications of international texts. The first act ratified the Declaration from certain European governments on the exploitation phase of the Ariane, Vega and Soyuz launchers at the GSC. This text, signed in March 2007, aims at defining a common legal framework for the exploitation of different launchers at the GSC. The second act ratified an amendment to the agreement between the French government and ESA concerning the GSC. Its goal is to adapt the utilisation rules of the GSC between the three main partners (CNES, ESA and the French government) following the adoption of the ESP.[122]

Following the adoption of the French Space Operation Act, CNES became responsible for the technical control of the space security and for safety at the GSC. To avoid conflicts of interest, it had to withdraw its participation from Arianespace. In 2009, the new contract between CNES and the French State will be prepared. It will cover the period 2010–2015 and will be structured along thematic lines. It will finally be more user-oriented, in order to better exploit the benefits space can offer to the scientific community, to enterprises and to citizens.[123]

A central goal of CNES's policy continued to be technology transfer to the private sector. As for the operational priorities, the French space agency concentrated its work on launchers. An integrated CNES/ESA team was created to pilot the activities related to Ariane 5, among others the Ariane 5ME project.

Furthermore, a report on the future of European launchers was presented to the French Prime Minister in May 2009. Other projects under development include the air-launched Perseus microsat launcher and the e-CORCE global imaging constellation. In addition, CNES should also support global initiatives in the field of climate change.[124] It is to be noted however, that financial hurdles limit CNES's ability to conduct missions on its own. In this respect, CNES will put emphasis on contributions to ESA missions in the field of space and Earth science.[125]

As the French national space policy is also considered as a means of foreign policy, France continued to pursue an active cooperation policy in 2008/2009. It focused not only on established space faring nations, such as Russia, the U.S. or Japan, but also on emerging space actors such as China, India or Brazil. In this respect, a new French-Indian agreement on space cooperation was signed in September 2008. Similarly, three new space cooperation programmes were launched with Brazil during the French-Brazilian Summit in December 2008.[126]

Finally, 2008/2009 was also marked by the boosting of French military space. In the White Paper on Defense and National Security published in June 2008, France affirmed its intention to enhance its military space capabilities.[127] Priorities include a missile-warning system, MUSIS, ELINT (Electronic signal Intelligence) and SIGINT (Signals Intelligence) satellites, military satcoms as well as the development of cooperation endeavours at the European level.

2.4.5.2. Germany

In 2008/2009, Germany continued to affirm itself as a leading space actor in Europe. It was the biggest contributor at the ESA Ministerial Council, committing itself to spend 2.7 billion euros. In terms of operational programmes, it focused on radar technologies, optical sensory and robotics. It also remained one of the main supporters for the use of the ISS, while putting emphasis on the economic and industrial benefits derived from space.

In line with its document "Mission Raumfahrt" released in August 2007, the Federal Ministry of Economics and Technology continued to view space technologies as key assets to stimulate growth and competitiveness. In this respect, the German government decided to strengthen the innovative industrial branch of robotics, in which Germany is one of the world leaders. It was decided to expand DLR's Institute for Robotics and Mechatronics to create a centre of excellence for automation, robotics and mechatronics.[128] Similarly, the first national conference on robotics took place in May 2009 in Berlin, with a focus on technology transfers and spill-overs.[129]

The German Aerospace Center (DLR) also continued to be involved in Public Private Partnerships (PPP). The five RapidEye EO satellites launched in August

2008 were developed in the framework of a PPP, and this option was chosen as well to develop the EnMAP EO satellite, to be launched in 2011, the TanDEM-X and the TerraSAR 2 satellites.[130]

At the operational level, the German Galileo Control Centre was inaugurated in September 2008 at the DLR site of Oberpfaffenhofen. When Galileo will be operational, it will be one of the largest Galileo Control Centres in Europe.[131]

Finally, Germany continued to strengthen its international ties with other space faring nations. In particular, a framework cooperation agreement on space science was signed in December 2008 between DLR and the China Manned Space Engineering Office (CMSEO).[132]

Germany's overall space budget for 2008 amounted 1.06 billion euros, encompassing 893 million euros for civilian programmes and 142 million euros for military programmes.

2.4.5.3. Italy

Italy remained a key European space actor in 2008/2009. However, the period was dominated by management changes at the Italian Space Agency (ASI) following national elections in April 2008. As a consequence, new budget orientations and a review of ASI programmes and priorities were initiated.

Enrico Saggese replaced Giovanni Bignami at the head of ASI in July 2008. The new ASI Commissioner will be in charge of working out the next National Space Plan, as well as to define the new nature and structure of the agency. Commissioner Saggese also announced that Italy will dedicate a larger part of its space budget to national activities. In general, the new Italian government intends to reduce public spending and a 20 million euros budget cut for ASI was adopted by the Italian Parliament in August 2008. These decisions will put several ESA projects in a difficult position, such as the ExoMars, GMES, ATV evolution and SSA programmes. Consequences were particular harsh for the two projects were Italy had a leading role: GMES and ExoMars.[133]

On the operational level, EO programmes remained at the core of ASI's activities in 2008/2009. Following the launch of COSMO-Skymed 3 in October 2008, the fourth satellite of the constellation is under construction and will be launched in 2010. In the field of launchers, Italy continued the development of the future European small launcher Vega, although the programme had to face some delays in 2009. Manned space was also an important priority, as ASI contributed to the ESA projects Columbus, ATV, Node-2, Node-3 and Cupola. ASI is also negotiating with NASA to definitely integrate the Multipurpose Logistics Module (MPLM) Donatello to the ISS as an Italian research and technology laboratory.[134] Furthermore, in November 2008 the Italian astronaut Paolo

Nespoli was assigned to mission 26/27 to the ISS scheduled for 2010–2011. In January 2009, astronaut Roberto Vittori started his training in preparation of the Shuttle mission he will take part to in 2010, and two Italians were selected among the new batch of six ESA astronauts.[135]

Italy also pursued an active policy of strategic international partnerships with other space faring nations. It first moved forward with the SIASGE radar satellite system, developed in cooperation with Argentina. The four COSMO-Skymed satellites will be integrated in a system for disaster prevention with the two Argentinean SAOCOM satellites.[136] ASI then signed a letter of intent for further cooperation with the Russian space agency Roscosmos in June 2009. Finally, a five-year bilateral cooperation agreement was signed between ASI and the Israeli Space Agency (ISA) in June 2009.[137]

In 2008, Italy had a budget of 879 million euros. 750 million euros were dedicated to civilian purposes, while 128 million euros were used for military space.

2.4.5.4. The United Kingdom

In 2008/2009, the United Kingdom continued to work amongst others on the issue of space exploration, building on its established strengths in robotic exploration. Indeed, it plays a major role in the ESA programmes Aurora and ExoMars and is planning the Moon Lightweight Interior and Telecom Experiment (MoonLITE) mission. The aim of it is to place a satellite in orbit around the Moon which will deploy four penetrators to deliver scientific instruments below the surface of the Moon. The launch of the mission is scheduled for 2014.

The British National Space Centre (BNSC) started to work on a study on space exploration at the beginning of 2009. It aims at reviewing the options for a British involvement in space exploration programmes. This initiative is in line with the U.K. Civil Space Strategy for 2008–2012, one of its cornerstones being a closer involvement in international initiatives on the future shape of space exploration to the Moon, Mars and beyond. The selection of a Briton in the new group of ESA astronauts could also indicate a shift in the British policy in the field of manned flights. Indeed, the U.K. has no tradition in this field and never participated in the ESA manned programmes.[138]

Further, an expert group composed of industry and government representatives was formed in June 2009 to prepare a report on future challenges and opportunities for the U.K. space industry. This Innovation Growth Team for Space (IGT) is intended to identify future innovation, technology and investment priorities for the sector. The overarching goal is to create a 20 years strategy for British leadership in space.[139]

The BNSC was also marked by a restructuration, as it moved to new offices in January 2009. The goal was to bring all the U.K. civil-space policy makers and funding partners to work together. The new BNSC premises also hosts its closest partners: the Science and Technology Facilities Council (STFC), the Natural Environment Research Council (NERC) and the Technology Strategy Board (TSB). Finally, an important event for the British space policy was the agreement on the establishment of a new ESA facility in the U.K. signed during the ESA Council in October 2008.[140]

The U.K. space budget for 2008 was around 543 million euros, including 341 million euros for civilian activities and 201 million euros for military activities.

2.5. The United States

2008 was a particular period for the United States, as it was an election year. This had important policy and budgetary implications for the federal agencies dealing with space, NASA and NOAA, as well as for the Department of Defence (DoD). At the policy level, no important decision could be made by these institutions in the second half of 2008, as it was likely that the newly elected Administration would set up new strategic orientations. At the budgetary level, this situation resulted in the adoption of a so-called continuing resolution by the House and the Senate in September 2008, keeping the funding level of the federal agencies at their 2008 levels until March 2009.[141] Only the adoption of the so-called Omnibus Appropriations Act, which was signed by President Obama on 11 March 2009, allowed setting up the FY 2009 budget of NASA and NOAA.[142]

The U.S. remained by far the main space power in terms of budget in 2008/2009. Its global space budget for 2008 amounted more than 66 billion U.S. dollars, encompassing 17.3 billion U.S. dollars for NASA and estimated 47.8 billion U.S. dollars for military space.[143] This trend is not likely to change, as the FY 2010 NASA budget amounted 18.68 billion U.S. dollars, a 5% increase from the 2009 budget.[144]

As a whole, 2008/2009 was a transition period for the main U.S. space-related agencies, marked by uncertainty about the programmes and funding levels. This situation was even worsened by the financial and economic crisis, as it further constrained budgets. On the other hand, NASA, NOAA and the DoD also benefited from the stimulus package decided by President Obama to reboost the U.S. economy.

The main issue for U.S. space policy during this period concerned the announced gap between the retirement of the Shuttle and the first launch of the successor system, the Orion capsule. Space security also remained an important topic in 2008/2009 and the FY 2009 budget for the DoD confirmed the ongoing trend of increased defence spending in the U.S.

NASA celebrated its 50th anniversary in 2008, and this year marked an important transition period towards its future programmes. Its administrator Mike Griffin was dismissed by President Obama in January 2009, and former astronaut Charles Bolden was nominated as its successor in May 2009, together with former NASA Associate Administrator Lori Garver as NASA Deputy Administrator. 2008/2009 was marked by strategic discussions regarding the future of manned flights after the retirement of the Shuttle. The crucial strategic asset of independent access to space is at stake in this question, underlying the importance and urgency of an adequate solution. Several issues are interrelated and a balance has to be found between different scenarios. Indeed, a Shuttle retirement in 2010 would imply relying solely on the Russian Soyuz spacecraft to access the ISS, which would have high economic and political costs. On the other hand, a Shuttle extension would mean delaying the Constellation programme and would bear significant safety risks. In addition, the Constellation programme was facing significant delays and funding difficulties. In order to explore in detail the different options and support the decision-making process, several reports were prepared during the period. The NASA first launched an internal report on Shuttle extension options in August 2008, followed in October by a study on possible ways to close the five-year gap between the Shuttle retirement and the operational launch of the Constellation programme.[145] Three other reports issued in 2009, focusing on the Constellation programme, have casted doubt on a 2015 first flight of Ares 1 and Orion.[146] Finally, a special committee chaired by former Lockheed Martin CEO Norman Augustine was appointed in May 2009. Its task is to prepare a broad and comprehensive report on the future of U.S. manned flights, to be delivered in August 2009, on time to translate its recommendations in the FY 2010 NASA budget request.[147]

First decisions have already been taken in 2008/2009 with regards to these questions. Concerning the Shuttle, the decision to retire it by 2010 was endorsed. NASA announced the schedule for the 10 remaining flights in July 2008, while an additional flight was added to the NASA FY 2010 budget by the Congress to deliver the Alpha Magnetic Spectrometer to the ISS.[148] It was also decided that NASA will rely on Russia to ship astronauts to the ISS with the Soyuz spacecraft, but that cargo services to the ISS after 2011 will be assured by U.S. private firms.[149] Finally, NASA postponed work geared towards returning humans to the Moon by 2020 in order to meet the deadline of an early 2015 first flight for Orion and Ares 1 without additional costs.[150] However, other important decisions concerning the future of the U.S. manned programme are still pending, waiting among others for the conclusions of the Augustine panel. Until the completion of the study, Altair lunar lander and Ares 5 cargo launch vehicle design contracts have been put on hold.[151]

At the operational level, NASA continued to work towards the completion of the ISS in 2008/2009, delivering the MPLM Leonardo to the station in November 2008 and the 4th set of solar arrays in March 2009. In line with the Constellation programme, it also concentrated its space exploration activities on the Moon and on Mars. Among others, the U.S. launched two unmanned missions in June 2009 that will prepare the return of humans on the Moon, the Lunar Reconnaissance Orbiter (LRO) and the Lunar CRater Observing and Sensing Satellite (LCROSS). NASA's exploration activities were also increasingly marked by the development of strategic partnerships, in particular with ESA for Mars missions but also more ambitious endeavours to Jupiter and Saturn. Finally, a marking event was the servicing mission to the Hubble telescope in May 2009, which will enable it to continue observations until 2014.[152]

As for military space, two programmatic documents were issued in January 2009. The updated military doctrine on space operations first, entitled "Space Operations", established a framework for the use of space capabilities and the integration of space operations into joint military operations.[153] The second document, the Regulation 900-1, issued by the Department of Army Space Policy, focused on operational aspects of space systems in the Army.[154] Another important event in the area of homeland security was the decision of Department for Homeland Security (DHS) Secretary Janet Napolitano to close the National Applications Office (NAO) in June 2009. This decision was taken due to concerns of civil liberties, as the NAO was intended among others to use spy satellites for domestic law enforcement purpose.[155]

2.6. Russia

In 2008/2009, Russia continued to rebuild its space capabilities with the stated objective to re-establish itself as a leading space actor. After a strong decrease of activities and funding in the 1990s, political and economic support to the Russian space policy started to increase again. In March 2008, a draft space policy for the period until 2020 was released, laying down ambitious policies in the areas of exploration, manned flights and space industry.[156] Throughout 2008 and 2009, Russia continued to implement the priorities identifies in the 2006–2015 Federal Space Programme, with a particular focus on the ISS, launch vehicles and rocket technology. Another striking feature was the revival of Russian exploration endeavours to the Moon and Mars.

A revised version of the Federal Space Programme is currently under review by the Russian government, in order to take into account new priorities. These include the completion of the civilian communication satellite network, comprising 24 Gonets M satellites, the development of the future Russian manned spacecraft, the development of the new generation launch vehicle Rus M and the implementation of Moon missions.[157] Russia also gave a new impetus to the GLONASS programme, as 6 additional satellites were launched in 2008 and the legal framework was precised through the adoption of Federal law on navigation activities in February 2009.[158] Besides the development of the Rus M launcher, dedicated to manned flights from the new Vostochny cosmodrome, Russia is also working on the Angara launcher, which will be Russia's next generation launch vehicle.[159]

Another feature of the new impetus given by the Russian government to its space policy in 2008/2009 is the announced shift in the industrial policy. Indeed, in October 2008, Prime Minister Putin announced heavy investments in the space sector for the period 2009–2011.[160] Similarly, a general reorganisation and rationalisation of the Russian space industrial base was initiated in 2008. The Russian government also launched the development phase of a new cosmodrome in the Far East in 2008. The Vostochny launch site is intended to reduce Russia's dependence from Baikonur, which is located on Kazakh territory.[161]

Finally, Russia continued to pursue an active policy of international cooperation in 2008/2009. In particular, it is engaged in the manned programmes of China and India, as well as in the development of the first South Korean rocket, the KSLV 1 programme. Besides, Russia also intended to strengthen its ties with emerging space faring nations, as it signed agreements on space with Nigeria and Egypt in June 2009.

Russia's space budget in 2008 represented around 1.04 billion euros, of which 943 million euros are used for civilian purposes and 104 million euros for military programmes.[162]

2.7. Japan

After several setbacks in the 1990s and at the beginning of the 2000s, the Japanese space policy is still in a reconstruction and redefinition phase. In 2008/2009, the implementation of the new Space law was still on top of the agenda. In this respect, national security concerns play a growing role in Japan's space policy. In parallel, Japan's ambitions to play a major role in space were testified by its growing involvement in the ISS and by the development of new space transportation systems.

Following the adoption of the Basic Law for Space Activities in May 2008, several steps towards the implementation of the space policy were undertaken. Seiko Noda was nominated as Space Minister, a Bureau of Space was set up within the Cabinet and a follow-up Council composed of Diet members from the main political parties was created to monitor government activities in space. Finally, the Strategic Headquarters for Space Policy (SHSP) was set up on 27 August 2008 with a mandate to formulate a Basic Space Plan.[163] The latter was released on 2 June 2009. National security was in the forefront of the policy, as the Plan opens the door for the development of space-based missile warning systems and other military satellites. In January 2009, a high-level MoD committee already recommended the development of other military space capabilities, such as SIGINT satellites, better reconnaissance satellites or SSA capabilities. The Basic Space Plan also recommends a sharp increase in the space budget, and details Japan's space utilisation until 2014. Besides the missile warning system, a focus will be put on the Japanese GPS augmentation system, the Quasi Zenith Satellite System.[164]

A highlight in 2008/2009 for Japan was the attaching of the JEM (Japanese Experiment Module) on the ISS, also named Kibo. The six components of JEM were attached to the ISS in the course of 2008 and 2009, and JAXA astronaut Wakata commissioned the Japanese module, allowing first experiments to take place in April 2009. In parallel, JAXA and Mitsubishi Heavy Industries (MHI) continued to work on the new heavy launcher H 2B, whose maiden flight is scheduled for September 2009. The new rocket will offer similar services as Ariane 5ECA, and will thus be able to compete with Arianespace and ILS on the commercial launch market. Finally, the HTV (H 2 Transfer Vehicle) is currently under development, paving the way for further steps in the Japanese exploration programme, as a human-rate version of the cargo spacecraft is also under preparation.[165]

Japan's space budget in 2008 was the second largest after the U.S., amounting to 2.1 billion euros. 1.79 billion euros are used for civilian programmes and 330 million euros for military purposes.[166]

2.8. China

China continued to present itself as a rising space-power in 2008/2009, with initiatives in the whole spectrum of space-related activities. In line with the strategic orientations set up in the 2006 issued White Paper on China's space activities, China continued to implement its space policy priorities in 2008/2009. The period was marked in particular with impressive achievements in the manned programme and ambitious developments for the future of the Chinese exploration policy. Finally, China continued to show increased commercial ambitions in the manufacturing and launching sectors.

The main event for China in 2008/2009 was its first Extra-Vehicular Activity (EVA) mission, which took place on 27 September 2008. China has developed a three-phased human spaceflight plan in 1992. Four unmanned missions were conducted during the first phase between 1999 and 2003. China is currently in the second phase consisting of manned flights, and prepares the third phase which foresees the completion of a space station. To support these developments, China is also currently working on the new Long March 5 launch vehicle, which will have performances comparable to Ariane 5. At the same time, the State Council and the Central Military Commission approved the construction of a new space launch centre in the southernmost province of Hainan at the end of 2008.[167]

China also continued to work towards its ambitious goal to send a taikonaut to the Moon by 2020. The country designed a 3-staged Moon programme, to master successfully the orbiting, landing and returning technologies. As part of the first phase, the Chang'e 1 lunar probe impacted the Moon in March 2009, and successfully concluded the mission. Three further Chang'e spacecrafts will be sent to the Moon in the coming years to test critical technologies.[168]

China launched a wide variety of satellites in 2008/2009, thus demonstrating its will to achieve a comprehensive network of space-based assets. In parallel to the modernisation of its meteorological satellites and the rapid developments of its GNSS system Compass/Beidou, China put a particular emphasis on EO satellites. A striking characteristic of most of its satellites is their alleged dual-use nature.

Finally, China continued to use its space activities as a foreign policy tool, with a particular emphasis on Africa, by that being in line with the overall geostrategic focus put on this continent in the recent years. In particular, China is currently building a new satellite for Nigeria, as a replacement for the lost Nigcomsat 1 and negotiated cooperative agreements with Kenya and Namibia for implanting tracking stations on their territory.[169] Similarly, China and Brazil agreed to share the data from the CBERS 02B satellite with African countries, making China an exporter of EO data for the first time.

China's space budget is difficult to assess, as there are no official figures released, but analysts estimated it around 1.43 billion euros in 2008.

2.9. India

India confirmed its status as a space power in 2008/2009. Indeed, a policy shift took place from a "need-based" approach centred on the use of space applications for the benefit of the citizens and to boost the country's development towards an approach more concerned with "high politics" issues. The new priorities laid out by the Indian government, in particular the focus on space exploration and on the commercial aspects of the Indian space programme, confirmed this trend.[170] As a consequence, India pursued two main orientations in 2008/2009: on the one hand the development and launch of operational mission and on the other hand an increase in space exploration activities.

In accordance with these ambitious goals, ISRO witnessed a sharp increase of its budget. In 2009/2010, it reached 924 million U.S. dollars, a 27% increase compared to 2008/2009.[171]

The most important achievement for India in 2008 was the launch of its Moon mission Chandraayan 1 in October. Two follow-up missions to the Moon are already under development. In February 2009, the Indian Plan Commission further adopted the manned flight programme on the basis of a five year plan at an estimated cost of 2.6 billion U.S. dollars. The objective is to send the first Indian astronaut in orbit by 2015. To support this ambitious exploration programme, India is currently developing an advanced version of its Geosynchronous Satellite Launch Vehicle (GSLV), which will also lessen India's dependency on foreign launchers.[172]

India is also becoming a growing actor on the commercial market since its first successful commercial launch in 2007. However, these perspectives were darkened by the in-orbit failure of the W 2M satellite, which was manufactured by ISRO and EADS and launched in December 2008. Finally, India is also increasingly acknowledging the benefits it could reap from a military use of space assets, in particular in light of the tensed security situation with its Pakistani neighbour. In this perspective, it launched a SAR satellite in April 2009, which will be used among others to monitor the Indian-Pakistani boarder.[173]

2.10. Emerging space powers

The recent years were marked by the increased space ambitions of a series of emerging countries. Amongst the marking events of 2008/2009 were the Iranian and North Korean rocket launches, both countries claiming to develop rocket technology for peaceful purposes despite high suspicion about potential military objectives among the international community. In Latin America, Brazil is trying to revitalise its space activities while Venezuela is the new rising space power. The Middle East witnessed a strong growth of commercial operators, while African countries concentrated on the benefits of space applications for development purposes. As a whole, more and more countries are engaging in space activities.

Asia was marked again by a buoyant activity in the space field in 2008/2009. North Korea launched a Taepodong 2 rocket in April 2009, to place a satellite on orbit according to the North Korean government. However, it was suspected that this failed launch was in reality a test for an ICBM capable of reaching the U.S. American coast. It also has to be noted that North Korea signed the Outer Space Treaty and the Moon Treaty on 12 March 2009, emphasising the peaceful use of outer space.[174] South Korea on its side continued the preparations for the maiden flight of its KSLV 1 launcher, while clearly stating its ambitions to become a leading space power in Asia. Other emerging space players in Asia include Thailand, which launched its Thai Earth Observation System (THEO) in October 2008 and designed a master plan for Thailand in space, Vietnam, which launched its first satcom in April 2008 and intends to master technologies for the production and launching of small satellites by 2020 and Indonesia, which is currently building a spaceport on the island of Biak with the help of Russia. Finally Kazakhstan also launched its own space initiatives in 2009, as it announced its intention to develop a national space industry and signed a strategic and technological partnership with Astrium to this end in May 2009.

In Africa, the most important developments are focusing on EO, as this application can support social, developmental, agricultural and environmental policies. The three most active countries in space remained the same in 2008/2009. While China is currently working on a replacement satellite for the lost Nigcomsat 1 satcom, Nigeria is developing its second EO satellite, Nigeriasat 2, which will be launched in 2009. Nigeria also established a National Geospatial Data Infrastructure (NGDI) in 2008, as a key asset for achieving the MDGs.[175] Algeria on its side, began the construction of the Centre for the development of satellites in December 2008. It will be responsible for the development, manufacturing and integration of future Algerian space systems.[176] Finally, South Africa created a South African National Space Agency in December 2008, whose task will be to implement the South African space programme. The South African National Space Policy was officially launched in March 2009, with a focus on capacity-building initiatives, the development of space applications and the fostering of international cooperation.[177] Besides these three actors, other African countries are showing interest for space applications, such as Angola which signed a contract with Russia in June 2009 for an Angolan satcom.

The Middle East was marked by the successful launch of an Iranian satellite in February 2009, which occurred after a failure in August 2008. Turkey on its side is currently developing a 1-m resolution reconnaissance and surveillance satellite system. However, most of the space projects in the Middle East are focusing on commercial activities. The region witnessed a booming of satcom operations with the emergence of new operators, in particular in the United Arab Emirates (UAE).

In South America, Venezuela intends to affirm itself as a growing space actor thanks to its oil revenues. It launched its first satellite in October 2008, the Simon Bolivar or Venesat 1 satcom. The country intends to acquire independent EO capabilities and the Venezuelan space agency launched a study for a national surveillance satellite. Brazil witnessed the harshest cut of its space budget since years in 2008/2009, but also has plans to expand its Alcântara aerospace base. Finally, the Brazilian-Ukrainian launching service company Alcântara Cyclone Space announced its first commercial launch from the Alcântara spaceport scheduled for 2010.

Australia worked towards the setting up of a comprehensive space policy in 2008 and 2009. A report was released in October 2008 by the Australian Senate Standing Committee on Economics, calling for the development of an Australian space policy including national security, economic, environmental and social aspects. Consequently, the Australian government announced new spending for space in May 2009, as 40 million Australian dollars were allocated to a new Australian Space Research Programme and 8.6 million Australian dollars for the establishment of a Space Policy Unit.[178]

3. Worldwide space budgets and revenues

The size of the global space sector can be estimated to total 157.3 billion euros, based on various sources as indicated in the following footnotes. The global space sector is composed of institutional spending and of commercial revenues. The institutional part amounts to 59.3 billion euros,[179] counting public expenditure in both the civilian and military space domain. Commercial space revenues make up 98 billion euros.[180] When comparing these numbers with other estimates, one should bear in mind possible difference in methodology and sources. The floating currency exchange rates should also be figured in.

3.1. Overview of institutional space budgets

The 59.3 billion euros of institutional space spending in 2008 represent a 37.7% share of the total financial space volume. This share is almost the same as the one estimated in the last edition of ESPI's Yearbook. Compared to 2007, institutional spending has remained robust with stable budget lines in the U.S. and Europe and positive budget developments in other world regions.[181]

Military and intelligence space spending constituted 59% of institutional space spending in 2008 worldwide (with 35.23 billion euros). Within the U.S. this share

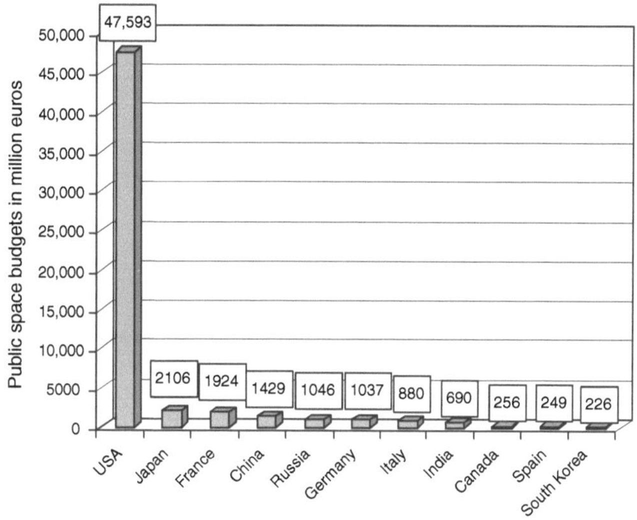

Fig. 3: *Estimate of the major space countries' public space spending in 2008.*

was even 72% (equalling 34.18 billion euros). This major role of military and intelligence space expenditure within governmental budgets is also in line with the previous years. Since not all relevant figures are publicly available, the actual value of the share is most probably even higher.

Civilian institutional space spending in 2008 amounted to 24.11 billion euros, with a fairly uneven distribution across the space faring nations and entities. Again, the U.S., accounted for the lion's share of this figure, spending 13.41 billion euros (equalling 56%). Civilian institutional spending outside the U.S. amounted to 10.69 billion euros.[182]

When looking at the total institutional space budgets of single countries as shown in Figure 3, the overwhelming dominance of U.S. spending becomes apparent. Only six countries worldwide spent more than 1 billion euro on space the U.S. Japan, France, China, Russia and Germany. Japan has overtaken France in comparison to last year. The expenditure of China can only be estimated. In any case, Asian countries figure prominently among the big spenders. On top, one should keep in mind the differing purchase parity in these countries.

It is also interesting to look at the single agencies and their spending in 2008, as illustrated in Figure 4.[183] As in the years before, the U.S. DoD was the largest spender in 2008, showing the strong role of the military in space. NASA is the institution with the second largest space spending and the biggest civilian space agency in the world. The third largest institutional spender is the U.S. National Reconnaissance Office (NRO). The U.S. Missile Defense Agency (MDA) is next.

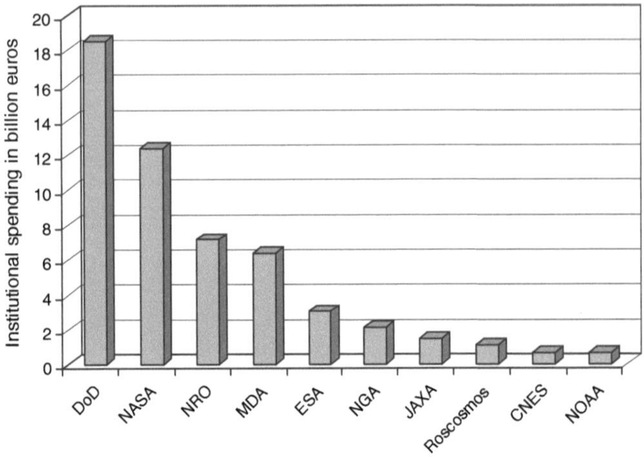

Fig. 4: *Estimate of the major institutions' spending on space in 2008.*

The European Space Agency follows, constituting the world's second biggest civilian space agency. After the U.S. National Geospatial-Intelligence Agency, there are three national agencies – JAXA, Roscosmos and CNES (the latter one's budget is counted without the French contribution to ESA here). All in all and not surprisingly, the dominance of U.S. institutions is apparent.

3.2. Overview of commercial space markets

The total revenues of the commercial space sector in 2008 are estimated to be 98.69 billion euros.[184] Table 1 lists the composition of this figure. Due to different data sources and used by previous editions of the ESPI Yearbook and variable currency

Tab. 1: *Estimated breakdown of global commercial space revenues in 2008.*

Type	Value in billion euros
Satellite services	60
Ground equipment	32.86
Satellite manufacturing	3.71
Launch industry	1.41
Insurance	0.68
Space tourism	0.03
Total	98.69

exchange rates, it is hard to compare the commercial space revenues to last year, but it is safe to assume a growth given the ever increasing demand for space-based applications and services. The lion's share of these commercial revenues was again constituted by satellite services together with ground equipment, as seen in the table. The different segments will be briefly discussed in the following.

Satellite services revenues amounted to 60 billion euros in 2008. This, much like last year, roughly represents a 60% share of total commercial space revenues. Satellite services have been economically driven by the emergence of new satellite operators, especially in developing countries, and booming demand in the Middle East and North Africa.[185] Satellite services comprise Direct Broadcast Services (DBS), Fixed Satellite Services (FSS), Mobile Satellite Services (MSS), and Remote Sensing (that is often included in FSS).

Direct Broadcast Services (DBS) refer to satellite broadcasts for "small dish reception". Direct-to-home (DTH) television is the best known, most established and by far the largest application. Satellite radio broadcasting is still minor but picking up due to the increasing number of car reception devices. DBS revenues in 2008 were 48.07 billion euros, corresponding to about 80% of all satellite service revenues.

Fixed Satellite Services (FSS) refer to satellite services (uplink and downlink) making use of fixed terrestrial terminals or terminals that do not change location very frequently. Sometimes satellite television is considered to be part of FSS. However within this estimate, FSS is understood to only comprise transponder agreements, managed network services and end-user broadband. FSS accounted for 10.36 billion euros of revenues in 2008.

Mobile Satellite Services (MSS) refer to satellite services making use of antennas that can be moved around, for example mounted on a vehicle or carried by a person. Satellite telephony is the best known example. MSS revenues in 2008 were 1.57 billion euros, representing only a small share of satellite service revenues. There is a perspective of growth with MSS in the process of converging with wireless and terrestrial solutions leading to integrated services and products.

Remote Sensing accounted for 0.5 billion euros of commercial space revenues in 2008,[186] again constituting only a minor share of satellite service revenues. However, the sector features a strong growth potential with an increasing number of available images, better quality and easy accessibility via Internet as well as a growing awareness of the value that Earth observation imagery holds for problems like cartography or resource monitoring.

The ground equipment sector comprises both network and consumer equipment. Its revenues grew strongly by 34% in 2008, reaching a total of 32.86 billion euros. This made ground equipment the fastest expanding sector in the space segment, currently accounting for one third of all commercial space revenues. The growth of the sector was led by consumer equipment, with GPS devices representing more than half of the ground equipment revenues.[187]

Satellite manufacturing contributed 3.71 billion euros to the total commercial space revenues. This again counts only the revenue share that originated from commercial customers (versus government and military orders). This commercial share within satellite manufacturing grew from 33% in 2007 to almost 50% in 2008. The total satellite manufacturing revenues (including government and military customers) decreased from 8.29 billion euros in 2007 to 7.5 billion euros in 2008. This largely reflects the reduced number of satellites launched in 2008.[188]

The commercial part of launch industry generated 1.41 billion euros of revenues in 2008, up from 1.1 billion euros in 2007.[189] Share between commercially contracted launches and government procured launches were almost evenly divided.[190] While the number of launches has more or less remained constant, launch costs have grown. This explains the increase in revenue. Regarding the share of commercial launch revenues, Europe leads the pack with 36%, followed by Russia with 29%. Multinational Sea Launch has returned to flight in 2008, holding the third rank with 24% (Figure 5).[191]

Insurance premiums of the space sector in 2008 are estimated to amount to 681 million euros. There was an upward pressure on premiums following several failures in 2007.[192] Although claims exceeded premiums in 2007, the space insurance market is considered healthy. It should be kept in mind that track records are not the primary source of impact on premiums, since the size of the space market is not sufficient to allow for differentiation.[193]

Space tourism revenues in 2008 amounted to 28 million euros, with a prospect of strong future growth.[194] The lion's share of these revenues was generated by Space Adventures, which offers orbital trips. For example, it flew Richard Gariott to the ISS for a one week stay. This visit may have been one of the last, since the ISS crew expanded to six persons in 2009 and is foreseen to continue in this configuration. Accordingly, there is no more room for tourists. As for suborbital

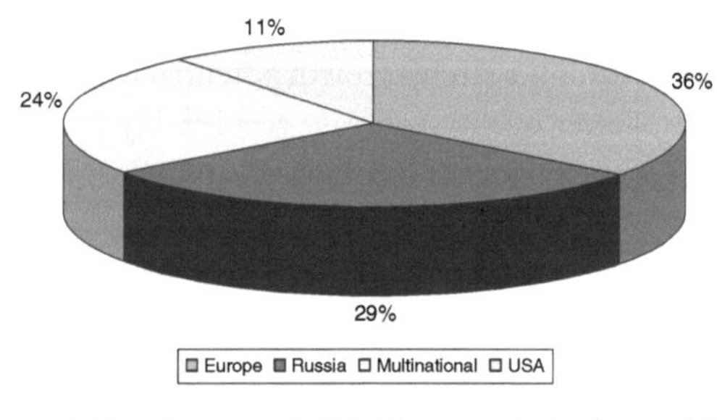

Fig. 5: *Share of commercial launch revenues in 2008 by country/entity (source: FAA).*

trips, Virgin Galactic, XCOR Aerospace and the joint venture between Armadillo Aerospace and Rocket Racing are already in the course of selling future flight opportunities.[195]

3.3. Evolution of the space industry

Space industry is understood in a broad way as the economic sector providing goods and services related to space. This definition, which goes beyond the narrow conception of space industry as encompassing only space hardware providers (launchers and satellites), seems to be more adequate to provide a general overview on the business trends in the space sector.

The most relevant trends for the industrial domain in 2008/2009 include the emergence of new satellite operators, a growing demand for FSS, a relatively steady amount of commercial satellite orders, efforts to cope with the consequences of the financial crisis, a resilient launch sector, evolving MSS (despite regulatory issues slowing down merger processes), a maturing market for remote sensing, and Ka-band internet access picking up momentum. The period was also marked by a series of restructurings.[196] Another notable trend was the booming market for small satellites.[197]

In general, the space industry remains dynamic. It has built up some momentum that is likely to carry it past the coming months despite the general financial crisis. Consolidations, takeovers, mergers, partnerships and alliances serve as a tool to achieve a higher degree of competitiveness on the global market. Key developments for selected geographical regions are highlighted in the following.

3.3.1. Industrial evolutions in Europe

Europe's space sector is consolidating, especially in countries whose space budget is not sufficient to ensure the growth of a space industrial base. Several mergers and takeovers took place both in the manufacturing sector and in the area of space operations and services providers.

As for the manufacturers, the Swiss Firm RUAG acquired Saab Space and Saab-owned Austrian Aerospace in July 2008 for a price of 335 million Swedish krona (56.3 million euros). RUAG wants to make civil and military space markets a strategic priority of the company by 2010, and the acquisition of Saab fits into this strategy. The augmented national constituency of Switzerland, Sweden and Austria will also strengthen RUAG's position vis-à-vis ESA due to the latter's principle of industrial return.[198] Furthermore,

RUAG bought the space division of the Swiss technology company Oerlikon (Oerlikon Space) in June 2009. Even though the amount of the deal was not made public, experts estimated it between 150 and 200 million Swiss franks (100–130 million euros). When the transaction will be completed by mid-2009, it will strengthen RUAG's position on the U.S. and European aerospace industry market.[199]

Also in July 2008, ELTA S.A., a Toulouse-based joint venture between energy conglomerate Areva (France) and satellite system manufacturer OHB Technology (Germany) purchased SMP S.A., a Toulouse-based manufacturer of satellite ground station radio frequency equipment. SMP has been involved in setting up the ground station for the German SAR-Lupe radar reconnaissance project. Financial details of the deal were not disclosed, but an OHB press release stated that the acquisition is completely financed from ELTA's own resources.[200]

In October 2008, Finmeccanica acquired DRS Technologies, a leading U.S. American supplier of integrated defence electronic products, services and support. The deal was worth 5.2 billion U.S. dollars and was aimed at strengthening Finmeccanica's position on the U.S. market, in particular the defence market.[201]

In December 2008, Dassault Aviation bought the 20.8% share that Alcatel-Lucent held in Thales, raising Dassault's share in Thales to 26%. Dassault paid 38 euros per share, for a total of 1.57 billion euros. The French government and EADS are other major shareholders of Thales. The deal is supposed to be finalised in spring 2009, subject to approval by the regulatory authorities. The closer link between Dassault and Thales is in line with the French industrial policy aiming at creating a national champion in the defence sector.[202]

Finally, EADS Astrium acquired a majority stake in SSTL by taking over the 85% SSTL stake held by the University of Surrey and the 10% share held by SpaceX. The deal took place in January 2009, following approval by the European Commission in December 2008. Provisions included in the sale make sure that SSTL remains independent and keeps its brand identity. The total sale price is estimated around 45 million British pounds (60.6 million U.S. dollars). SSTL is specialised in small and low-cost satellites. It has increasingly been competing with Astrium in the past years, which has developed its own product line of small satellites. However, SSTL had difficulties in earning contracts from ESA, with the notable exception of GIOVE A, the test satellite for Galileo. According to the European Commission, the new entity will not be a dominant actor in the Earth observation satellite market.[203]

Mergers and takeovers also occurred in the area of space operations and service providers. In July 2008, SES announced plans to merge its subsidiaries SES New

Skies and SES Americom into a single division in order to unleash synergies. For regulatory reasons, the two subsidiaries will remain the licensees for their particular satellite fleets. SES Americom also considered a possible restructuring of its IP-Prime service, which distributes more than 200 television channels to local telephone networks and telecommunication operators. Eventually, though, the service was shut down in January 2009 due to a lack of consumer uptake, two years after it arrived on the market.[204] Apart from that, SES Astra and Eutelsat Communications founded the joint venture Solaris Mobile in 2008, to operate the S-band payload on the W2A satellite to provide video, radio, data and two-way communications to mobile devices.[205] The joint venture was selected by the EC to provide MSS in Europe (see Sect. 2.4.1).

Also in July 2008, hedge fund investor Harbinger Capital Partners and MSS provider Inmarsat opened discussions on the possibility of Harbinger to purchase Inmarsat. Harbinger, which already holds a 28% share of Inmarsat, did not indicate a potential bidding amount. Instead, the talks focused on regulatory issues, which will be hard to tackle given the fact that U.S. authorities are the most important customer of Inmarsat, which also provides service to U.K. and Australian armed forces. Given the prior need for official U.S. approval, the talks ended in late July. Harbinger then secured the U.S.-based satellite operator SkyTerra/MSV (Mobile Satellite Ventures), with a view to comforting U.S. authorities and preparing the ground for a later Inmarsat takeover.[206]

Another relevant event in July 2008 was the announcement by EADS Astrium that it has purchased the 41% share that CNES held in Spot Image, which brings the Astrium share in Spot Image to 81%. Other shareholders of Spot Image are Telespazio (7.7%), SSC (6.7%) and IGN (2.7%). Spot Image will now work side by side with Infoterra in the Astrium Services' Earth observation division.[207]

In the same month, Intelsat announced the finalisation of its debt restructuring process, which followed the offering by its subsidiary Intelsat Corporation of 658.119 million U.S. dollars of senior notes.[208]

Abertis Telecom raised its stake in Hispasat to 33.4% in October 2008. It bought the 5% share previously held by EADS/CASA for a price of 35 million euros. Abertis already holds a 32% share of Eutelsat, which in turn is Hispasat's second largest stakeholder with an ownership percentage of 27.7%.[209]

In January 2009, the Swedish Space Corporation (SSC) acquired 100% of its long-time joint-venture partner in satellite operations, Universal Space Network (USN). The financial details of the bargain have not been made public, but the deal received all necessary approvals from the U.S. government in May 2009. The two companies had created the PrioraNet in the late 1990s, a global satellite tracking ground station network. U.S. government entities like NASA or the DoD represent 50–60% of USN business.[210]

Finally, the Israeli company Elbit Systems Ltd., acquired more than 90% of the shares of privately-owned Shiron Satellite Communications Ltd., for 16 million U.S. dollars in February 2009. Elbit will use Shiron's satellite broadband services to augment its offer for remote locations to defence customers.[211]

3.3.2. Industrial evolutions in the United States

In the area of manufacturing, mergers and takeovers continued to occur throughout 2008 in the U.S. as well. In July 2008, Balaton Group, a private equity firm from Toronto, acquired mobile satellite antenna manufacturer MotoSAT of Salt Lake City. The amount of the deal was not disclosed.[212] In October 2008, however, MotoSAT issued a statement saying that Balaton "no longer have the support of their financial committee to continue with the offer to purchase MotoSAT" due to the situation of the economy.[213]

Telecom equipment producer Comtech Telecommunications and Radyne merged in August 2008, after Comtech had acquired Radyne in May 2008 for 223.6 million U.S. dollars. Comtech aspires to augment its offer of satellite Earth station products and to integrate Radyne's shared bandwidth satellite networking solution (referred to as SkyWire) through this deal.[214]

In September 2008, Northrop Grumman purchased 3001 International Inc., a geospatial intelligence firm providing products and analysis to various U.S. and other governmental and military entities. 3001 International will become part of Northrop Grumman's Information Technology Sector. Financial details of the deal were not made public. In December 2008, Northrop Grumman merged its Space Technology and Integrated Systems sectors to create a 10 billion U.S. dollars Aerospace Systems division.[215]

In October 2008, Sierra Nevada Corporation (SNC), a high tech electronics, engineering and manufacturing company, acquired small satellite manufacturer SpaceDev. It had already bought another small satellite manufacturer, Microsat, in January 2008. The transaction price was stated as 38 million U.S. dollars in cash. More than 70% of SpaceDev's revenue in 2007 was constituted by U.S. government contracts.[216] Another deal in October 2008 was the 22.7 million U.S. dollar agreement between Globalstar and Ericsson Federal for the latter to supply a core network system for Globalstar's satellite gateway ground stations.[217]

Another notable event was the major investment in SpaceX made by Steve Jurvetson and his DFJ partners in June 2009, as the total sum invested could amount 60 million U.S. dollars.[218]

Mergers and takeover also occurred in the field of space operations and service provisions. In July 2008, Antenna System Provider EMS Technologies purchased Sky Connect for approximately 15.5 million U.S. dollars. Sky Connect operates more than 2000 tracking and voice systems onboard general aviation aircrafts. With this deal, EMS hopes to increase the level of Iridium-based business for civilian and military aviation markets.[219] Furthermore, EMS acquired Satamatics Global Inc., a provider of Inmarsat machine-to-machine services, in December 2008 and for 46 million U.S. dollars.

Also in July 2008, the U.S. Federal Communications Commission (FCC) approved the merger between Sirius Satellite Radio and XM satellite Radio into a new entity, Sirius XM Radio. The regulatory approval took more than 17 months altogether.[220] In the further course of events, Liberty Media granted 530 million U.S. dollars to Sirius XM in February 2009, receiving seats in the Sirius XM board and a 40% share of the company's equity stake in return. The investment followed a Sirius warning that the company might face bankruptcy due to imminent debt repayments.[221]

In September 2009, MSS provider Iridium merged with GHL Acquisition, an affiliate of Greenhill & Co. Inc. to create a new company named Iridium Communications. According to the deal, Greenhill will provide 500 million U.S. dollars in cash to Iridium Holdings. The transaction made Iridium debt-free and enabled the company to start producing IridiumNEXT, a new generation of satellites.[222]

Finally, DBSD North America filed a voluntary petition under Chapter 11 of the Bankruptcy Code in the U.S. in May 2009. The aim of this action is to complete a restructuring of its 750 million U.S. dollars of convertible notes due in August 2009.[223]

2008/2009 was also marked by policy developments relevant to the aerospace sector. The Defense Authorization Act passed the U.S. Senate on 17 September 2008 and the U.S. House of Representatives on 24 September 2008. It was signed into law by the President on 14 October 2008. The bill foresees reviews by the Secretary of Defense "to determine whether there are any security risks associated with participation of covered contractors in certain space activities of the People's Republic of China".[224] This has potential consequences for companies like Thales Alenia Space, producing satellites free from ITAR (International Traffic in Arms Regulations) components that can accordingly be launched on Chinese rockets. In the same line, a coordinated action of the FBI and other U.S. governmental entities to counter illegal aerospace technology transfer and weapons proliferation to Iran and China was launched in November 2008. Around 150 managers, engineers and companies were swept, partly due to claims of benefiting China's space programme and Iran's plans for ballistic missile capacity. Charges included exporting restricted U.S. technology to these countries.[225]

3.3.3. Industrial evolutions in Russia

After hard times in the 1990s, marked by a collapse of the space budget and a sharp decline in the workforce, the Russian space industry had to struggle with large internal reforms. In the recent years, however, the Russian space industry is resurgent.[226] Russia now plans to invest heavily in the space industry. According to a statement made in October 2008 by the Russian Prime Minister Vladimir Putin, Russia foresees allocating 200 billion roubles (7.68 billion euros) from the federal budget to the space industry for the period 2009–2011. Putin also called the development of a national satellite navigation system a priority, other areas of interest being geological research from space, ecological monitoring of farming, forest and water resources.[227]

Russia also decided to create a State corporation for the rocket and space industry in 2009. According to the head of Roscosmos, it would help to create the conditions for a stable economic development of the space industry by uniting and coordinating scientific and production operations.[228] Such considerations show that the Russian aerospace industry undergoes controlled structural changes rather than being subject to market induced transformations, typical for the Western aerospace industry.[229]

In parallel with these developments, Russia strives to limit foreign investments in strategic sectors of its economy, including defence and aerospace. A bill had already been passed by the Duma in May 2008, listing 42 strategic sectors where foreign investments require prior approval by the Russian government. Similarly, companies controlled by foreign governments cannot take control stake in any Russian strategic-sector enterprise.[230]

In the framework of the construction of a new cosmodrome in Vostochny, the Russian government plans to implement an overall shift of its space industry to the Far East. Among others, this involves a design bureau for small satellite development that could be opened in Uglegorsk. Another candidate is the Moscow Aviation Institute (MAI), a leading supplier of graduates to the Russian space industry, which might open a branch in the Amur region. Some rocket producers have also considered the possibility of moving their work to the region, including the Gagarin aviation plant.[231]

Following three failures of the Proton launcher in 2006, 2007 and 2008, its prime contractor Khrunichev launched a quality improvement programme. In parallel, it tried to gain control over the manufacturing cycle, which had been spread over dozens of independent companies. Meanwhile, 37 of those have been consolidated, meaning that Khrunichev is now controlling roughly 80% of the supply chain. This move was in line with the overall goal of implementing the concentration and streamlining programmes of the Russian space industry. From

2009 on, the Russian government will entirely switch to Khrunichev's Breeze M upper stages and stop using the Block DM stage.[232]

3.3.4. Industrial evolutions in Japan

The Japanese aerospace industry is still relatively small in comparison to others. Yearly sales of the space system industry have been low for years, while those of the service industry have been growing. Japanese satellite communication providers have relied on satellites from abroad and the Japanese satellite industry has lost ground following the Japan–U.S. Satellite Procurement Agreement of 1990. JAXA and its predecessor NASDA (National Development Space Agency of Japan) have only been performing satellite development for scientific research and technology demonstration since. All of this has led to calls for increased industrial competitiveness.[233]

The new Basic Space Plan for Japan will come into effect in May 2009. One of its five pillars is the development of industries in the space area. The Basic Space Law, enacted in May 2008, also aims at fostering and energising the industry, strengthening its international competitiveness and technological capabilities. In this respect, JAXA will pursue R&D on fundamental technology and pass new technology to the industry. In the same vein, private investments will be facilitated by tax and finance measures.[234] Another aim of JAXA is to enable spin-in of technologies from the private sector into the space domain, in addition to the traditionally pursued technology spin-off beyond the space industry.[235] In general, the new Basic Space Law ushered a paradigm shift, favouring industry users over technology development for its own sake.[236]

An example of cooperation between JAXA and the industry is constituted by the development of the new H 2B rocket to be launched in 2010. This project is implemented as a PPP between JAXA and MHI, allowing for an effective utilisation of the small allocated budget. The approach also facilitates transfer of publicly developed technology to the private sector, which is an aim of JAXA, as laid down above. MHI had already taken over H 2A launch operations from JAXA to start commercial launch services in 2007.[237]

3.3.5. Industrial evolutions in China

The space industry in China is performing well, building upon stable growth of the economy and increasing wealth. This trend has been supported by the U.S. removing the Great Wall Industry Corp. (CGWIC) from its blacklist of the Office of Foreign Assets Control in June 2008. The removal followed the CGWIC announcement to expand its internal and external control procedures. CGWIC had formerly been accused of providing technology for ballistic missiles to Iran.[238]

Chinese space organisations have undergone structural changes in 2008. The Commission for Science, Technology and Industry (COSTIND), working in matters of national defence, has lost its rank as a ministry. It became the State Defense Science and Technology Bureau (SDSTB), a department of the civil ministry for Industry and Information. The big five companies of the Chinese Military Industrial Complex are China National Nuclear Corp. (CNNC), China Aerospace Corp. (CASC), Aviation Industries of China (AVIC), China State Shipbuilding Corp. (CSSC) and China Ordnance Corp. (Norinco). All of them are now under direct control of the State Council.[239]

Furthermore, a new centre for aerospace industry is being created in Shenzhen. The city will become one of the most important industrial sites for aerospace in China, including research, development, and manufacturing activities. A new satellite manufacturing company, Shenzhen Aerospace Spacesat Co. Ltd., was created there in March 2009. In addition, the research institute of CASTC (China Aerospace Science and Technology Corporation) will move to Shenzhen, along with CASTC's core business of space manufacturing and aerospace technology applications.[240]

3.4. Industrial overview

In order to get a more detailed overview on the main developments of the space industry in 2008, a segmental appraisal will be conducted in the following section. Three areas will be treated: the launch sector, the satellite manufacturing sector and satellite operators. These three segments make up the two main components of the overall space industrial business, namely the launch sector and the satellite industry. The two strands of the business are closely interrelated, as none of these industry branches can prosper without the other. Indeed, satellite manufacturers and satellite operators need a guaranteed and stable access to space, whereas launch providers rely on orders from the satellite industry to sustain their activities.

It is important to clarify some central concepts which will be at the centre of the analysis in the following sections, in particular the notions of commercial launch and commercial payload. Indeed, since commercial space business is growing in significance and progressively replacing the traditional forms of government-operated space activities, it became more difficult to define and interpret what commercial launches and commercial payloads encompass. For the following sections, the definitions that have been worked out by the Federal Aviation

Administration (FAA) will be used.[241] A commercial launch is defined as having one or more of the following characteristics:

a) it is licensed by the FAA,
b) the primary payload's[242] launch was internationally competed, meaning that the launch contract is one in which the launch opportunity was available in principle to any capable launch service provider,
c) the launch is privately financed without government support.

A commercial payload[243] is described as having one or both of the following characteristics:

a) the payload is operated by a private company,
b) the payload is funded by the government, but provides satellite service partially or totally through a private or semi-private company.

3.4.1. Launch sector

Despite its crucial importance for the satellite industry, the launch sector is an enabler rather than a significant economic activity. The revenues it generates are far less important than the ones originating from the satellite manufacturing business and from satellite services.

2008 was again a very active year for the launch sector, with 68 launches conducted by launch providers from Russia, the United States, Europe, China, India, Japan, Ukraine and the multinational consortium Sea Launch. This represented the same figure than in 2007. Two launches failed, one being a competed launch (a Proton rocket carrying an SES satellite in March 2008), and one being a non-competed launch (a test flight of the Falcon 1 launcher).

Fifty-eight percent of the total launches conducted in 2008 were non-competed, representing 40 launches, and 42% of them were competed launches, representing 28 launches. Only four actors performed competed launches, whereas seven actors performed non-competed launches. As a whole, there was a significant increase of competed launches in 2008, compared to 2007. This was mainly due to the return to service of the Sea Launch vehicle Zenit 3SL, which performed 6 launches in 2008 and only one in 2007. After a hard year, GEO (Geostationary Orbit) launches were again the top commercial activity, and the whole space transportation market was largely driven by the demand for GEO satcoms. This is likely to remain the same in the near term. Competed launches were particularly important for Europe and Sea Launch. U.S. launch services in contrast, continued to rely heavily on the governmental market, with only 6 out of 15 U.S. American launches being competed. In Russia as well, the relatively important domestic institutional demand continued to support the launch sector, as 15 out of 24 payloads launched by Russia were of domestic institutional nature. India and Japan focused on non-competed launches. Russian launchers conducted 9 competed launches, followed by U.S. American vehicles and the Sea Launch rocket conducting six competed flights each, and the European Ariane 5 performing five competed launches.

Arianespace was again dominating the market in 2008, as its Ariane 5 vehicle launched more than 37 metric-tons in GTO (Geostationary Transfer Orbit), representing almost 40% of the total commercial mass launched into GTO. It placed 9 payloads into orbit in 5 launches, and with 28 successes in a row now, Ariane 5 confirmed its technical maturity. Arianespace attributed the lack of a 6th launch in 2008 to late-arriving satellites. This illustrates a core feature of the company: with the ability to carry two satellites at a time, the use of Ariane 5 is maximised, but it also makes the company more vulnerable to satellite schedule slips.[244]

Sea Launch launched about 27 metric-tons in GTO (more than 28% of the total commercial mass launched in GTO) in 6 launches. The company also conducted the first Land Launch mission from Baikonur in April 2008. Following a harsh year 2007, marked by the grounding of its Zenit 3SL vehicle, shortages in the supply chain, and a meager near-term launch manifest, the Sea Launch board dismissed its President Robert A. Peckham and replaced him with Chief Financial Officer Kjell Karlsen in July 2008.

International Launch Services (ILS) launched 5 Proton rockets, totaling 23 metric-tons in GTO (around 25% of the total commercial mass launched into GTO in 2008). Following a failure in March 2008, the Proton M vehicle was grounded until August.[245]

As for the U.S. American launch providers, Boeing Launch Services (BLS) conducted 2 launches, placing two satellites on LEO, while Lockheed Martin Commercial Launch Services (LMCLS) launched one 6.5 tons heavy satellite on GTO, representing 7% of the total mass launched in GTO.[246] The two companies traditionally don't compete in the commercial launch market, as their launch vehicles wouldn't be cost-competitive for such launches, and as they can count on a steady governmental demand.[247]

Other launch providers also conducted competed launches in 2008, although not launching mass in GTO. The Ukrainian launch provider ISC Kosmotras, exploiting the Dnepr 1 launcher-based on the technology of the SS 18 ICBM, carried two launches in 2008 from Yasny. AKO Polyot launched three Kosmos 3 M rockets from Baikonur and Kapustin Yar. Orbital Science Corp. performed one competed launch in 2008 using a Pegasus XL launcher. The low-cost start-up

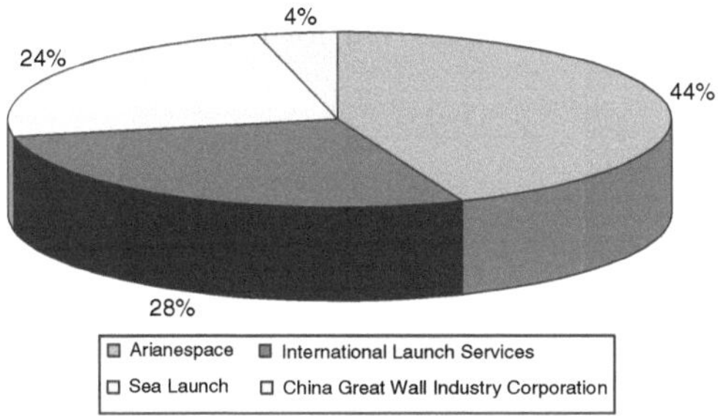

Fig. 6: *Share of launch contracts for GEO satellites in 2008 by launch service provider.*

SpaceX had a turbulent year in 2008: after experiencing three failures, the Falcon 1 vehicle was launched successfully in September 2008.

An estimated 25 contracts were signed in 2008 for geostationary communication satellites. The three main actors in the sector were the same than in 2007, namely Arianespace, ILS and Sea Launch[248] (Figure 6).

Arianespace had a very solid year in terms of contracts signed, as the company won 13 out of 18 contracts open to competition. Among these "Services and Solutions" contracts are 11 GEO satcoms, one EO satellite for Chile (SSOT) and a meteorological satellite for EUMETSAT (MSG 4). Furthermore, Arianespace signed a contract with Roscosmos in September 2008 for 10 Soyuz vehicles, to be launched from the second half of 2009 onwards either from Baikonur or from Kourou.[249] The company also signed a contract with Astrium for the production of 35 Ariane 5ECA in February 2009, worth 4 billion euros. With this contract, Arianespace has a total of 49 Ariane 5 under production.[250]

ILS signed 7 launch contracts for GEO satcoms in 2008. Sea Launch could win 6 contracts, including a five-launch contract with Intelsat, which should give a solid basis to the company in the near-term future. 3 Sea Launches and 4 Land Launches are expected for 2009, even if Land Launch still needs to make its presence felt on the market. Its targeted market segment is the small satcoms market (weighing less than 3000 kg), typically the class of secondary payloads on Ariane 5.[251] Sea Launch also negotiated 9–10 vehicles a year for both Sea Launch and Land Launch operations with its suppliers, for launches starting from 2010 or 2011.[252] However, Sea Launch announced that it went bankrupt in June 2009 and placed itself under the protection of Chapter 11 on bankruptcy. With debts estimated between 500 million and 1 billion U.S. dollars, the launch provider will have to undergo a severe restructuration process.[253]

Among the other actors in the launch sectors, China Great Wall Industry Corporation won a contract for the Pakistani satcom Paksat 1 R, to be launched by the end of 2011 to replace Paksat 1. ULA also signed a commercial contract, although not for a GEO satellites, but for the Italian EO satellite COSMO-Skymed 4.

3.4.2. Satellite manufacturing sector

Satellite services represents the most mature and lucrative market in the space sector. Indeed, space-based communications is the core business for satellite service providers as well as for satellite manufacturers. Therefore, looking at the market share of satellites launched and ordered in a given year is not only a good indication of the vitality of domestic space industries, but it also helps assessing the global trends of the space industry.

One hundred and two payloads were launched in 2008, excluding the failures and the 4 Shuttle flights.[254] Almost 40% of the payloads were commercial. Twenty-two percent of the launched payloads were U.S. American, 21% were manufactured by Russia, 14% by China and 9% by Germany. Aggregated Europe accounted for 21% of the payloads launched.[255]

Ninety satellites were launched in 2008.[256] Europe was the leader with 28 satellites launched (31% of the total satellites launched), followed by the United States (22 satellites representing 24.5% of the total number), Russia (15 satellites accounting for 16.5% of the total number) and China (13 satellites and 14.5% of the total number). Thirty satellites were launched to GEO, 60 satellites into other orbits. Out of 90 satellites launched, 41 were commercial. Eighteen of the commercial satellites were European, representing 44% of the total number of commercial satellites launched, whereas 16 commercial satellites were U.S. American, accounting for 39% of the total number of launched satellites. Five commercial satellites were Russian (12% market share), Japan and Israel had one commercial satellite launched in orbit each (2.5% market share). Twenty-two of the commercial satellites were launched into GEO, 19 into other orbits. When looking at GEO satellites, the European domination is even more evident: 50% of the GEO commercial satellites launched in 2008 were European (6 for EADS Astrium, 5 for Thales Alenia). 36.5% of the GEO satellites launched were U.S. American, and Japan, Russian and Israeli GEO satellites each accounted for 4.5% of the total number of GEO satellites launched.

When looking at the performances by satellite manufacturers, the Russian company ISS Reshetnev launched the most satellites with 12 units, including 6 GLONASS navigation satellites. German satellite manufacturer OHB Systems had a very active year, as 9 of its satellites were launched in orbit. EADS Astrium and Thales Alenia also had a very good year, with respectively 8 and 7 satellites launched. Chinese manufacturers built a total of 13 satellites, but none was a

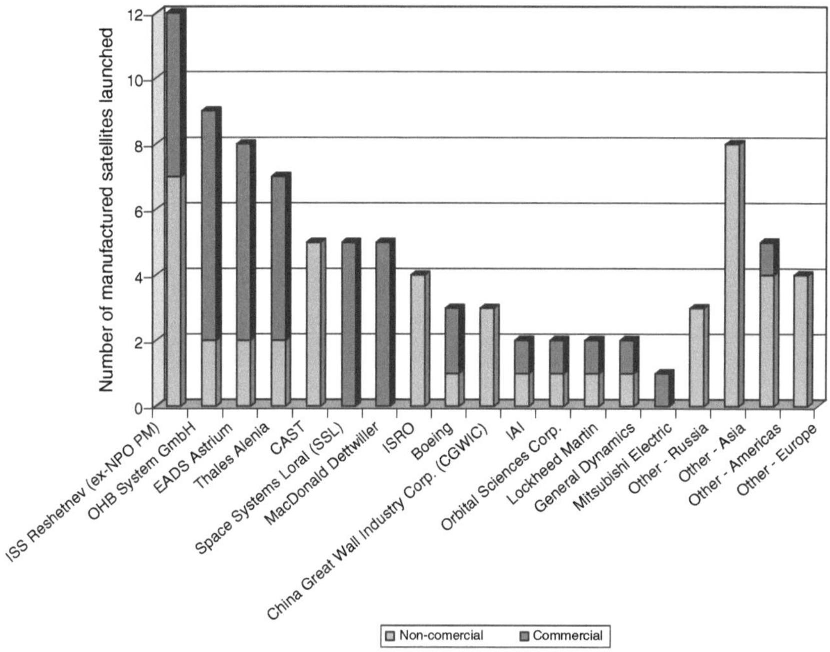

Fig. 7: *Satellites launched in 2008 by manufacturer and commercial status.*[257]

commercial satellite. The two top U.S. satellite manufacturers were Space Systems/Loral (SSL) and MacDonald Dettwiler with 5 satellites each, followed by Boeing, with 3 of its satellites having been launched (Figure 7).

2008 was a solid year again in terms of satellite contracts awarded. Twenty-three commercial GEO satellites were ordered, 3 more spacecrafts than in 2007. Manufacturers from the U. S. won 13 contracts, whereas European manufacturers signed 9 contracts. There was also a satellite order with two co-prime contractors: the Korean Koreasat 6, which will be built by Thales Alenia and Orbital Sciences together (Figure 8). As a whole, 80 GEO satcoms and 48 LEO were on the order lists of the industry at the end of 2008, witnessing the good health of the satcom market.[258] The current situation is marked by overcapacity regarding the launch systems. Indeed, the launch providers are offering 25–29 launch opportunities a year, but the market is limited to some 20 satellites.[259]

Looking at the satellite manufacturers in more detail, EADS won 3 contracts, whereas Thales Alenia had 6 orders, including a co-order with Orbital Sciences. The order for Inmarsat's EuropSat is firm, but still subjected to regulatory approval by the EC for the company's S-Band mobile communication project.

As for the U.S. satellite manufacturers, the striking feature in 2008 was the surprising absence of Boeing, despite favorable U.S. dollar exchange rates

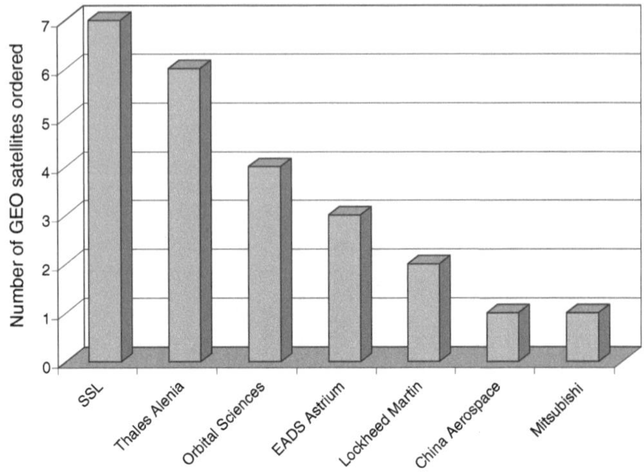

Fig. 8: *GEO satellite orders in 2008 by manufacturer.*

throughout the year. Orbital Sciences won 4 contracts. Loral Space and Communications has injected 350 million U.S. dollars in the past two years into its SSL division. This should enable the company to build 9 large satcoms a year. Loral is the only satellite manufacturer which lives by commercial work alone, and as a consequence, it is also more vulnerable in the commercial market than other manufacturers.

Commercial contracts for other spacecrafts (e.g. non-GEO satellites) were also signed in 2008. Chinese authorities announced that they will replace the Nigerian satellite which experienced an in-orbit failure. Similarly, Khrunichev is building a KazSat 2 satellite for Kazakhstan after the failure of KazSat 1, even though it's unclear if it is a response to the failure or a provision in the initial contract. In the non-GEO segment, Thales Alenia won a contract for 16 LEO satellites dedicated to internet access for poor nations, to be operated by O3B Networks. Thales Alenia is also the only big satellite manufacturer which sells telecommunication payloads independently from its satellite platform business.

With the exception of Loral, no satellite manufacturer relies on commercial orders alone. Institutional orders therefore, constitute an important segment for the satellite manufacturing branch, particularly in the United States. Boeing signed a contract for a 6th Wideband Global satcom for the U.S. Air Force. Lockheed Martin also won a U.S. Air Force contract (GPS3 navigation satellites), as well as a contract for the Geostationary Environmental Satellite series-R. In Europe, EADS Astrium got orders from the Chilean and Spanish governments for Earth Observation satellites and from ESA for the BepiColombo science mission and the Earth Care and Sentinel 2 Earth Observation spacecrafts. Thales

Alenia on its side won the ESA contract for the Sentinel 3 Earth Observation satellite.

In terms of future perspectives, 20–25 annual satellites are expected to be launched in the next few years. 10–12 of these satellites will have a power superior to 8 kW, 4–5 spacecrafts will have a power ranging between 3 and 6 kW, whereas the market of small GEO satcoms has a tendency to shrink. The current "high cycle" of the manufacturing market should last until 2013 and corresponds to the replacement of older satellites. It will be followed by a low cycle. Seventy percent of the expected new orders will concern traditional applications (HDTV, 3DTV and telecom), while two applications are particularly dynamic: mobile and large band.[260]

3.4.3. Satellite operators sector

In 2008, the business of selling commercial transponder capacity remained strong. Most satellite operators had solid performances despite the global economic and financial crisis. The growth was driven among others by satellite TV. New operators continued to enter in the market, therefore testifying the attractiveness of the satcom business.[261] The list of the top 25 FSS operators in 2008 (Table 2)

Tab. 2: *Top 10 FSS operators in 2008.*[262]

Rank	Company	Country	2008 revenues in million $	Satellites in orbit	Satellites on order
1	Intelsat	Bermuda	2360	51	9
2	SES	Luxembourg	2300	41	8
3	Eutelsat	France	1280	27	5
4	Telesat Canada	Canada	581.9	13	1
5	JSAT Corp.	Japan	399	12	3
6	Russian Satellite Communications Co.	Russia	228.1	11	1
7	Hispasat	Spain	193.73	3	3
8	Star One	Brazil	177.76	7	0
9	SingTel Opus	Australia	163.8	6	3
10	Arabsat	Saudi Arabia	157.6	4	4

testifies these trends and also the booming of new markets in Asia and the Middle East. Arabsat from Saudi Arabia entered the top 10, while AsiaSat from Hong Kong and ISRO/Antrix from India progressed in the rating. Three European FSS operators are among the top 10, accounting for 25% of all FSS revenues in 2008 (14.5 billion U.S. dollars/10.36 billion euros). Out of the top 25 FSS operators, 14 are from Asia.

As for the hierarchy of the top 10 FSS operators, Intelsat, SES and Eutelsat remained the leaders, but Intelsat took the first place, which was occupied by SES in 2007. Behind the three major operators, which account for 41% of all FSS revenues in 2008, there is a big gap in terms of revenues and satellite owned.

4. The security dimension

Security-related space activities, and in particular their classical military kernel, form an important subset of general space activities. This chapter lists key features, developments and activities of central players in this domain in order to provide an overview. The information contained is stated as found in publicly accessible sources.

4.1. The global space military context

According to the Stockholm International Peace Research Institute (SIPRI), global defence spending in 2008 rose to 1043 billion euros in 2008, representing an increase of 4% compared to 2007.[263] The U.S. is the biggest spender, followed by China and France. Regional players like Brazil also invest heavily in enhanced military capabilities. It should be kept in mind that spending is not always clearly allocable to defence categories. The standard confidentiality and secrecy along with potential opaqueness can further complicate matters. Floating currency exchange rates, different purchasing powers and work force costs add another degree of ambiguity. Accordingly, numerical budget values should be handled and analysed with care.

In any case, the sheer order of magnitude of global military spending puts the worldwide institutional space expenditures of 59.3 billion euros in 2008 into perspective (see Sect. 3.1). Defence spending has been largely unimpacted by the financial crisis in 2008, although smaller countries might be inclined to cancel or delay procurement efforts, and larger multinational endeavours might face difficulties as well. It should be kept in mind, though, that military expenditures often represent long-standing commitments or are part of a long-term strategy.

The general problem of non-transparency in assessing military spending is even more acute for space-related defence budgets. Indeed, there is rarely a specific budget line dedicated to space in national defence budgets. Moreover, military space applications are often based on spill-over from a wide number of areas, such as R&D or missile technology. Country-specific features make military budgets even harder to assess, as in the particular cases of China and Russia, characterised by opaqueness in military matters. The U.S. as well is marked by a certain amount of secrecy in this field, as there are undisclosed lines in its defence budget.

Despite these uncertainties, it is estimated that the U.S. concentrate 97% (34.18 billion euros) of the worldwide institutional spending in space-related defence. In 2008/2009, it therefore remained the clear leader in military space activities. The next two most active military space actors, Russia and China, are far from catching up with the U.S., both in terms of funding and capabilities. Russia had a very active year 2008 in military space, in line with its will to give a new impetus to its space policy and more generally to raise its international profile as an important power. It is the country with the most complete panoply of military capabilities in space after the U.S. China also continued to build up its military forces in general and to diversify its mission-array in space. It is difficult to asses the military capabilities of the newly launched Chinese spacecrafts though, as they are all officially labelled civilian. In Europe, the main countries pursuing military space activities are France, Germany, the U.K., Italy and Spain, as well as Greece, Belgium and Sweden to a lesser extent. A number of other countries are also investing in space-related security activities, mainly but not exclusively for reconnaissance satellites. These countries include Japan, South Korea, North Korea, Canada, Brazil, India, Iran, Pakistan, Israel, Turkey and Ukraine.

As for the military payloads launched in 2008, there is an ambiguity in clearly defining and identifying them. Three broad categories of spacecrafts can be distinguished. The first and most obvious category encompasses satellites explicitly and specifically dedicated to military use. These include the early warning systems, the military communications satellites, or the technology demonstrators. The second category refers to officially recognised dual-use satellites, such as the navigation or reconnaissance spacecrafts. The last category, also the most debatable, is composed of officially civilian satellites, but which are highly suspected to serve military goals as well. All the Chinese spacecrafts fall into this category.

By adding all the spacecrafts belonging to one of the three categories, a total of 28 military or dual-use payloads were launched in 2008 (Figure 9). In terms of type of spacecraft, reconnaissance satellites (9 launches) and navigation satellites (8 launches) were the most frequently launched. Most of the reconnaissance satellites

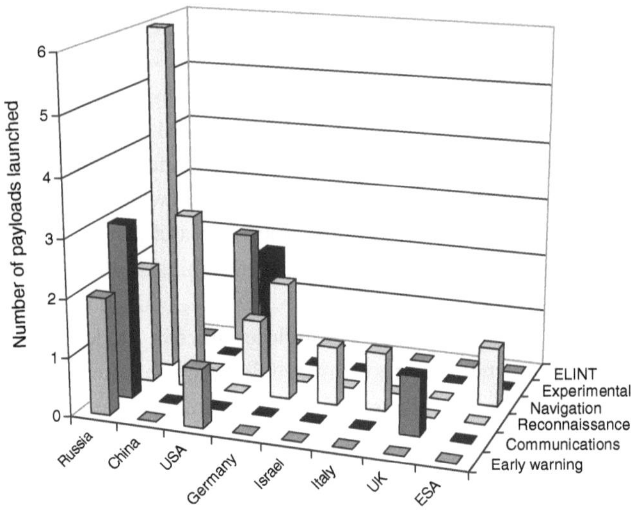

Fig. 9: *Military/dual use spacecrafts launched in 2008 by country/entity and by mission.*

were equipped with Radar (Israeli TecSAR 1, German SAR-Lupe 4 and 5, Italian COSMO-Skymed 3), but there were also optical satellites, in particular the Russian and Chinese spacecrafts. The high number of navigation satellites launched is explained by the fact that Russia intensively worked towards the completion of its GLONASS navigation constellation by launching 6 new navigation satellites. Other types of military spacecrafts launched in 2008 included communications satellites (4), early warning (3), experimental satellites (2) and ELINT (2).

In terms of countries, Russia was the most active actor in 2008, with 13 relevant spacecrafts launched. It was followed by China, with 5 satellites launched and the U.S., with 4 security-related spacecrafts launched. Europe taken as a whole launched 5 military/dual use satellites (one for ESA, the U.K. and Italy, 2 for Germany), while Israel launched one.

4.2. Europe

Overall, Europe has continued to acknowledge and to exploit the link between space and security within the reporting period. This development is flanked by statements at the political level, like the 5th European Space Council that identified space and security as one of the priorities of the European Space Policy. However, only a small number of European countries is pursuing military space activities, European institutions are still at an early stage of involvement, and European spending on military space is considerably lower than in the U.S. Moreover, European actors keep on using civilian space assets for security and military purposes, albeit not to the extent the U.S. does.

4.2.1. National initiatives

The number of European countries actively engaged in military space is very limited. France, Germany, the U.K., Italy and Spain, the main actors, will be discussed in the following. Beyond that, some smaller countries try to access military space systems through cooperation mechanisms. In general, space systems of security or military relevance at the national level are designed and optimised to primarily serve national interests. There are bilateral agreements outside the official European structures to facilitate utilisation of complementary capabilities. However, there is no truly coordinated and coherent European approach to setting up national military space systems in a harmonised way. This is supposed to change with new cooperation models like MUSIS that is described below. Notwithstanding these trends, European States insist that space must not be weaponised (see Sect. 4.2.2).

France as the biggest space player in Europe has put forth its intention to enhance its military space capabilities in the White Paper on Defence and National Security of 2008. This expansion is seen as part of an effort to increase general reconnaissance and intelligence capacities. The Helios 2B optical and infrared satellite had initially been scheduled for launch in late 2008. Following technical problems with the Ariane 5 GS vehicle, the launch was postponed to December 2009. This is not seen as a major problem, though. Given the fact that Helios 1A and 2A work perfectly and the exact nature of the Helios 3 system is not decided upon yet,[264] the delayed launch of Helios 2B can be considered as optimised timing. Helios 2 also involves Belgium, Greece, Italy and Spain. Germany and Italy have certain rights to use Helios in return for offering data from their own space-based surveillance systems (see below).[265] France also pursues its plans for the optical Earth observation system Pleiades. The first satellite is foreseen to be launched in 2010, the second one in 2011.[266]

On 12 February 2009, the military test satellites Spirale A and B were launched from Kourou by an Ariane 5 ECA. They are demonstrators for a distinct French space-based infrared early warning system and they are planned to verify associated technologies and procedures. The early warning system is foreseen to detect ballistic missiles in their boost phase. While EADS Astrium was the system's prime contractor and will be in charge of operation, Spirale A and B were integrated and tested by Alcatel Space. The actual system is planned to be operational by 2019, and France is ready to shoulder the estimated cost of

1 billion euros on its own.[267] France had been active in trying to persuade other European countries to join in the development of a joint missile warning system, but responses have not been encouraging.[268]

Continuing the trend towards utilisation of space systems for security purposes, Germany has proceeded with building up dedicated military capacities in the area of Earth Observation and Communication. It completed its military X-band Radar space-based reconnaissance satellite system SAR-Lupe by launching the last constellation member, SAR-Lupe 5, from Plesetsk by a Kosmos 3 M rocket on 22 July 2008. The satellite, as its predecessors, had been built by OHB. The total system costs are around 900 million euros with an expected lifetime of around 10 years.

The satellite COMSATBw 1 was launched in October 2009. COMSATBw 2 is scheduled for launch in 2010. As dedicated assets, they make up the second stage of the satellite-based communication system SATCOM of the German Armed Forces. Stages 0 and 1 had been virtual ones, consisting of leased satellite capacity. While accommodating different zones of interest, both new satellites will operate in the Ultra High Frequency (UHF) and Super High Frequency (SHF) band. The total system costs are estimated at roughly 950 million euros with an expected lifetime of around 15 years. The satellites are operated by DLR. MilSat Services will furnish additional satellite capacity from Intelsat to support operations.[269]

The United Kingdom has not launched military space assets in the reporting period. Traditionally, it has been using its close relationship with the U.S. to gain access to military space capabilities. However, it is active by itself in the area of satellite communication, as evidenced by the Skynet system. The full Skynet constellation provides X-band and UHF coverage. Skynet 5, representing the latest Skynet stage, consists of three satellites, the last one of which (Skynet 5C) was launched in 2008. There is a purchase option for a fourth satellite Skynet 5D. The three present satellites are military hardened, and they comply with NATO standards. They were designed, built and launched by EADS Astrium. Skynet 5 are the first military spacecraft that are not owned by a government or an international organisation. The system is operated by Paradigm Secure Communications, an EADS Astrium subsidiary, through a Public Financing Initiative (PFI).[270]

Italy launched its COSMO-Skymed3 satellite on 25 October 2008 from Vandenberg aboard a Delta 2 7420. The COSMO-Skymed family in its full constellation will feature four satellites, all of them built by Thales Alenia Space. The fourth satellite is scheduled for launch in 2010. Employing Radar in the X-band, COSMO-Skymed has been set up as dual-use system from the outset.[271] It will deliver services for various fields of applications and it is planned to complement the French Pleiades system within Optical and Radar Federated

Earth Observation (ORFEO) project. The Italian SICRAL 1B satellite for military communication was launched in April 2009.

Spain did not launch military space assets in the reporting period either. It disposes of its own military communication satellite Spainsat operating in the X-band and Ka-band. As a back-up, the Spanish Ministry of defence uses capacities of the X-band satellite XTAR-EUR that is operated by XTAR-LLC. Spain is also pursuing a Spanish National Earth Observation Programme. This programme consists of the Radar satellite Paz for primarily military purposes and the optical satellite Ingenio for primarily civil purposes. For both, Astrium España will be the prime contractor. The two satellites are planned to enter into service in 2012.[272]

As mentioned above, there are various cooperation mechanisms and bilateral agreements at the national level. France and Germany exchange data from their Helios and SAR-Lupe systems. Within ORFEO, France and Italy will do the same for their Pleiades and COSMO-Skymed systems. France and Italy are also in the course of setting up the Athena-Fidus programme for broadband telecommunications services for armed forces and civil protection. It is planned to launch a geostationary satellite in 2013. CNES and ASI are in charge of developing the associated space segment. Later on in the operational phase, the system will also be available to other nations.[273] Another project between France and Italy is the military communication satellite SICRAL 2, that is envisaged to be launched in 2011. Both systems would be primers in being designed and constructed commonly.[274]

Multilateral efforts are going on as well. The MUSIS project joints France, Germany, Italy, Spain, Belgium, Greece and Sweden as an observer country. Its aim is to define a future European system for space-based surveillance and reconnaissance. This comprises an architecture answering the capacity needs in the field of Earth observation as a basis for future optical and Radar satellite programmes within a coordinated and coherent European approach. Furthermore, the project comprises the definition of a multi-sensor user ground segment allowing each participating country to access all satellites of the future system through a unique entry point. Similar coordination efforts are going on for the next generation of military communication satellites to be procured. Moreover, countries form consortia to offer their respective communication satellite capacities to NATO.[275]

4.2.2. European Union level

The European Union's development into a significant space actor is reflected in the domain of space and security. The European Union primarily handles security and defence matters in its second pillar, which is the domain of the Member States.

4.2.2.1. Council/Space Council

At its fifth meeting on 26 September 2008, the Space Council labelled space a crucial asset for CFSP and ESDP. Europe was called upon to develop an autonomous capacity for the monitoring and surveillance of space and space infrastructures. Priorities in the field of security in space were specified as improving the coordination between civilian and military space programmes, developing a comprehensive SSA system and securing Europe's independence regarding critical space technologies and programmes. The area of space exploration, finally, was described as both a political and global endeavour. Europe was summoned to develop the necessary key technologies for space exploration while acknowledging the fact that international cooperation is indispensable in this realm.

In addition, the Space Council commented on the role of the military in Galileo and GMES. It stated that Galileo will be usable by European military forces, although it is not yet clear to what extent. At the same time, it was made clear that military users of the flagship programmes must comply with the principle that the latter are civil systems under civil control. The Council also welcomed ESA's intention to enhance its dealings with EDA.[276]

At its sixth meeting on 29 May 2009, the Space Council recognised the progress with the enhanced coordination of space and security-related activities between key European space policy actors. It also called for rapid implementation of the proposed EU Regulation on the GMES Initial Operations Programme, and it stressed the need for priority funding of the GMES space component.[277]

At its meeting from 8/9 December 2008, the Council of the European Union adopted a draft Code of Conduct for outer space activities.[278] Its main intention, as laid down in article 1, is "to enhance the safety, security and predictability of outer space activities for all". To achieve this goal, general principles were specified in article 2. They included the "freedom of access to, exploration and use of outer space" as well as the right of self-defence. Measures regarding space operations and debris as well as notification, registration, information and consultations were also listed.

At its meeting on 11/12 December 2008, the European Council issued conclusions on the European Security and Defence Policy.[279] The conclusions comprised a declaration on the enhancement of the ESDP. It stated that Europe has established itself as a political player and that threats specified in the European Security Strategy of 2003 still exist, along with new risks likely to endanger the security of the EU. To live up to these challenges, Europe was called upon to improve its civilian and military capabilities. A special focus was set on specialisation, pooling and sharing of resources within the priority areas of planning, crisis management, space and maritime security.

In line with the global framework above, the European Council also issued at declaration on strengthening capabilities in order to meet the associated requirements.[280] As for space, it decided to strengthen information-gathering and space-based intelligence. This includes providing COSMO-Skymed and Helios 2 images to EUSC (European Union Satellite Centre), the letter of intent for a similar agreement for SAR-Lupe signed on 10 November 2008, the preparation of a new generation of satellites for Earth observation within the MUSIS project and accounting for military requirements in space surveillance.

4.2.2.2. European Commission

On 11 September 2008, the European Commission issued the European Space Policy Progress Report. As for security and defence, it stated that Commission and EU Council General Secretariat are currently working on the identification of relevant user requirements for GMES security services, and that both have set up a structured dialogue involving also EDA and EUSC in order to exchange information and to optimise synergies.

4.2.2.3. Agencies

The European Union Satellite Centre (EUSC) is the only operational space entity of the European Union. Its goal is to support decision-making by providing geospatial intelligence data.[281] Over the last year, its tasking orders doubled and the amount of imagery used tripled. In late 2008, it communicated that France, Germany and Italy have agreed to provide the EU with their military Earth observation satellite data. Previously, the EUSC had depended on commercially available imagery data, predominantly from outside Europe.[282]

Throughout 2008, EUSC has continued to support EU missions in the frame of ESDP. This includes EUFOR Chad/RCA, the Atalanta EU naval taskforce in the Gulf of Aden and the EU Monitoring Mission in Georgia. Furthermore, it continued its cooperation with institutions like the United Nations in the framework of the UN mission in the Democratic Republic of Congo (MONUC). EUSC also played an important role in the GMES initiative.[283]

EDA is increasingly involved in space matters. It already has provided the European Commission with military requirements for GMES. It will also provide military requirements for the planned European SSA system in the course of

2009. Besides that, EDA is interested in military satellite communications and satellite data relay. Another intention of EDA is to ensure that the next generation of security-related Earth observation satellites are devised as a network rather than independent national systems as it is the case today.[284] In this regard, a Letter of Intention has been signed by five Member States to express their willingness to cooperate within the MUSIS project, involving EDA.[285]

A workshop on critical space technologies for European strategic non-dependence was held in Brussels on 9 September 2008. It was organised by ESA, EDA and the European Commission and it gathered around a 100 stakeholders from 20 countries. The topic was considered as being of central importance for Europe to stay a global space actor. It was decided to take concrete steps by the three institutions, in coordination with the Member States. This includes the creation of an ESA/EC/EDA task force involving European industry and R&D actors with the goal of developing a list of critical technologies.[286]

4.2.2.4. European Parliament

On 10 July 2008, the European Parliament adopted a resolution on space and security-based on the "Von Wogau-Report".[287] The report stressed the necessity of the European Space Policy to guarantee the autonomy and independence of the EU and it reiterated the need to build up a comprehensive European space-based architecture for security and defence. It also emphasised that Europe should not contribute to the militarisation and weaponisation of space. Furthermore, it called for the development of European capabilities in the fields of autonomous threat assessment, Earth observation and reconnaissance, navigation, positioning and timing, telecommunications, space surveillance, space-based early warning against ballistic missiles, signal intelligence and autonomous access to space. Questions of governance, financing and international cooperation were highlighted as well.

4.2.3. European Space Agency

The European Space Agency is bound by its convention to follow exclusively peaceful purposes. In the course of a wider interpretation of its mandate, ESA has started to embark on activities that can be of military relevance, under the condition that they are not aggressive. Such activities comprise ESA's involvement in the European flagship projects Galileo and GMES, which feature a link to security. Other projects pursued by ESA and relevant for security are GIANUS (Global Integrated Architecture for iNnovative Utilisation of space for Security) and the upcoming European Data Relay Satellite (EDRS).

A major activity in the area of space and security is constituted by the set-up of a European system for SSA. The ESA Ministerial Council of November 2008 decided to start an associated preparatory programme. Its aim is to assist in defining an operational system that is foreseen to link existing and new assets, both at national and European level. Such a system will also be of interest to military and civilian security entities throughout Europe. Their requirements are being consolidated by the European Defence Agency to be integrated into system specification at a later stage.

ESA is also engaging in the space structured dialogue on security and related activities. The 4th Space Council's resolution on the European Space Policy had already called for a structured dialogue between relevant European institutions. The 5th Space Council followed up by adopting the resolution on "Taking forward the European Space Policy". The agenda of the Structured Dialogue meetings includes the EU Code of Conduct, SSA and EDRS. Within the dialogue, ESA contributes to security on Earth as well as to security in space. Among other things, ESA's activities could consist of supporting the identification and federation of user requirements for the associated systems.

4.2.4. Other European institutions

The Assembly of the Western European Union (WEU) has taken on the role of a discussion platform for security-related issues. At its 55th plenary session on 2–4 December 2008, the Assembly adopted the report "A common security and defence strategy for Europe – reply to the annual report of the Council" and a corresponding recommendation "On the revision of the European Security Strategy – reply to the annual report of the Council".[288] There, the Assembly called for speeding up the introduction of an open and competitive European Defence Equipment Market (EDEM). To that end, it demanded to draw up a space operationality concept for the ESDP. According to the recommendation, this concept should be based on space assets and resources available in Europe and it should envisage the development of new capabilities, for example in the area of space-based early-warning systems.

Another space-related report adopted at the same session was entitled "Multinational Space-based Imaging System (MUSIS): European space cooperation for security and defence".[289] The report recalled the Assembly's early awareness of the need for a European space-based observation system and it addressed the lessons learnt from the past, the obstacles to implementation and its further course. It also acknowledged the threat of an arms race for space control. The Assembly stated that EDA has a role to play in the MUSIS project and it recommended that EUSC should get full access to the data it needs. Moreover, it warned that the project must not be delayed by the entry of possible new participants.

4.3. The United States

In 2008, the United States were not particularly active in terms of military launches, as they accounted for only 14% of the total military satellites launched. However, it was an important year for U.S. military space, as a series of important developments, taking into account new threats and new needs, were on top of the agenda. As for the launcher policy, the main priority remained the guaranteed access to space for government payloads, but there was a growing trend towards low-cost and responsive launch solutions. New satellite systems were also developed, in the traditional areas of communications, navigation or early warning, while the deployment of nanosatellites was considered for various missions. The inclusion of the cyber-dimension into military space was another important development, while the topic of export control continued to fuel the debate. Finally, the U.S. engaged in several cooperation or collaboration endeavours in areas linked with space security.

Concerning the launch systems, the U.S. Air Force unveiled an Evolved Expandable Launch Vehicle (EELV) New Entrant Evaluation Plan in February 2009. The document contained a set of requirements for companies seeking to compete for government launch contracts. To fulfil these conditions, the full spectrum of launch needs should be made available. The proposed launcher should be able to put medium, intermediary and heavy payloads into GTO. Currently, the EELV launches are performed by United Launch Alliances (ULA), using the Delta 4 and Atlas 5 rocket families. ULA has 11 EELV launches planned for 2009. SpaceX is the only launch provider that could match the requirements of the plan, even though it only tested its Falcon 1 launcher so far.[290] In general, the Air Force Space Command hopes to reduce the launching costs in the next decade by implementing new technologies and methods. In particular, GPS tracks for rockets after their launch will replace the current ground-based sensors by 2011, and automated destruction capabilities for vehicles which veer off course will be introduced by 2018.[291]

In a move to both reduce costs and to offer responsive capabilities, the DoD conducted research in the field of small satellites and plug-and-play satellite technologies. The U.S. Army studied the potential of a constellation of nanosatellites for imagery dubbed the Kestrel Eye concept. A study was conducted in the second half of 2008, but no funding for the development of such a programme has been planned so far. Still, much of the initial design work has already been done by the Defense Advanced Research Projects Agency (DARPA). As a consequence, the Army Space and Missile Defense Command (SMDC) could begin fielding such satellites within two years. Each spacecraft would cost around 710,000 euros, would weigh 10 kg and have a 1-m resolution. They could be used on a lower orbit as expendable assets as well, for punctual missions requiring a better resolution. The SMDC is also developing the Operational Nanosatellite Effect, a swarm of small communications satellites. As a whole, such small spacecrafts could help stimulating the market for new, low-cost and quick-reaction launchers.[292] Plug-and-play satellite technologies then, have been developed by the U.S. Air Force Research Laboratory (AFRL). This technology will help to rapidly develop, build and launch spacecrafts to augment or replace existing assets. Indeed, it will be able to change or add components on a spacebus shortly before launch. The concept was already tested in laboratory, but the TacSat 5 satellite will demonstrate the technology in space for the first time.[293]

The specific question of export control was again emphasised when the National Research Council published a report on this topic in January 2009. It called for changes in the U.S. export control rules, as they are harming the U.S. economic competitiveness. As these rules hurt the space-industrial basis, they are also threatening national security in the long run. U.S. export control rules were set up during the Cold War to prevent the dissemination of U.S. science and industrial innovations. As the world has changed, a reform has become vital. The report recommended an annual review of items placed on the export control list, as well as the formation of two independent panels, one to process export control applications, the other one to handle appeals.[294]

Protection against cyber-attacks was another rising concern, as the vulnerability of mission-critical systems to such threats became more obvious. Space assets, in particular, could be primary objectives for cyber-attacks, with potentially disastrous consequences given the dependence of ground troops on space systems. The U.S. Strategic Command was therefore seeking for people able to counter and launch cyber-attacks, while the Pentagon is currently creating a military cyberspace command. Such a structure should be able to conduct both defensive and offensive duties in cyberspace, but its structures and attributions are still unclear.[295]

In various areas, the United States engaged in international cooperation or collaboration to serve its interests in military space. This was the case in three specific areas in particular: for the development of concrete military programmes, on space surveillance and on Radar imaging satellites.

The DoD Operationally Responsive Space (ORS) office first called for more collaboration between the U.S. and its allies in military space programmes. The imitation of what is already being done in civilian space programmes (NASA partnerships with other agencies) and on programmes for military aircrafts (F-16 or F-35) could bring economic and diplomatic benefits for the U.S. The ORS is therefore currently developing a strategy for international partnerships. Since its foundation in 2007, it has already worked with Canada on its Radarsat 2 satellite and with the U.K. on its TopSat imaging satellite.[296]

The USAF announced its intention to widen access to its space surveillance data. The necessity of such a move was highlighted by the collision between two satellites which took place in February 2009. The USAF agreed to provide access to its data catalogue for commercial operators and other governments, and a new policy in this regard should be announced in June 2009. So far, the satellite operators have access to the data, but the system is not working efficiently enough to drastically reduce collision risks. The dilemma for the USAF is to help coordinating a global space traffic management while being reluctant to disclose sensitive military information.[297]

The DoD also revealed its plans to use international Radar satellites to satisfy its persistent needs in this field. The Pentagon considered purchasing Radar satellite imagery from international and commercial systems currently in orbit. These would include the Canadian Radarsat 2, the German SAR-Lupe constellation and the Israeli TecSAR satellite. Funding was requested for this purpose, and data acquisition could start in 2009. The DoD so far failed to develop its own space-based radar system, and in 2008, a joint project of the USAF and NRO for a constellation of radar satellites was abandoned.[298]

Another issue of concern were the disagreements which often occur between military and intelligence space users for operational requirements and utilisation of space-based assets. In this regard, a report published in August 2008 recommended merging classified and unclassified national security space capabilities and procurements to create a single authority, the National Security Space Authority (NSSA). This structure would take over the activities currently pursued by the NRO, the Air Force's Space and Missile Systems Center (SMC) and the Air Force Research Laboratory's Space Vehicle directorate. The National Space Council would serve as the highest authority which could take a decision in case of disagreement between the military and intelligence users.[299]

As a whole, 2009 will be an important year, as the DoD and the USAF will have to make major decisions on the future of space capabilities. In particular, the purchase of major satellite programmes, the updates of the GPS constellation and the development of new infrared systems will be on top of the agenda. In the field of telecommunications, the DoD will purchase the WGS 7, 8 and 9 satellites, while the WGS 2 and 3 satellites will be launched. A decision on the long-delayed Transformation Communications Satellite Systems (TSAT) is also due. This satellite network will provide the USAF with next-generation data communications via laser, and the launch is scheduled for 2018. As for navigation, the Navstar 2RM satellite was launched in March 2009. A spy satellite probably able to perform both SIGINT and ELINT missions, NRO 26, was also launched in January. Other planned launches for 2009 include the military meteorological satellite DMSP 5D3, the TacSat 1, 3 and 4 small research satellites for modular buses, the Space Tracking Surveillance System (STSS) 1 and 2 for early warning and the Space Based Surveillance System (SBSS 1) able to detect and track space objects. In comparison with 2008, where only 4 military satellites were launched, 2009 will be a very active year for the United States military space community.

4.4. Russia

Although the budget for defence and other military tasks has constantly been increased over the past years, due to the financial crisis and its consequences for Russia's economic growth, Russia has announced that it will cut its defence budget by roughly 15% from about 28.6 billion euros to approximately 24.3 billion euros in 2009.[300] This will probably affect spending on military space activities, too. Still, military usage of space is considered a priority. Overall, Russia is still following the guidelines from the long-standing official Russian Federal Space Programme (2006–2015) and the Global Navigation System Federal Programme.

The Georgian war in 2008 revealed some shortcomings in Russian military space capabilities. The satellite based GLONASS navigation system still suffers from the consequences of insufficient financial and technical care in the period of the late Soviet Union. Satellite imaging is another area of weakness, for example in regard to timeliness.[301]

Accordingly, Russia keeps up its high rate of launching satellites. In 2008, it launched 13 military satellites (6 GLONASS, 2 early warning, 2 optical reconnaissance and 3 communication satellites), representing a 46.6% share of all military launches worldwide. Latest activities include the launch of the military communication and relay satellite Raduga 18 on 28 February 2009 from Baikonur. Furthermore, GLONASS is constantly being completed. GLONASS in its final configuration will consist of 30 satellites. In 2010, Russia will start launching the GLONASS K satellites.

According to official claims, the destruction of satellites by China in 2007 and by the U.S. in 2008 has forced Russia to develop anti-satellite weapons on its own. Furthermore, Russia has announced to build up an air and space defence system until 2020, as a consequence of the military conflict with Georgia in summer 2008 and as a reaction to the former U.S. government's plans to set up a missile defence shield in Eastern Europe.[302] On top, Russia is pursuing plans for a missile warning system in space.[303]

On 10 February 2009 a Russian military communication satellite, Kosmos 2251, Type Strela 2 M, collided with an U.S. Iridium communication satellite. The Russian satellite had been out of service since 1995, two years after its launch. Russia did not comment on claims that the satellite was out of control. The incident has caused a lot of space debris, also including bigger fragments which might imperil other satellites or space operations.

4.5. Japan

The Basic Law of Outer Space, endorsed in May 2008 by the Japanese Diet, was enacted on 27 August 2008. Since then, two new bodies have been created to deal with the topic. The Ministry of Defence (MoD) has established the Committee on the Promotion of Outer Space Development and Use. The Committee is chaired by the Senior vice-minister of defence. It held its first session on 11 September 2008. Another cabinet-level body set up in parallel is the Strategic Headquarters for Space Policy (SHSP). It is presided by the Prime Minister himself, and for the first time, it met on 12 September 2008.[304] The Minister of Defence is also member of this Headquarter. Both institutions are formulating their guidelines, the former from the MoD's point of view, the latter compiling the opinion of the Japanese Government. The MoD has presented its guidelines on 15 January 2009. The focus is on a broad range of space-based capabilities like missile warning, ground surveillance and navigation. Although the report is not very detailed, it states a number of space capabilities Japan is aiming for.

These capabilities include more and higher-resolution imaging satellites to complement the nation's existing fleet of four Information Gathering Satellites, a dedicated military communications satellite, a missile warning satellite – or a missile warning payload hosted aboard another satellite – to support the nation's Aegis Ballistic Missile Defence system, the development of small, low-cost satellites and rockets that can be launched on short-notice, a signals intelligence satellite, an independent navigation and positioning capability and satellite protection and space situational awareness capabilities. The Japanese government could also launch an early warning satellite that can detect missile launches abroad by 2013. Moreover, it wants to increase the number of reconnaissance satellites from the current number of three to four.

The SHSP has published its first draft of the five-year space policy plan on 27 April 2009. After incorporating public opinion, the five-year space policy plan will be finalised at the end of May 2009. As the next step, it is planned to review the existing National Defence Programme Guidelines (NDPG) against the background of both Basic Plan and Guidelines and to then set up a new five-year Mid-Term Defence Build-Up Programme, which is slated for completion in December 2009. Future military space activities of Japan might include the launch of two more Earth observation satellites – one optical, one Radar – as part of the IGS (Integrated Geo Systems) along with infrared missile warning and signals intelligence satellites around 2016.[305]

The SHSP policy plan urges the nation to actively utilise space in various areas including national security, industrial promotion and people's lives. The draft points out that space is taking on a greater role in the security field, and satellites necessary for defence will be discussed while forming the government's new defence programme outline.[306] The budget for the new objectives is not specified yet. It will be subject to ongoing discussions with the Ministry of Finance. This will be a crucial point, as discussions on introducing an early warning satellite failed to make progress due to the huge cost of some 500 billion yen (3.7 billion euros) per satellite. In 2008, Japan launched no military satellite.

4.6. China

China's officially communicated policy is that it explores and develops technology in outer space for exclusively peaceful purposes. China maintains that outer space should be free of weapons and that preventing an arms race in outer space complies with the interest of all States. It also states its willingness to cooperate with other countries in this regard.[307] However, some of the activities pursued in space feature a strong dual-use, if not a primarily military potential. China increased its defence spending threefold in real terms during the past decade. On the other hand, the military budget is still estimated to represent a moderate 2.1% share of the Chinese GDP. In 2008, China launched five satellites that can be considered as military, although they are officially designated as civilian. It is suspected that the three EO satellites launched by China (Shijan 3, Yaogan 4 and 5) are dual-use satellites, while two science satellites (Shijan 6 3A and 6 3B) would perform ELINT missions.[308] Together, the five assets constitute 17.9% of the worldwide military launches in 2008.

In March 2009, China announced plans for a space laboratory station of its own, called Tiangong 1. The launch of this station is planned for late 2010. The project is being led by the General Armaments Department (GAD) of the People's Liberation Army (PLA). China acknowledges that the Tiangong space station will also involve military space activity and technology development. The design, as shown on Chinese television, includes a large module with a docking system and a service module section with solar arrays and propellant tanks. The concept features similarities to man-rated concepts for Europe's Automated Transfer Vehicle.[309] The module will mainly operate autonomously, and it will be visited by Chinese taikonauts periodically, possibly to retrieve sensor data or reconnaissance imagery.

One can argue, on the other hand, that the manned space station is not primarily military and that its defence-related utilisation is just one element along with scientific, engineering and commercial tasks as with comparable missions of the U.S. or Russia. Specifically, the fact that the project is handled by the military does not necessarily imply a predominantly military nature, and the deployment of military astronauts is commonplace with other space actors as well.[310]

4.7. India

In June 2008, the Indian Minister of Defence announced a raise of its military budget by 50% to almost 28 billion euros making military expenditure 3% of the annual GDP. The budget is needed to modernise the Indian Armed Forces, to purchase new military equipment and for national security purposes.[311] On 22 October 2008, India launched its first unmanned Moon mission Chandrayaan 1. One mission objective is to look for Helium-3, an isotope thought to be valuable for use in nuclear fusion. It is rare on Earth, but could be more common on the Moon.[312]

India has already launched satellites for countries around the globe and it is developing a new launcher for heavier payloads. These space activities are to a certain degree founded in conflicts with China and Pakistan and have a military aspect. India wants to keep up with other Asian countries in a so called "friendly contest" that has serious technological and military implications. China's ASAT test in 2007 had raised fears about Chinese military ambitions in space. According to high ranking officials, India needs to optimise the use of space for military applications to counter China's progress in this realm. The optimisation includes space-based applications like surveillance, intelligence, communications, navigation and precision guidance. The alleged handover of Cartosat 2A , carrying an Israeli-made SAR instrument, to the Indian military fits into this picture.[313]

The Indian Army is in the process of expanding its knowledge about space applications and of enhancing the efficient use of space-based capabilities. To this end, a Space vision 2020 is being prepared. Furthermore, India has set up a special cell to counter threats to space assets. The cell functions under the auspices of the Integrated Defence Services Head Quarters, and it will act as a single window for integration among the armed forces, the Department of Space and ISRO.[314] It could also be an element on the way to setting up a dedicated aerospace command.

4.8. Other selected space actors

Apart from the more traditional actors in the domain of military space, there are also countries whose corresponding endeavours are still in the process of development. Australia is a marked example that will be treated in more detail here. Australia officially owns 10 satellites with none of them being military. There is one defence-owned payload on the SingTel/Optus C1 satellite.[315] Consequently, Australia totally depends on external satellites for its defence and other security-related space applications. More Australian involvement in space technologies, even if aimed at civil purposes, will probably lead to more military or dual use applications, too. In fact, most of the recent Australian space activities have been carried out by the Defence Science and Technology Organisation (DSTO). DSTO has long been involved in applied defence space support research and technology innovation in the operation of systems to access and exploit satellite communications, remote sensing and position, navigation and timing products.[316]

Beyond that, a U.S. study recently stated that the military use of space power is ideally suited to Australia's geopolitical environment and showed that the Australian Defence Force (ADF) can use space assets to effectively increase the utility of existing terrestrial based forces. It says that Australia has been slow to accept the military use of space and its untapped potential. Accordingly, the future challenge for the ADF is to formulate a viable military space policy and doctrine that provides a vision for the future employment, organisation and integration of space assets into the ADF in order to provide the optimum defence structure for Australia's continued security.[317]

Australian activities in the area of military space include an MoU that the Department of Defence has signed with the U.S. Department of Defence in 2008. The value of the MoU is 662 million euros, and it refers to Australia's contribution to the U.S. Wideband Global Satcom system. The U.S. is in charge of five WGS (Wideband Global Satcom) satellites. The Australian Department of Defence is funding the capital cost of the sixth WGS satellite, the costs of launching the satellite and the incremental costs associated with managing the construction of the satellite, the launch services and the operation of the satellite. Nevertheless, it will be a U.S. manufactured and U.S. launched satellite, and it will be controlled from U.S. facilities.

Moreover, in December 2008 the Defence Material Organisation (DMO) has signed a contract with BAE Systems Australia Ltd. for the joint supply of five maritime satellite communications terminals. The maritime satellite communications terminals, known as MASTIS, provide a major enhancement to the Royal Australian Navy's operations and life at sea through the delivery of vastly increased bandwidth. Furthermore, the revitalisation of the Woomera Rocket Range, which was very intensely used in the 1960s, is still a topic under discussion.[318]

Another actor shaping its posture in the military space domain is NATO. It obviously relies heavily on space and its applications, but until recently, it had no dedicated policy document to govern the overall approach. Instead, space is fragmented into narrow functional areas, there is partial duplication of systems and capabilities and a general lack of interoperability. A first step towards more coherence had been taken by the Joint Air Power Competence Center (JAPCC) through releasing the "NATO Space Operations Assessment" in 2008. One of the recommendations therein was to develop a NATO space policy as a foundation. This policy was suggested to be followed by a strategy, a doctrine and appropriate guidance.[319]

A publicly releasable revision to the Assessment was issued in January 2009, calling again for a holistic approach to space matters. It clarified and expanded several sections. Among others, it featured new annexes that covered the military applications of space. The most significant update, though, was the recommendation to establish a Space Office at NATO Headquarters. This Space Office was suggested to consist of eight persons, civilian and military, in two branches. One branch was envisaged to be in charge of planning, programming and architecture, while the other one was foreseen to focus on operations and the integration of capabilities. The NATO Space Office was also seen as a communication port for interaction with other organisations like ESA or the EU.[320]

4.9. Threats to the space environment

Space assets face various threats, both natural and man-made. Naturally caused threats include adverse space-weather effects. Man-made threats, like jamming or debris creation, can be accidental or deliberate. In the domain of man-made threats, several events and incidents occurred.

There were reports that the London-based Al-Hiwar satellite TV channel fell victim to jamming efforts in May 2009. The channel is transmitted by the Hotbird satellite. Supposedly, jamming took place during a special feature on "Democratic change and scenarios of change in Egypt" from the Arab Cultural Centre in London. Suspension of the transmission via Hotbird also led to suspension on the Arabsat and the Atlantic bird satellites as well as on the internet. No specific party was accused of undertaking the jamming endeavour.[321]

On 10 February 2009, as a first event of its kind, an accidental collision between two intact spacecraft occurred. The U.S. communication satellite Iridium 33 and the Russian decommissioned communication satellite Kosmos 2251 collided at a velocity of more than 11 km per second with almost perpendicular orbit planes. The event created numerous debris objects. By March 2009, 823 pieces of larger debris have been identified and catalogued by the U.S. Space Surveillance Network (SSN).[322]

Thus, the event has caused a significant increase in the amount of space debris. The debris population had already been enhanced dramatically by the destruction of the meteorological satellite Fengyun 1C by China in early 2007. Twenty-five percent of all catalogued objects in LEO stem from the event. In total, more than 2300 fragments of a diameter larger than 5 cm have been catalogued by SSN as a consequence, and the number of pieces larger than 1 cm is estimated to exceed 150,000.[323] Table 3 lists catalogued orbital debris per country as of April 2009.

Tab. 3: *Orbital debris per major space country as of 1 April 2009, catalogued by the U.S. Space Surveillance Network (source: NASA).*

Country/ organisation	Payloads	Rocket bodies and debris	Total
China	77	2855	2932
CIS	1381	3637	5018
ESA	39	36	75
France	49	410	459
India	36	115	151
Japan	114	71	185
USA	1101	3449	4550
Other	429	98	527
Total	3226	10,671	13,897

[1] International Monetary Fund. World Economic Outlook. April 2009. Crisis and Recovery. Washington DC: International Monetary Fund, 2009.
[2] Ibid.
[3] These were the central banks from the U.S., Great Britain, Canada, Sweden, Switzerland and the European Central Bank (ECB).
[4] "Declaration Summit on Financial Markets and the World Economy." 15 Nov. 2008. G20. 31 July 2009. http://www.g20.org/Documents/g20_summit_declaration.pdf.
[5] "Statement of G7 Finance Ministers and Central Bank Governors." 14 Feb. 2009. G8 Information Centre. 31 July 2009. http://www.g7.utoronto.ca/finance/fm090214.pdf.
[6] "The Global Plan for Recovery and Reform." 2 Apr. 2009. G20. 31 July 2009. http://www.g20.org/Documents/final-communique.pdf.

[7] For further details, see: Council of the European Union. Brussels European Council. 11 and 12 December 2008. Presidency Conclusions. Doc. 17271/1/08 of 13 February 2009. Brussels: European Union; Council of the European Union. Council Resolution. The Contribution of Space to Innovation and Competitiveness in the Context of the European Economic Recovery Plan and Further Steps. Doc. 10500/09 of 29 May 2009. Brussels: European Union.
[8] These six issue-areas were defined and conceptualised in the frame of a conference organised by ESPI: European Space Policy Institute. Threats, Risks and Sustainability – Answers by Space, 10-11 Dec. 2007, Vienna, Austria. following publication: Schrogl, Kai-Uwe; Mathieu, Charlotte and Lukaszszyk, Agnieszka, eds. Threats, Risks and Sustainability – Answers by Space. Studies in Space Policy, vol. 2. Vienna: SpringerWienNewYork, 2009.
[9] These include among others France and Germany.
[10] The "near abroad" is a term used in Russia to describe the Post-Soviet States (except Russia itself).
[11] The negotiations are conducted by two privately constituted bodies led by their respective governments: the Straits Exchange Foundation (SEF) for the Republic of China/Taiwan, and the Association for Relations Across the Taiwan Straits (ARATS) for the People's Republic of China.
[12] Council of the European Union. Declaration on Strengthening Capabilities. Doc. 16840/08 of 11 December 2008. Brussels: European Union.
[13] "Declaration on Alliance Security Issued by the Heads of State and Government Participating in the Meeting of the North Atlantic Council in Strasbourg/Kehl on 4 April 2009." 4 Apr. 2009. NATO Press Release (2009)043. 31 July 2009. http://www.nato.int/cps/en/natolive/news_52838.htm?mode =pressrelease.
[14] The five permanent members of the UN Security Council (USA, Russia, China, France, Great Britain) and Germany.
[15] The IPCC Assessment Reports are comprehensive assessments of climate change, each one having played a crucial role for negotiations under the UNFCCC.
[16] IPCC website. 3 Aug. 2009. http://www.ipcc.ch.
[17] UNFCC website. 3 Aug. 2009. http://unfccc.int.
[18] Lardier, Christian. "Contrat Etat-CNES: Priorité aux Bénéficiaires." Air & Cosmos 2179 3 July 2009: 34.
[19] "The Global Plan for Recovery and Reform." 2 Apr. 2009. G20. 31 July 2009. http://www.g20.org/Documents/final-communique.pdf.
[20] Council of the European Union. Council Adopts Climate-Energy Legislative Package. Doc. 8434/09 (Presse 77) of 6 April 2009. Brussels: European Union.
[21] Commission of the European Communities. Communication from the Commission to the European Parliament, the Council, the European Economic and Social Committee and the Committee of the Regions. Towards a Comprehensive Climate Change Agreement in Copenhagen. COM (2009) 39 final of 28 January 2009. Brussels: European Union.
[22] Commission of the European Communities. White Paper. Adapting to Climate Change: Towards a European Framework for Action. COM (2009) 147 final of 1 April 2009. Brussels: European Union.
[23] Council of the European Union/Commission of the European Communities. Climate Change and Security: Recommendations of the High Representative on Follow-up to the High Representative and Commission Report on Climate Change and International Security. Doc. S412/08 of 18 December 2008. Brussels: European Union.
[24] "International Energy Agency." 2008. OECD/IEA. 3 Aug. 2009. http://www.iea.org/about/docs/iea2008.pdf.
[25] The value of the OPEC basket price was 131,22 U.S. dollars in July 2008 and fell down to 38.60 U.S. dollars in December 2008.
[26] "Commission Demands New Power in Gas Crisis." 18 June 2009. Euractiv.com. 3 Aug. 2009. http://www.euractiv.com/en/energy/commission-demands-new-powers-gas-crises/article-183290.
[27] "Towards a Sustainable Energy Future. IEA Programme of Work on Climate Change, Clean Energy and Sustainable Development." 2008. OECD/IEA. 3 Aug. 2009. http://www.iea.org/G8/2008/G8_Towards_Sustainable_Future.pdf.

[28] The EU and the U.S. couldn't agree on a compromise to reduce their agriculture subsidies, whereas emerging economies wanted to introduce special safeguard mechanisms to protect their farmers.
[29] "Agricultural outlook 2009–2018." 2009. OECD/FAO. 4 Aug. 2009. http://www.oecd.org/dataoecd/2/31/43040036.pdf.
[30] These are the Food and Agricultural Organisation (FAO), the World Food Programme (WFP) and the International Fund for Agricultural Development (IFAD).
[31] "G8 Leaders Statement on Global Food Security." 9 July 2008. Ministry of Foreign Affairs of Japan. 4 Aug. 2009. http://www.mofa.go.jp/POLICY/economy/summit/2008/doc/doc080709_04_en.html.
[32] "Agriculture and Food Security at the Core of the International Agenda." 10 Apr. 2009. Agriculture G8 Final Declaration. 4 Aug. 2009. http://www.g8agricultureministersmeeting.mipaaf.com/en/index.php?pL1=news&newsId=8895ca36c193f25a5c2a63beb291a259&page=1.
[33] "Statistics: Graphs & Maps." 2009. UNEP. 4 Aug. 2009. http://www.unwater.org/statistics.html.
[34] "The 2008 Zaragoza Charter." 14 Sept. 2008. 2008 Zaragoza International Exhibition. 4 Aug. 2009. http://www.expozaragoza2008.es/docs/repositorio/TribunaDeAgua/docs_pdf/cartazgz-en.pdf.
[35] "Water for a sustainable Europe – our vision for 2030." 30 June 2008. European Water Partnership. 4 Aug. 2009. http://www.ewp.eu/wp-content/uploads/2009/03/water-vision-for-europe.pdf.
[36] Council of the European Union. Council Resolution. The Contribution of Space to Innovation and Competitiveness in the Context of the European Economic Recovery Plan and Further Steps. Doc. 10500/09 of 29 May 2009. Brussels: European Union.
[37] Council of the European Union. Brussels European Council. 11 and 12 December 2008. Presidency Conclusions. Doc. 17271/1/08 of 13 February 2009. Brussels: European Union.
[38] Commission of the European Communities. A more research-intensive and integrated European Research Area. Science, Technology and Competitiveness Key Figures Report 2008/2009. Brussels: European Communities, 2008.
[39] For an overview of IHY's activities, see Thompson, Barbara J.; Gopalswamy, Natchimuthuk Davila, Joseph M. and Hans J. Haubold, eds. Putting the "I" in IHY. The United Nations Report for the International Heliophysical Year 2007. Studies in Space Policy, vol. 3. Vienna: SpringerWien NewYork, 2009.
[40] These are Resolution 1816 (2008) of the UN SC adopted on 2 June 2008, and Resolution 1851 (2008) of the UN SC adopted on 16 December 2008.
[41] "International Shipping and World Trade. Facts and Figures." Nov. 2008. International Maritime Organisation. Maritime Knowledge Center. 4 Aug. 2009. http://www.imo.org/includes/blastDataOnly.asp/data_id%3D23754/InternationalShippingandWorldTrade-factsandfigures.pdf.
[42] "ACI Global Traffic Forecast Report 2008-2027". 10 Sept. 2008. International Civil Aviation Organisation. 4 Aug. 2009. http://www.icao.int/ceans/Docs/Ceans_Wp_066_en.pdf.
[43] Commission of the European Communities. Communication from the Commission. A Sustainable Future for Transport: Towards an Integrated, Technology-lead and User Friendly System. COM (2009) 279/4 of 17 June 2009. Brussels: European Union.
[44] Commission of the European Communities. Communication from the Commission to the European Parliament, the Council, the European Economic and Social Committee and the Committee of the Regions. Single European Sky II: Towards More Sustainable and Better Performing Aviation. COM (2008) 389 final of 25 June 2008. Brussels: European Union.
[45] Godin, Benoît. "Science, Accounting and Statistics: The Input-Output Framework." Research Policy 36:9 (2007): 1388–1403.
[46] Commission of the European Communities. A more research-intensive and integrated European Research Area. Science, Technology and Competitiveness Key Figures Report 2008/2009. Brussels: European Communities, 2008.
[47] R&D intensity is the ratio between the Gross Domestic Expenditure on R&D (GERD) and GDP in current euros.
[48] Eurostat Statistics. 5 Aug. 2009. http://epp.eurostat.ec.europa.eu/portal/page/portal/science_technology_innovation/data/main_tables.
[49] Ibid.

[50] Commission of the European Communities. A more research-intensive and integrated European Research Area. Science, Technology and Competitiveness Key Figures Report 2008/2009. Brussels: European Communities, 2008.
[51] Eurostat Statistics. 5 Aug. 2009. http://epp.eurostat.ec.europa.eu/portal/page/portal/science_technology_innovation/data/main_tables.
[52] Commission of the European Communities. A more research-intensive and integrated European Research Area. Science, Technology and Competitiveness Key Figures Report 2008/2009. Brussels: European Communities, 2008.
[53] Ibid.
[54] Commission of the European Communities. A more research-intensive and integrated European Research Area. Science, Technology and Competitiveness Key Figures Report 2008/2009. Brussels: European Communities, 2008.
[55] Ibid.
[56] Commission of the European Communities. Green Paper. The European Research Area: New Perspectives. COM (2007) 161 final of 4 April 2007. Brussels: European Union.
[57] Commission of the European Communities. A more research-intensive and integrated European Research Area. Science, Technology and Competitiveness Key Figures Report 2008/2009. Brussels: European Communities, 2008.
[58] The PTC is an international treaty administered by the World Intellectual Property Organisation (WIPO) which makes it possible to seek patent protection for an invention simultaneously in each of the signatory countries by filling a single international patent application.
[59] Commission of the European Communities. A more research-intensive and integrated European Research Area. Science, Technology and Competitiveness Key Figures Report 2008/2009. Brussels: European Communities, 2008.
[60] Schrogl, Kai-Uwe. "Raumfahrt im VN-System. Anwendungen – Weltraumrecht – Verkehrskontrolle." Presentation. Vorlesungsreihe "Interplanetare Raumfahrtmissionen". Technical University of Dresden, Dresden, Germany. 23 April 2009.
[61] United Nations General Assembly. Resolution on the Prevention of an Arms Race in Outer Space. UN Doc. A/RES/63/40 of 12 January 2009. New York: United Nations.
[62] United Nations General Assembly. Resolution on Transparency and Confidence-Building Measures in Outer Space Activities. UN Doc. A/RES/63/68 of 12 January 2009. New York: United Nations.
[63] United Nations General Assembly. Resolution on International Cooperation in the Peaceful Uses of Outer Space. UN Doc. A/RES/63/90 of 18 December 2008. New York: United Nations.
[64] United Nations General Assembly. Sixty-Third General Assembly. First Committee. 12th Meeting (AM). UN Doc. GA/DIS/3371 of 20 October 2008. New York: United Nations.
[65] United Nations General Assembly. Sixty-Third General Assembly. Fourth Committee. 16th Meeting (AM). UN Doc. GA/SPD/410 of 24 October 2008. New York: United Nations.
[66] United Nations General Assembly. Outer Space Scientific and Technical Committee Concludes its 46th Session in Vienna. UN Doc. UNIS/OS/377 of 23 February 2009. Vienna: United Nations.
[67] United Nations General Assembly. Outer Space Legal Committee Concludes its 48th Session in Vienna. UN Doc. UNIS/OS/379 of 3 April 2009. Vienna: United Nations.
[68] United Nations General Assembly. Provisional Agenda for the Fifty-second Session of the Committee on the Peaceful Uses of Outer Space. UN Doc. A/AC.105/L.273. Vienna: United Nations.
[69] The other three advisory missions are running in Namibia, Burkina Faso and Ghana.
[70] United Nations General Assembly. Report on Activities Carried out in 2008 in the Framework of the United Nations Platform for Space-based Information for Disaster Management and Emergency Response. UN Doc. A/AC.105/929 of 14 January 2009. Vienna: United Nations.
[71] "United Nations Programme on Space Applications Activities Schedule." 2009. United Nations Office for Outer Space Affairs. 6 Aug. 2009. http://www.oosa.unvienna.org/oosa/en/SAP/sched/index.html.
[72] United Nations General Assembly. International Committee on Global Navigation Satellite Systems Concludes Third Meeting. UN Doc. UNIS/OS/374 of 30 December 2008. Vienna: United Nations.

[73] United Nations Inter-Agency Meeting on Outer Space Activities. Draft Report of the Secretary General on the Coordination of Space-Related Activities Within the United Nations System: Directions and Anticipated Results for the Period 2009-2010. UN Doc. IAM/2009/CRP.5 of 3 March 2009. Vienna: United Nations.
[74] UNESCO website. 6 Aug. 2009. http://portal.unesco.org/en/.
[75] UNGIWG website. 6 Aug. 2009. http://www.ungiwg.org/unsdi.htm.
[76] ITU website. 6 Aug. 2009. http://www.itu.int/net/home/index.aspx.
[77] UNIDIR website. 6 Aug. 2009. http://www.unidir.org/html/en/home.html.
[78] "CD Report 2009." 2009. Reaching Critical Will. 6 Aug. 2009. http://www.reachingcriticalwill.org/political/cd/speeches09/reports.html.
[79] Group on Earth Observations website. 6 Aug. 2009. http://earthobservations.org/index.html.
[80] UNOOSA website. 6 Aug. 2009. http://www.unoosa.org/.
[81] These are: Bangladesh, China, Pakistan, Iran, Mongolia, Peru, Indonesia and Turkey.
[82] "Asia-Pacific Space Cooperation Organization Starts Operation." 16 Dec. 2008. www.chinaview.cn. 6 Aug. 2009. http://news.xinhuanet.com/english/2008-12/16/content_10514901.htm.
[83] APRSAF website. 6 Aug. 2009. http://www.aprsaf.org/.
[84] Kedjar, A. "3rd African Leadership Conference on Space Science and Technology for Sustainable Development." Technical Presentation. Committee on the Peaceful Uses of Outer Space, Fifty-Second Session. United Nations, Vienna, Austria. 11 June 2009.
[85] "Department of Trade and Industry (DTI) Minister Mandisi Mpahlwa, launch of the South African National Space Policy." 6 Mar. 2009. South African Government Information. 6 Aug. 2009. http://www.search.gov.za/info/previewDocument.jsp; Mostert, S. "The African Resource Management (ARM) Satellite Constellation." African Skies/Cieux Africains 12 (2008): 53–56.
[86] "Ministers Meet to Define the Role of Space in Delivering Europe's Global Objectives." 18 Nov. 2008. European Space Agency. 11 Aug. 2009. http://www.esa.int/esaCP/SEMUPQ4DHNF_index_2.html.
[87] Lardier, Christian. "Près de 10 Md€ pour l'ESA." Air & Cosmos 2149 28 November 2008: 53.
[88] "Ministers Meet to Define the Role of Space in Delivering Europe's Global Objectives." 18 Nov. 2008. European Space Agency. 11 Aug. 2009. http://www.esa.int/esaCP/SEMUPQ4DHNF_index_2.html.
[89] Bigot, Bernard/d'Escatha, Yannick and Laurent Collet-Billon. "L'Enjeu d'une Politique Européenne de Lanceurs: Assurer Durablement à l'Europe un Accès Autonome à l'Espace. " 18 May 2008. 11 Aug. 2009. http://lesrapports.ladocumentationfrancaise.fr/BRP/094000223/0000.pdf.
[90] ESA website. 11 Aug. 2009. http://www.esa.int/.
[91] Council of the European Union. Council Resolution Taking Forward the European Space Policy. Doc. 13569/08 of 29 September 2008. Brussels: European Union.
[92] Wouters, Jan. "Space in the Treaty of Lisbon." Yearbook on Space Policy 2007/2008: From Policies to Programmes. Eds. Schrogl, Kai-Uwe; Mathieu, Charlotte and Peter Nicolas. Vienna: SpringerWienNewYork, 2008. 116-124.
[93] "Réunion Informelle des Ministres Européens Chargés de l'Espace à Kourou: Communiqué Final. 22 July 2008. Ministère de l'Enseignement Supérieur et de la Recherche. 12 Aug. 2009. http://www.enseignementsup-recherche.gouv.fr/cid21740/reunion-informelle-des-ministres-europeens-charges-de-l-espace-a-kourou-communique-final.html.
[94] Commission of the European Communities. Commission Working Document. European Space Policy Progress Report. COM (2008) 561 final of 11 September 2008. Brussels: European Union.
[95] Institut Français des Relations Internationales (Ifri), ed. The French Presidency of the EU and the Dynamics of European Space. Report of the Conference Held on July 2, 2008, Paris, France. Paris: Ifri, 2008.
[96] Commission of the European Communities. Elements for a European Strategy for International Relations in Space. Annex to COM (2008)561 final of 11 September 2008. Brussels: European Union.
[97] "Critical Space Technologies for European Strategic Non-Dependance." 19 Sept. 2008. European Space Agency. 12 Aug. 2009. http://www.congrex.nl/08C37/.

[98] European Parliament and Council of the European Union. Regulation of the European Parliament and of the Council on the Further Implementation of the European Satellite Navigation Programmes (EGNOS and Galileo). Doc. (EC) No. 683/2008 of 9 July 2008. Brussels: European Union.
[99] "Galileo Global Nav System Examined by European Court of Auditors." 1 July 2009. Satnews.com. 12 Aug. 2009. http://www.satnews.com/cgi-bin/story.cgi?number=1653440132.
[100] de Selding, Peter B. "Galileo Cost Overrun Passes European Commission Audit." Space News 29 June 2009: 15.
[101] Commission of the European Communities. Communication from the Commission to the European Parliament, the Council, the European Economic and Social Committee and the Committee of the Regions. Global Monitoring for Environment and Security (GMES): We Care for a Safer Planet. COM (2008)748 final of 12 November 2008. Brussels: European Union.
[102] Council of the European Union. Council Conclusions on Global Monitoring for Environment and Security (GMES): "Towards a GMES Programme." Doc. 16267/08 of 26 November 2008. Brussels: European Union.
[103] Commission of the European Communities. Proposal for a Regulation of the European Parliament and the Council on the European Earth Observation Programme (GMES) and its Initial Operations (2011–2013). COM (2009)223 final of 20 May 2009. Brussels: European Union.
[104] Council of the European Union. Press Release. 2907th Council Meeting. Transport, Telecommunications and Energy. Doc. 16326/1/08 REV 1 (Presse 345) of 27 November 2008. Brussels: European Union.
[105] European Parliament and Council of the European Union. Decision of the European Parliament and of the Council on the Selection and Authorisation of Systems Providing Mobile Satellite Services (MSS). Decision 626/2008 EC of 30 June 2008. Brussels: European Union.
[106] Solaris Mobile is a joint venture between SES and Eutelsat.
[107] Council of the European Union. Brussels European Council. 11 and 12 December 2008. Presidency Conclusions. Doc. 17271/1/08 of 13 February 2009. Brussels: European Union.
[108] Council of the European Union. Council Resolution. The Contribution of Space to Innovation and Competitiveness in the Context of the European Economic Recovery Plan and Further Steps. Doc. 10500/09 of 29 May 2009. Brussels: European Union.
[109] European Parliament. European Parliament Resolution on Space and Security. Doc. 2008/2030 (INI) of 10 July 2008. Brussels: European Union.
[110] "Declaration on Strengthening Capabilities." 11 Dec. 2008. Council of the European Union. 12 Aug. 2009. http://www.consilium.europa.eu/uedocs/cms_data/docs/pressdata/en/esdp/104676.pdf.
[111] Council of the European Union. Council Conclusions and Draft Code of Conduct for Outer Space Activities. Doc. 17175/08 of 17 December 2008. Brussels: European Union.
[112] WEU Assembly website. 13 Aug. 2009. http://www.assembly-weu.org/.
[113] "Conclusions Xth European Interparliamentary Space Conference." 14 Oct. 2008. EISC. 13 Aug. 2009 http://www.eisc-europa.eu/images/stories/documents/10theisc.pdf.
[114] Bénéteau, Alain/Laporterie-Déjean, Florence/Baron, Hélène and Christine Guidice. NEREUS. "Network of European Regions Using Space Technologies." Presentation. Eurisy. Dublin, Ireland. 20–21 May 2008; "Le Bulletin de Spheris N.36." March 2009. Spheris. 13 Aug. 2009. http://www.spheris-fr.com/Info/index.php?RID=5&catdoc_idt=7.
[115] "EUMETSAT website. 13 Aug. 2009. http://www.eumetsat.int/.
[116] Council of the European Union. Council Resolution. The Contribution of Space to Innovation and Competitiveness in the Context of the European Economic Recovery Plan and Further Steps. Doc. 10500/09 of 29 May 2009. Brussels: European Union.
[117] "EUMETSAT website. 13 Aug. 2009. http://www.eumetsat.int/.
[118] Ibid.
[119] Taverna, Michael A. "Squaring the Circle." Aviation Week & Space Technology 2 Feb. 2009: 38.
[120] Lardier, Christian. "L'Europe Mise sur les Applications." Air & Cosmos 2150. 5 Dec. 2008: 38.
[121] For details see Schmidt-Tedd, Bernhard and Isabelle Arnold. "The French Act Related to Space Activities. From International Law Idealism to National Industrial Pragmatism." Aug. 2008. ESPI

Perspectives 11. 7 Jan. 2010. http://www.espi.or.at/images/stories/dokumente/Perspectives/espi-perspectives_11.pdf.

[122] Loncle, François. Rapport N° 1470 Fait au Nom de la Commission des Affaires Etrangères. 17 Feb. 2009. Paris: Assemblée Nationale.

[123] Lardier, Christian. "Objectif 2009 du Cnes: Préparer 2010–2015." Air & Cosmos 2158. 6 Feb. 2009: 38+; Lardier, Christian. "Contrat Etat-Cnes: Priorité aux Bénéficiaires." Air & Cosmos 2179. 3 July 2009: 34+.

[124] Ibid.

[125] de Selding, Peter B. "Financial Hurdles Limit CNES's Ability to Fly Solo." Space News 20 April 2009: 10.

[126] French Foreign Ministry website. 14 Aug. 2009. http://www.diplomatie.gouv.fr/fr/.

[127] Défense et Sécurité Nationale. Le Livre Blanc. Paris: Odile Jacob/La Documentation Française, 2008.

[128] "German Federal Minister of Economics Michael Glos and Bavarian Prime Minister Günther Beckstein visit DLR in Oberpfaffenhofen." 17 July 2008. DLR. 14 Aug. 2009. http://www.dlr.de/en/desktopdefault.aspx/tabid-3432/7418_read-13067/7418_page-5/.

[129] "Bundeswirtschaftsminister zu Guttenberg: «Raumfahrt-Robotik Bringt einen Technologieschub für Deutschland»." 13 May 2009. Pressemitteilung Bundesministerium für Wirtschaft und Technologie. 14 Aug. 2009. http://www.bmwi.de/BMWi/Navigation/Presse/pressemitteilungen,did=301028.html.

[130] ESD Partners. European Space Directory 2009. 24th Edition. Paris: ESD Partners Publication, 2009: 87.

[131] "Inauguration of the Galileo Control Centre at DLR." 8 Sept. 2008. DLR. 14 Aug. 2009. http://www.dlr.de/en/desktopdefault.aspx/tabid-3432/7418_read-13448/7418_page-4/.

[132] "Germany and China Step up Cooperation in Space." 16 Dec. 2008. DLR. 14 Aug. 2009. http://www.dlr.de/en/desktopdefault.aspx/tabid-667/7411_read-14864/.

[133] Taverna, Michel A. and Douglas Barrie. "Double Whammy." Aviation Week & Space Technology 22 Sept. 2008: 35.; Lardier, Christian. "Le Rôle Futur de l'Europe dans l'Exploration." Air & Cosmos 2144 S 24 Oct. 2008: 60+; de Selding, Peter B. "Italian Funding Shift Casts Shadow on ESA Initiatives." Space News 6 Oct. 2008: 8.

[134] MPLM is a large pressurised container used to transfer cargo to and from the ISS. Modifications would be necessary to transform the MPLM into a Permanent Logistics Module (PLM).

[135] "Italy Space Activities Year 2008." ASI. 14 Aug. 2009. http://www.asi.it/files/COPUOS%20 %20Italian%20Space%20Activities%20Report%202008.pdf.; "Gelmini: «Italy is a Leader Player in Space.»" 18 June 2009. ASI. 14 Aug. 2009. http://www.asi.it/en/news_en/gelmini_italy_is_a_leading_player_in_space.

[136] "Italy and Argentina Move Forward in Space Cooperation." 22 Apr. 2009. ASI. 14 Aug. 2009. http://www.asi.it/en/news_en/italy_and_argentina_move_forward_in_space_cooperation__0.

[137] "Gelmini: «Italy is a Leader Player in Space»." 18 June 2009. ASI. 14 Aug. 2009. http://www.asi.it/en/news_en/gelmini_italy_is_a_leading_player_in_space.

[138] BNSC website. 14 Aug. 2009. http://www.bnsc.gov.uk/.

[139] Ibid.

[140] BNSC. UK in Space 2009. London: BNSC, 2009.

[141] Congress of the United States of America. Consolidated Security Disaster Assistance and Continuing Appropriations Act for 2009, H.R. 2638. 3 Jan. 2008.

[142] "H.R. 1105: Omnibus Appropriations Act, 2009." Govtrack.us. 17 Aug. 2009. http://www.govtrack.us/congress/bill.xpd?bill=h111-1105.

[143] Space Foundation. The Space Report 2009. Colorado Springs, CO: Space Foundation, 2009.

[144] Lardier, Christian. "Un Budget Nasa en Hausse de 5,1%." Air & Cosmos 2172 15 May 2009: 39.

[145] Iannotta, Becky. "Shuttle Extension Options Have Common Denominator: High Cost." Space News 5 Jan. 2009: 1+.

[146] Iannotta, Becky. "NASA Postponing Moon Work to Keep Orion on Track." Space News 27 Apr. 2009: 13.
[147] Lardier, Christian. "Un Budget Nasa en Hausse de 5,1%." Air & Cosmos 2172 15 May 2009: 39.
[148] Hsu, Jeremy. "NASA Sets Schedule for Remaining Shuttle Flights." Space News 14 July 2009: 5.
[149] Berger, Brian. "Despite Waiver, NASA To Stop Using Russian Cargo Vehicle." Space News 6 Oct. 2008: 4.
[150] Iannotta, Becky. "NASA Postponing Moon Work to Keep Orion on Track." Space News 27 Apr. 2009: 13.
[151] Coppinger, Rob. "Altair and Ares V Design Contracts on Hold Till Third Quarter." 5 July 2009. Flightglobal.com. 23 Nov. 2009. http://www.flightglobal.com/articles/2009/05/07/326213/altair-and-ares-v-design-contracts-on-hold-till-third.html.
[152] NASA website. 17 Aug. 2009. http://www.nasa.gov/.
[153] "Space Operations. Joint Publication 3:14." 6 Jan. 2009. Joint Chiefs of Staff. 17 Aug. 2009. http://www.dtic.mil/doctrine/jel/new_pubs/jp3_14.pdf.
[154] "Army Regulation 900-1." 23 Jan. 2009. Department of the Army Space Policy. 17 Aug. 2009. http://www.fas.org/irp/doddir/army/ar900-1.pdf.
[155] Klamper, Amy. "Homeland Security Kills Nascent National Applications Office." Space News 29 June 2009: 5.
[156] Mathieu, Charlotte. "Assessing Russia's Space Cooperation with China and India. Opportunities and Challenges for Europe." 12 June 2008. ESPI Report 12. 17 Aug. 2009. http://www.espi.or.at/images/stories/dokumente/studies/espi%20final%20report%20ric.pdf.
[157] "Roscosmos Priorities." 16 June 2009. Roscosmos. 17 Aug. 2009. http://www.roscosmos.ru/NewsDoSele.asp?NEWSID=6491.
[158] Glonass Information-Analytical Center website. 17 Aug. 2009. http://www.glonass-ianc.rsa.ru/pls/htmldb/f?p=202:1:1244260643211211752::NO:::.
[159] Russianspaceweb website. 17 Aug. 2009. http://www.russianspaceweb.com/ppts_lv.html.
[160] "Russia Set to Invest Heavily in Space Industry." 22 Oct. 2008. New Scientist Space and Reuters. 18 Aug. 2009. http://www.newscientist.com/article/dn15014-russia-set-to-invest-heavily-in-space-industry.html.
[161] Russian space web website. 18 Aug. 2009. http://www.russianspaceweb.com/svobodny.html.
[162] Euroconsult.
[163] Suzuki, Kazuto. "A Brand New Space Policy or Just Papering Over a Political Glitch? Japan's New Space Law in the Making." Space Policy 24.4 (2008): 171–174.
[164] Kellender-Umezu, Paul. "Missile Warning System at Forefront of Japan's New Space Policy." Space News 8 June 2009: 8.
[165] JAXA website. 18 Aug. 2009. http://www.jaxa.jp/.[166] Euroconsult.
[167] "ChangZhen 5 (Long March 5) Launch Vehicle." 20 Feb. 2009. Sinodefence.com. 18 Aug. 2009. http://www.sinodefence.com/space/launcher/changzheng5.asp; "New Space Launch Center to Be Built in China's Southernmost Hainan." 3 Mar. 2009. Space Daily. 18 Aug. 2009. http://www.spacedaily.com/reports/New_Space_Launch_Center_To_Be_Built_In_China_Southernmost_Hainan_999.html.
[168] "China Able to Send Man to Moon Around 2020." 12 Mar. 2009. Space Daily. 18 Aug. 2009. http://www.spacedaily.com/reports/China_Able_To_Send_Man_To_Moon_Around_2020_999.html.
[169] National Institute for Defense Studies. East Asian Strategic Review 2008. Tokyo: National Institute for Defense Studies, 2008.
[170] Gopalaswamy, Bharath. "Indian Space Policy: Aiming Higher." Space News 21 July 2008: 21.
[171] Ibid. Lardier, Christian. "L'Inde Augmente son Budget Spatial de 27%." Air & Cosmos 2161 27 Feb. 2009: 42+.
[172] Ibid.
[173] "India Launches Spy Satellite." 20 Apr. 2009. Space War. 18 Aug. 2009. http://www.spacewar.com/reports/India_launches_spy_satellite_999.html.

[174] Shin, Jenny. "Chronology of North Korea's Missile Flight Tests." 13 July 2009. 21 Aug. 2009. http://www.cdi.org/pdfs/NKMissileTestTimeline7.16.09.pdf.
[175] Agbaje, Ganyi I. "Current Trends in Nigeria's Space Development Programme to Facilitate Geospatial Information (GI) Sharing and Implementation of the NGDI." 2008. 18 Aug. 2009. http://www.state.gov/documents/organization/110820.pdf.
[176] ASAL (Agence Spatiale Algérienne) website. 18 Aug. 2009. http://www.asal-dz.org/.
[177] South African Space Portal. 18 Aug. 2009. http://www.space.gov.za/.
[178] "Super Science – Space and Astronomy." 12 May 2009. Australian Government. Ministers for Innovation, Industry, Science and Research. 19 Aug. 2009. http://minister.innovation.gov.au/Carr/Pages/SUPERSCIENCESPACEANDASTRONOMY.aspx.
[179] Space Foundation. The Space Report 2009. Colorado Springs, CO: Space Foundation, 2009.
[180] The figure stems from an ESPI analysis of the aforementioned "Space Report 2009" and other data, e.g. from the Futron Corporation. Exact sources will be listed for the specific figures stated in the following.
[181] Space Foundation. The Space Report 2008. Colorado Springs, CO: Space Foundation, 2008.
[182] Space Foundation. The Space Report 2008. Colorado Springs, CO: Space Foundation, 2008.
[183] Ibid.
[184] The figure stems from an ESPI analysis of the aforementioned "Space Report 2009" and other data, e.g. from the Satellite Industry Association/Futron. Exact sources will be listed for the specific figures stated in the following.
[185] "Space Policies, Issues and Trends 2008/2009." May 2009. ESPI. 16 Dec. 2009. http://www.espi.or.at/images/stories/dokumente/studies/espi%20report%2018.pdf.
[186] Satellite Industries Association/Futron. State of the Satellite Industry Report. Diego: SIA; Bethesda. Futron, June 2009.
[187] Ibid.
[188] Ibid.
[189] Space Foundation. The Space Report 2008. Colorado Springs, CO: Space Foundation, 2008.
[190] Satellite Industries Association/Futron. State of the Satellite Industry Report. Diego: SIA; Bethesda. Futron, June 2009.
[191] "Space Policies, Issues and Trends 2008/2009." May 2009. ESPI. 16 Dec. 2009. http://www.espi.or.at/images/stories/dokumente/studies/espi%20report%2018.pdf.
[192] Space Foundation. The Space Report 2008. Colorado Springs, CO: Space Foundation, 2008.
[193] "Space Policies, Issues and Trends 2008/2009." May 2009. ESPI. 16 Dec. 2009. http://www.espi.or.at/images/stories/dokumente/studies/espi%20report%2018.pdf.
[194] Space Foundation. The Space Report 2008. Colorado Springs, CO: Space Foundation, 2008.
[195] "Space Policies, Issues and Trends 2008/2009." May 2009. ESPI. 16 Dec. 2009. http://www.espi.or.at/images/stories/dokumente/studies/espi%20report%2018.pdf.
[196] "Space Business Review." Dec. 2008. Milbank. 8 Jan. 2010. http://www.milbank.com/NR/rdonlyres/B34A51C3-5450-463C-B8C1-008D261CD07A/0/December_2008_Space_Business_Review.pdf"Space Business Review." Dec. 2009. Milbank. 8 Jan. 2010. http://www.milbank.com/NR/rdonlyres/4429D43E-4A1A-4F8C-B129-067607E6530A/0/December_2009_Space_Business_Review.pdf.
[197] Lardier, Christian. "Le Marché des Petits Satellites en Plein Essor." Air & Cosmos 2143. 17 Oct. 2008: 36+.
[198] de Selding, Peter B. "Swiss Firm Align Self for Role in Civil and Civil and Military Space." Space News 21 July 2008: 6; Sherrer, Peter. "YIR: RUAG Aerospace." Dec. 2008. SatMagazine. 8 Jan. 2010. http://www.satmagazine.com/cgi-bin/display_article.cgi?number=663514835&method=print#.
[199] "Oerlikon Verkauft Geschäftsbereich Space an RUAG." 2 June 2009. Common Press Release by Oerlikon and RUAG. 8 Jan. 2010. http://www.ruag.com/de/media_relations/Medienmitteilungen_PDF/Medienmitteilungen_PDF_2009/02.06.09_Oerlikon_RUAG_MediaRelease_de.pdf; "OC Oerlikon Verkauft Raumfahrtsparte an Ruag." 2 June 2009. News.ch. 8 Jan. 2010. http://www.news.ch/OC+Oerlikon+verkauft+Raumfahrtsparte+an+Ruag/389059/detail.htm.

200 "ELTA S.A. Acquires Maker of Ground Station Equipment." Space News 4 Aug. 2008: 9; "OHB Shareholding Elta S.A. Extends its Space Activities by Acquiring SMP S.A., Toulouse." 28 July 2008. OHB Technology AG Press Release. 8 Jan. 2010. http://www.finanznachrichten.de/nachrichten-2008-07/11379387-press-release-ohb-technology-ag-ohb-shareholding-elta-s-a-extends-its-space-activities-by-acquiring-smp-s-a-toulouse-015.htm.

201 "Finmeccanica Completes Acquisition of DRS Technologies for 5.2 billion U.S. dollars." 22 Oct. 2008. Finmeccanica Press Release. 8 Jan. 2010. http://www.drs.com/MediaCenter/1022press.aspx.

202 Lagorce, Aude. "Dassault Buys Alcatel-Lucent's Thales Stake." 19 Dec. 2008. Market Watch. 8 Jan. 2010. http://www.marketwatch.com/story/dassault-aviation-buys-alcatel-lucents-thales-stake; Tran, Pierre. "Dassault Buys $2.2B Stake in Thales." 19 Dec. 2008. DefenseNexs. 8 Jan. 2010. http://www.defensenews.com/story.php?i=3871176.

203 de Selding, Peter B. "European Union Approves Astrium's Acquisition of SSTL." Space News 5 Jan. 2009: 13; de Selding, Peter B. "Acquisition by Astrium Seen Boosting SSTL's Competitiveness." Space News 19 Jan. 2009: 6; "Astrium Acquires Surrey Satellite Technology Limited." 13 Jan. 2009. EADS Astrium Press Release. 8 Jan. 2010. http://www.astrium.eads.net/en/press-center/press-releases/2009/astrium-acquires-surrey-satellite-technology-limited?set_language=en.

204 "SES Merging New Skies and Americom Units." Space News 14 July 2008: 15. de Selding, Peter B. "SES Consolidating Its New Skies and Americom Subsidiaries." Space News 8 Sep. 2008: 6; de Selding, Peter B. "SES is Shutting Down IP-Prime Satellite Television Service." Space News 5 Jan. 2009: 7.

205 Solarismobile website. 8 Jan. 2010. www.solarismobile.com/about-us/.

206 "Inmarsat Shareholder May Buy the Company." Space News 14 July 2008: 19; Rousseau, Claude. "MSS Industry Sees XM-Sirius Syndrome." Space News 8 Sep. 2008: 19; Peel, Lilly. "Harbinger Talks on Inmarsat Takeover Back On." 26 July 2008. Times Online. 8 Jan. 2010. http://business.timesonline.co.uk/tol/business/industry_sectors/telecoms/article4402240.ece; Galani, Una. "Clear Hand." 25 July 2008. Reuters Breakingviews. 8 Jan. 2010. http://www.breakingviews.com/2008/07/25/Harbinger-Inmarsat.aspx.

207 "EADS-Astrium Fait Passer sa Participation dans Spot Image à 81%." Air & Cosmos 2134 18 July 2008: 6.

208 "Space Business Review." July 2008. Milbank. 8 Jan. 2010. http://www.milbank.com/NR/rdonlyres/951D8C88-6390-4D08-B18B-829B0C78D7DC/0/July_2008_Space_Business_Review.pdf.

209 "Abertis Telecom Raises Its Stake in Hispasat to 33.4%." Space News 27 Oct. 2008: 3; " Abertis Strengthens Hispasat Position." 22 Oct. 2008. Satellite Today. 8 Jan. 2010. http://www.satellitetoday.com/enterprise/headlines/Abertis-Strengthens-Hispasat-Position_25075.html.

210 de Selding, Peter B. "Swedish Space Corp. Pays Cash for Universal Space Network." Space News 26 Jan. 2009:6; "Swedish Space Completes Acquisition of Universal Space Network." 4 May 2009. Space Mart. 8 Jan. 2010. http://www.spacemart.com/reports/Swedish_Space_Completes_Acquisition_Of_Universal_Space_Network_999.html.

211 "Elbit Takes Majority Stake In Israeli Satcom Firm." Space News 12 Jan. 2009: 3.

212 "Balaton Group Acquires Satellite Antenna Maker." Space News 14 July 2008: 8.

213 MotoSAT website. 12 Nov. 2009. http://www.motosat.com/.

214 "Comtech to Buy Radyne for $11.50/shr." 12 May 2008. Reuters. 8 Jan. 2010. http://www.reuters.com/article/mergersNews/idUSBNG2464720080512; Shrivastava, Anshu. "Comtech to Acquire Radyne in an All-Cash Accretive Transation." 12 May 2008. TMCnet.com. 8 Jan. 2010. http://voipservices.tmcnet.com/feature/articles/27728-comtech-acquire-radyne-an-all-cash-accretive-transaction.htm.

215 "Northrop Grumman Buys Geospatial Intel Company." Space News 15 Sep. 2008: 8; "Northrop Reorg Merges Space and Aviation Sectors." Space News 12 Jan. 2009: 3.

216 "Sierra Nevada Corp. Acquires SpaceDev." Space News 27 Oct. 2008: 8.

217 "Globalstar Announces Annual and Fourth Quarter Results for 2008." 31 Mar. 2009. Globe Newswire. 8 Jan. 2010. http://phx.corporate-ir.net/phoenix.zhtml?c=203507&p=irol-newsArticle&ID=1271841&highlight=.

[218] "New Round of Financing Sallies Forth For Space X." 29 June 2009. Satnews.com. 8 Jan. 2010. http://www.satnews.com/cgi-bin/story.cgi?number=196942570.
[219] "Acquisition Enhances EMS Aircraft Telecom Business." Space News 15 Sep. 2008: 8.
[220] Iannotta, Becky and Ferster Warren. "Sirius XM Cites $400 Million Savings." Space News 4 Aug. 2008: 4.
[221] "Sirius XM Avoids Bankruptcy With Liberty Media Loan." Space News 23 Feb. 2009: 8.
[222] de Selding, Peter B. "Iridium Gets New Owner, Cash to Start Next Constellation." Space News 29 Sept. 2008: 6; Wylie, Peter. "Iridium and GHL Acquisitions to Merge." 23 Sept. 2008. Fierce Telecom. 8 Jan. 2010. http://www.fiercetelecom.com/story/iridium-and-ghl-acquisitions-merge/2008-09-23.
[223] "DBSD North America Files Pre-Negotiated Plan to Restructure Debt." 18 May 2009. Space Mart 8 Jan. 2010. http://www.spacemart.com/reports/DBSD_North_America_Files_Pre_Negotiated_Plan_To_Restructure_Debt_999.html.
[224] "Duncan Hunter National Defense Authorization Act For Fiscal Year 2009." 14 Oct. 2008. GPO 8 Jan. 2010. http://www.dod.mil/dodgc/olc/docs/2009NDAA_PL110-417.pdf.
[225] Covault, Craig and James Ott. "Caught in the Net." Aviation Week & Space Technology 3 Nov. 2008: 34.
[226] Coppinger, Rob. "Russian Space Industry Rises Again." 1 Oct. 2007. Flightglobal. 8 Jan. 2010. http://www.flightglobal.com/articles/2007/10/01/217401/russian-space-industry-rises-again.html.
[227] "Russia Set to Invest Heavily in Space Industry." 22 Oct. 2008. NewScientist Space. 8 Jan. 2010. http://www.newscientist.com/article/dn15014-russia-set-to-invest-heavily-in-space-industry.html?full=true&print=true.
[228] "Russia to Decide on State Space Corporation in 2009 – Ivanov." 21 Oct. 2008. RIA Novosti. 8 Jan. 2010. http://en.rian.ru/russia/20081021/117856949.html.
[229] Makarov, Yuri and Dmitry Payson. "Russian Space Programmes and Industry: Defining the New Institutions for New Conditions." Space Policy 25.2 (2009): 90–98.
[230] "Russia Moves To Secure Hold on Strategic Sector." Space News 21 July 2008: 8.
[231] "Centers: Svobodny." 31 Dec. 2009. Russianspaceweb.com. 11 Jan. 2010. http://russianspaceweb.com/svobodny.html.
[232] de Selding, Peter B. "Commercial Launch Providers Looking Ahead to Busier 2009." Space News 5 Jan. 2009: 11.
[233] Hanawa, Yuji. "Japanese Aerospace Industries' Requests to the Government." 6 Oct. 2008. Presentation. JAXA 12 Jan. 2010. http://eeepitnl.tksc.jaxa.jp/en/event/MEWS/21th/data/1 6.pdf.
[234] Myoken, Yumiko. "The Bill of Basic Space Law." Apr. 2008. Presentation. 12 Jan. 2010. http://ukinjapan.fco.gov.uk/resources/en/pdf/5606907/5633988/The_Bill_of_Basic_Space_Law.pdf.
[235] Tachikawa, Keiji. "2009: A New Era for Japan's Space Program." JAXA. 12 Jan. 2010. http://www.jaxa.jp/article/interview/vol44/p3_e.html.
[236] Suzuki, Kazuto. "A Brand New Space Policy or Just Papering Over a Political Glitch? Japan's New Space Law in the Making." Space Policy 24.4 (2008): 171–174.
[237] JAXA website. 6 Aug. 2009. http://www.jaxa.jp/article/special/transportation/kouchiyama_e.html.
[238] Covault, Craig. "Reopening China."Aviation Week & Space Technology 8 Sep. 2008: 28.
[239] "Restructuration de l'Organisation Spatiale Chinoise." Air & Cosmos 2132 4 July 2008: 32.
[240] "Long March 5 Will Have World's Second Largest Carrying Capacity." 4 Mar. 2009. Space Daily. 12 Jan. 2010. http://www.spacedaily.com/reports/Long_March_5_Will_Have_World_Second_Largest_Carrying_Capacity_999.html.
[241] "Commercial Space Transportation: 2008 Year in Review." Jan. 2009. FAA. 12 Jan. 2010. http://www.faa.gov/about/office_org/headquarters_offices/ast/media/2008%20Year%20in%20Review.pdf.
[242] A primary payload is defined by the FAA as the payload with the greatest mass on a launch vehicle for a given launch.
[243] In the following sections, to avoid confusion, and in order to be in line with the definitions used in ESA documents, the term competed launch will be used instead of commercial launch.

[244] de Selding, Peter B. "Commercial Launch Providers Looking Ahead to Busier 2009." Space News 5 Jan. 2009: 11.
[245] It may be noted that the Proton K version already flew in June 2008, carrying an early warning satellite for the Russian Ministry of Defence.
[246] The two firms created a joint venture, United Launch Alliances (ULA) to provide Delta and Atlas launch vehicles to U.S. governmental customers.
[247] Foust, Jeff. "Satellites, Launches and the Recession." 30 Mar. 2009. The Space Review. 12 Jan. 2010. http://www.thespacereview.com/article/1341/1.
[248] de Selding, Peter B. "Commercial Launch Providers Looking Ahead to Busier 2009." Space News 5 Jan. 2009: 11.
[249] "Arianespace Signs Contract With Russian Space Agency for 10 Soyuz Launchers." 20 Sept. 2008. Arianespace Press Release. 12 Jan. 2010. http://www.arianespace.com/news-press-release/2008/09-20-08-Soyuz-Launcher-Contract.asp.
[250] "Arianespace Orders 35 Ariane 5ECA Launchers From Astrium." 2 Feb. 2009. Arianespace Press Release. 12 Jan. 2010. http://www.arianespace.com/news-press-release/2009/02-02-09-launcher-order-from-astrium.asp.
[251] de Selding, Peter B. "Commercial Launch Providers Looking Ahead to Busier 2009." Space News 5 Jan. 2009: 11.
[252] de Selding, Peter B. "Chief Executive Replacement Signals Shift at Sea Launch." Space News 14 July 2008: 13.
[253] "Sea Launch Dépose le Bilan." Air & Cosmos 2178 26 June 2009: 41.
[254] Russian and Chinese manned flights are considered payloads since the spacecrafts are carried by separate launch vehicles.
[255] Payloads are assigned to the nation that commissioned them, not according to the nationality of the manufacturer.
[256] This figure was obtained by subtracting the manned flights, the cargo spacecrafts and the hardware for the ISS to the payload figure.
[257] Sources include the FAA and the Futron satellite manufacturing report from January 2009.
[258] "L'Europe Rivalise Avec les Etats-Unis". Air & Cosmos 2137 5 Sept. 2008: 16.
[259] "Satcoms: une Surcapacité de Lancement." Air & Cosmos 2138 12 Sept. 2008: 32.
[260] "L'Europe Rivalise avec les Etats-Unis." Air & Cosmos 2137 5 Sept. 2008: 16. "Satcoms: une Surcapacité de Lancement." Air & Cosmos 2138 12 Sept. 2008: 32.
[261] de Selding, Peter B. "The List: Top Fixed Satellite Service Operators." Space News 6 July 2009: 12.
[262] Ibid.
[263] "SIPRI Yearbook 2009. Armaments, Disarmament and International Security. Summary." 2009. SIPRI. 3 Jan. 2010. http://www.sipri.org/yearbook/2009/files/SIPRIYB09summary.pdf.
[264] Helios 3 could be developed nationally or be part of a wider European system.
[265] de Selding, Peter B. "EU Satellite Centre Gains Access to European Recon Data." Space News 8 Dec. 2008: 10.
[266] "Pléiades". CNES. 12 Jan. 2010. http://www.cnes.fr/web/CNES-fr/3227-pleiades.php.
[267] de Selding, Peter B. "France Prepared To Go It Alone on Missile Warning System." Space News 16 Feb. 2009: A1.
[268] de Selding, Peter B. "Rest of Europe Cool to French Missile Warning Ambitions." Space News 27 Oct. 2008: 13.
[269] Bundeswehr website. 12 Jan. 2010. http://www.bundeswehr.de/.
[270] Skynet 5 Presentation. Paradigm Services. 12 Jan. 2010. http://www.paradigmservices.com/our_services/skynet5.
[271] Cosmo-SkyMed website. 12 Jan. 2010. http://www.cosmo-skymed.it/en/index.htm.
[272] Paz Satellite Presentation. EADS Astrium. 12 Jan. 2010. http://www.astrium.eads.net/en/news/the-paz-satellite-a-defence-and-security-solution-provided-by-astrium-espana.html.
[273] Athena-Fidus Presentation. CNES. 12 Jan. 2010. http://www.cnes.fr/web/CNES-en/5912-athena-fidus.php.

[274] de Selding, Peter B. "Two Firms offer to Help Finance Italian Military Comsat." Space News.com. 12 Jan. 2010. http://www.space.com/spacenews/archive07/sicral_0702.html.
[275] de Selding, Peter B. "Europe Studies Military Demand for Commercial Satcom." Space News 10 Nov. 2008: A2.
[276] de Selding, Peter B. "European Ministers Endorse Military Role for Galileo." Space News 29 Sept. 2008: 8.
[277] "Space Ministers Highlight Contribution of Space to Leading Economic Recovery Through Innovation." 29 May 2009. ESA. 4 Jan. 2010. http://www.esa.int/esaCP/SEM3VJ1OWUF_index_0.html.
[278] "Council Conclusions and Draft Code of Conduct for Outer Space Activities." 17 Dec. 2008. EU Council General Secretariat. 12 Jan. 2010. http://register.consilium.europa.eu/pdf/en/08/st17/st17175.en08.pdf.
[279] "Conclusions of the Brussels European Council of 11 and 12 December 2008 on the European Security and Defence Policy." EU Council. 12 Jan. 2010. http://www.consilium.europa.eu/uedocs/cms_data/docs/pressdata/en/esdp/104699.pdf.
[280] "Declaration on Strengthening Capabilities." 11 Dec. 2008. EU Council. 12 Jan. 2010. http://www.consilium.europa.eu/uedocs/cms_data/docs/pressdata/en/esdp/104676.pdf.
[281] EUSC Presentation. EUSC website. 12 Jan. 2010. http://www.eusc.europa.eu/index.php?option=com_content&task=view2&Itemid=10.
[282] de Selding, Peter B. "EU Satellite Center Gains Access to European Recon Data."Space News 8 Dec. 2008: 10.
[283] "Presidency Report on ESDP." 9 Dec. 2008. EU Council. 12 Jan. 2010. http://register.consilium.europa.eu/pdf/en/08/st16/st16686.en08.pdf.
[284] de Selding, Peter B. "European Defence Agency Has Strong Growing Interest in military Space." Space News 6 Oct. 2008: 12.
[285] "Presidency Report on ESDP." 9 Dec. 2008. EU Council. 12 Jan. 2010. http://register.consilium.europa.eu/pdf/en/08/st16/st16686.en08.pdf.
[286] "Critical Technologies for European Strategic Non-Dependence." 9 Sept.2008. ESA Congrex. 12 Jan. 2010. http://www.congrex.nl/08C37/.
[287] European Parliament. Resolution on Space and Security. Doc. INI/2008/2030 of 10 July 2008. Brussels: European Union.
[288] "A Common Security and Defence Strategy for Europe-Reply to the Annual Report of the Council." 2 Dec. 2008. WEU Assembly. 12 Jan. 2010. http://www.assembly-weu.org/en/documents/sessions_ordinaires/rpt/2008/2028.pdf.
[289] "Multinational Space-Based Imaging System (MUSIS): European Space Cooperation for Security and Defence." 3 Dec. 2008. WEU Assembly. 12 Jan. 2010. http://www.assembly-weu.org/en/documents/sessions_ordinaires/rpt/2008/2025.pdf.
[290] Brinton, Turner. "Pentagon Lays Framework for New EELV Participants." Space News 16 Feb. 2009: A2.
[291] Singer, Jeremy. "Space Command to Push for Launch Range Overhaul." Space News 21 July 2008: A5.
[292] Singer, Jeremy. "U.S. Army Examining Potential of Low-Cost Imagery Satellites." Space News 21 July 2008: A1.
[293] Brinton, Turner. "Air Force Solicits Bids for Plug-N-Play Satellite Technology." Space News 16 Feb. 2009: A2.
[294] Werner, Debra. "NRC Report Says U.S. Export Control System Is «Broken»." Space News 19 Jan. 2009: 13; Brinton, Turner. "Gen. Chilton: Export Controls Pose Threat to National Security." Space News 23 Mar. 2009: 20.
[295] "No Separate Command for AFCYBER." 10 Oct. 2008. CDI Space Security Update #7. 12 Jan. 2010. http://www.cdi.org/program/document.cfm?DocumentID=4396&StartRow=1&ListRows=10&appendURL=&Orderby=D.DateLastUpdated&ProgramID=68&from_page=index.cfm; "U.S. Seeking out Cyber Warriors". Ibid.

[296] Brinton, Turner. "ORS Director Urges More Milspace Collaboration with U.S. Allies." Space News 16 Feb. 2009: A3.
[297] de Selding, Peter B. "U.S. Air Force to Widen Access to Detailed Space Surveillance Data." Space News 9 Mar. 2009: 1.
[298] Opall-Rome, Barbara. "DoD makes Plans to Leverage International Radar Satellites." Space News 1 Sept. 2008: 4.
[299] "Allard Commission Calls for Combining Black, White Space." 5 Sept. 2008. CDI Space Security Update #6. 12 Jan. 2010. http://www.cdi.org/program/document.cfm?documentid=4367&programID=68&from_page=./friendlyversion/printversion.cfm.
[300] Schaerlaeckens, Leander. "Russian Budget Cuts Could Impact EU Defence Market." 23 Feb. 2009. UPI. 12 Jan. 2010. http://www.upi.com/Security_Industry/2009/02/23/Russian-budget-cuts-could-impact-EU-defense-market/UPI-79231235426701/.
[301] Saradzhyan, Simon and Nabi Abdullaev. "Georgian War Highlights Gaps in Russian Space Capabilities." Space News 1 Dec. 2008: 16.
[302] "Russia to Build Space Defence System." 27 Sept. 2008. DAWN. 12 Jan. 2010. http://www.dawn.com/2008/09/27/int6.htm.
[303] Saradzhyan, Simon and Nabi Abdullaev. "Georgian War Highlights Gaps in Russian Space Capabilities." Space News 1 Dec. 2008: 16.
[304] "Space-Related Defence Policies and Future Topics for Consideration." Nov. 2008. Japanese Ministry of Defence. 12 Jan. 2010. http://www.mod.go.jp/e/d_policy/pdf/space2008.pdf.
[305] Kallender-Umezu, Paul. "North Korean Missile Test Puts Focus on Japanese Space Policy." Space News 13 Apr. 2009: 7.
[306] "Space Policy." 28 Apr. 2009. APAN Executive Summary. 12 Jan. 2010. http://www1.apan-info.net/vic/Home/VICAdmin/ProductFullView/tabid/1755/ArticleID/11833/CBModuleId/5208/Default.aspx.
[307] "China military Space Projects." Globalsecurity.org. 12 Jan. 2010. http://www.globalsecurity.org/space/world/china/military.htm.
[308] Gunter's Space Page. 13 Jan. 2010. http://space.skyrocket.de/.
[309] Covault, Graig. "China Readies Military Space Station-Launch Coincides with Shuttle Phase-Out." 2 Mar. 2009. Space Flight Now. 13 Jan. 2010. http://www.spaceflightnow.com/news/n0903/02chinastation/.
[310] Chen, Lan. "Is the Chinese manned Space Program a Military Program?" 30 Mar. 2009. The Space Review. 13 Jan. 2010. http://www.thespacereview.com/article/1340/1.
[311] "India to Boost Military Spending by 50%." 11 June 2008. RIA Novosti. 13 Jan. 2010. http://en.rian.ru/world/20080611/110101921.html.
[312] "Indian Space Ambitions Soar With Lunar Launch." 22 Oct. 2008. New Scientist. 13 Jan. 2010. http://www.newscientist.com/article/dn15008.
[313] Jayaraman, K.S. "India's Space Cell Leverages ISRO Technology for Armed Forces." Space News 16 Feb. 2009: A4.
[314] "India to Scale-Up Military use of Space: Army Chief." 17 June 2008. Thaindian News. 13 Jan. 2010. http://www.thaindian.com/newsportal/sci-tech/india-to-scale-up-military-use-of-space-army-chief_10060839.html.
[315] Rule, Christopher. "Wideband Global Satellite Communications System." Australian Department of Defence Magazine. 13 Jan. 2010. http://www.defence.gov.au/defencemagazine/editions/200708_08/cdg_comms.pdf.
[316] The Australian Senate Standing Committee on Economics. "Lost in Space? Setting a New Direction for Australia's Space Science and Industry Sector." Canberra: Australian Senate Printing Unit, Nov. 2008.
[317] Drover, Kenneth J. "Space Power: Military Imperatives in Australia's Environment." USAF Air War College, Mar. 1989.
[318] Australian Department of Defence. "Joint Purchase of Maritime Satellite Communications Terminals for Warships." Media Release. 16 Dec. 2008.

[319] Single, Thomas. "Considerations for a NATO Space Policy." Sept. 2008.ESPI Perspectives 12. 14 Jan. 2010. http://www.espi.or.at/images/stories/dokumente/studies/espi_perspectives_12.pdf.
[320] NATO Joint Air Power Competence Centre. "NATO Space Operations Assessment." Kalkar. Jan. 2009.
[321] "Al-Hiwar Satellite TV Off-Air After «Deliberate Jamming»." 8 May 2009. Radio Netherlands Worldwide. 13 Jan. 2010. http://blogs.rnw.nl/medianetwork/al-hiwar-satellite-tv-off-air-after-deliberate-jamming-paper.
[322] "Satellite Collision Leaves Significant Debris Clouds." NASA Orbital Debris Quarterly News 13.2 Apr. 2009. 13 Jan. 2010. http://www.orbitaldebris.jsc.nasa.gov/newsletter/pdfs/ODQNv13i2.pdf:1-2.
[323] "Fengyun-1 C Debris." NASA Orbital Debris Quarterly News 13. 1 Jan. 2009. 13 Jan. 2010. http://www.orbitaldebris.jsc.nasa.gov/newsletter/pdfs/ODQNv13i1.pdf:2.

Developments in space policies, programmes and technologies throughout the world and in Europe

Wolfgang Rathgeber and Christophe Venet

1. Space policies and programmes

Seen from an abstract level, there are two co-existing development lines in space polices and programmes. One is consolidation of projects and efforts already set out and pursued before. The other one is the search and decision for new goals and strategic objectives. In both cases, the trend towards international cooperation is being reinforced. Cooperation schemes often include emerging space actors.

1.1. Highlights in policies and programmes

Single space actors continue to aim at serving specific requirements and to follow the meta-goal of figuring and progressing within the international space system. One of the most critical areas in this regard is space transportation and independent access to space. As a subset, this includes human spaceflight efforts. Most space actors – pubic and private ones – are in the continuous process of planning, refining or re-designing launch vehicles and the associated infrastructure. A more recent trend is the consideration of public actors to involve commercial entities in space transportation efforts, as evidenced in the findings of the so-called Augustine report (see Sect. 2.3.1).

In the area of science and exploration, efforts around the ISS continued with addition of new hardware, logistic missions like the European ATV, conduct of experiments and the start of regular six person crew operation. China is pursuing plans for its own space station. The Moon is an important destination of space science and exploration. Currently, the focus is on orbiters and probes, but landing and rover projects as well as sample return missions are also foreseen. In parallel, several robotic Mars missions are being geared up, with first preparations for a human component like the long term simulation experiment by ESA.

In the area of applications, communication keeps on being regarded as crucial. This is due both to its commercial success as well as to its importance for security. Military and crisis management operations are very demanding in terms of bandwidth. This need can not be satisfied by public means. Apart from procuring and launching new dedicated hardware, public entities are also active in leasing commercial capacity.

As for positioning, navigation and timing, the U.S. and Russia are modernising and completing their own systems. China, India and Japan are building or planning their own systems as well, just like Europe that has lost time in the course of reconfiguring the set up process for its Galileo system. Originally conceived as purely civilian, the latter is no longer precluded from military utilisation.

Earth observation is also commercially successful and relevant for security in all its dimensions. Accordingly, many countries try to gain access to Earth observation data, many of them through their own assets. Europe is increasingly exploiting the inherent dual use nature that Earth observation share with other space applications. In the process, it tries to improve data sharing and exchange mechanisms for relevant missions. This is done both inside and outside European Union structures.

1.2. Highlights in partnerships

International cooperation in space activities is steadily increasing. It is being sought for various reasons. The motivation can be purely technological or scientific reasons, taking advantage of complementary capabilities and capacities or exploiting common mission objectives. However, cooperation in space activities can also be designed to follow more general high-level political goals. This can occur when actors seek to build or reinforce strategic partnerships or alliances with other actors. The same holds true inversely – sometimes cooperation in space matters is not being sought because the potential partners are avoided or even opposed for political reasons.

Space cooperation allows monitoring and influencing other actors' behaviour. Beyond that, it has the potential to relieve the budgetary pressure, although most of the time cooperation is performed on a "no exchange of funds" basis. International space cooperation schemes can be categorised along various lines. One can discern bilateral or multilateral patterns, with the first one sometimes being a precursor of the second. One can also discern cooperation between public or private entities, or a mixture of both. In the following, real-world space cooperation highlight endeavours are discussed along geographical regions. Details will be given in subsequent chapters.

Europe is the only space actor that has cooperation schemes with all other space actors, and with non-space faring actors. Accordingly, European international space relations will be considered in more detail in the following. Typical examples of space cooperation will be given. When already listed for one specific partner region, the relevant projects will not be mentioned again for the other partner region or regions.

ESA with its rich foreign relations expertise and experience plays a central role in international European space efforts. In the reporting period, there were four States having signed a European Cooperating State Agreement with ESA: The Czech Republic, Hungary, Romania and Poland. The Czech Republic formally joined ESA as the 18th Member State in November 2008. ESA and Canada have very close relations due to the status of Canada as an associated ESA member. Canada holds a special interest in ESA activities like the ATV and the EDRS. The ESA-Canada cooperation agreement is set to be extended beyond 2010. Relevant talks are under way.

The International Space Station is an important area of cooperation between Europe and the U.S. (and the other ISS partners). Apart from human spaceflight activities, ESA and NASA ran the Ulysses solar mission together. They also presented their Mars Exploration Joint Initiative (MEJI) in June 2009. This might lead to ExoMars being launched with an Atlas 5 rocket and other joint activities in the framework of ExoMars. ESA and NASA also eye a common mission to the Saturnian moons Titan and Enceladus, and they pursue the Europa Jupiter System Mission (EJSM – known in Europe as "Laplace") together. In addition, the so called James Webb telescope is to be launched as a cooperation between ESA, NASA and CSA (Canadian Space Agency).

Cooperation between ESA and Russia includes joint efforts in the domain of space transportation and of human spaceflight. Preparations for the first Soyuz launch from Europe's spaceport in French Guiana, scheduled for 2010, are going on. Integration of the Russian ground equipment began in 2008, to be followed by a qualification phase. The launch infrastructure is designed in a way that allows for adaptation to human spaceflight. ESA has also run a 105 day isolation experiment together with the Russian Institute for Biomedical Problems. The simulation was part of ESA's Mars 500 programme.

ESA's cooperation with Japan encompasses the BepiColombo mission to Mercury, where Japan will supply one spacecraft. Japan has also shown interest in upcoming Cosmic Vision missions, and ESA is in charge of the European data node of JAXA's ALOS (Advanced Land Observing Satellite), distributing its data to a variety of scientific and operational users in Europe and Africa. ESA supports ALOS as a third party mission, offering assistance through its expertise and ground infrastructure.

ESA is also working together with China, which has shown interest in Cosmic Vision missions as well. The Chinese lunar mission Chang'e 1 had been supported by ESA through spacecraft and ground operation services. The Dragon Programme to jointly exploit data from European and Chinese Earth observation satellites has been extended, just like the joint Double Star Programme for studying the Earth's magnetosphere.

ESA cooperates with India on the Chandrayaan 1 lunar mission. It supplied three out of the eleven instruments onboard, and it provides assistance in flight dynamics, data archiving and processing. ESA and ISRO will share the data produced by the instruments. A cooperation agreement between ESA and ISRO had already been signed in 1978. Its latest renewal took place in 2007, adding another five years to its validity.

Apart from this, ESA has initiated the GMES Africa initiative. In addition, it leads the Tiger initiative of the Committee on Earth Observation Satellites (CEOS), which has been extended to a second phase. A cooperation agreement between ESA and Argentina has been renewed in 2008, and relations with other South American States have been strengthened. Moreover, ESA holds an observer status in COPUOS.

Beyond ESA, other European entities run international cooperation agreements that are relevant for space as well. EUMETSAT is a key partner in the Jason 2 programme that forms the space component of the Ocean Surface Topography Mission (OSTM) and that is run together with NOAA, CNES and NASA. Besides, EUMETSAT has formal cooperation agreements with ISRO, the China Meteorological Administration (CMA), the Japan Meteorological Agency (JMA), the Korea Meteorological Administration (KMA), the Meteorological Service of Canada (MSC) and the Russian Federal Service for Hydrometeorology and Environmental Monitoring (Roshydromet).

The European Union has established space dialogues with both the U.S. and Russia, treating selected topics and issues. The EU-U.S. dialogue on civil space applications was launched in 2006. It aims to enhance space cooperation in areas like Earth observation, satellite navigation, satellite communication, space science and exploration.[324] The European Commission and the Russian Federal Space Agency have established a structured dialogue on space cooperation that includes also ESA. The fourth meeting of the steering board took place in March 2009.[325] Chinese participation in the EU/ESA-run Galileo navigation system remains to be defined.

As evidenced by the examples above, the U.S. has traditionally kept close ties to Europe in space matters, cooperating primarily in the areas of human spaceflight and space exploration. The U.S. is also involved in multinational endeavours like the ISS or the Global Exploration Strategy. In general, ITAR are an impediment to a wider U.S. approach to international cooperation – both by restricting U.S.

technology transfer and by enhancing foreign ambitions to develop domestic "ITAR-free" systems.

Space transportation systems are important assets for space cooperation, which is increasingly being sought by Russia – primarily with the U.S. and Europe. Russian space transportation systems are already crucial for ISS operation. They will even become indispensable during the U.S. capability gap following the Shuttle retirement. The Soyuz system will soon complement European launch capabilities.

As for Russian cooperation with other space actors, the Chinese National Space Administration (CNSA) opened an office in Moscow in December 2008 after Roscosmos had opened a representation in Beijing before. Russia puts an emphasis on cooperation with India, among other things regarding its GLONASS navigation system. Moreover it provides assistance to South Korea in the launch sector, and it maintains contacts with various emerging space actors.

Japan has a number of space cooperation schemes running, including Europe as mentioned above. A major component of Japanese space cooperation is APRSAF. This forum gathers countries from the region and aims to enhance the development of the members' space programmes and to facilitate the exchange of views on future cooperation opportunities. The 15th session was held in Hanoi in December 2008, and the 16th session will take place in Bangkok in January 2010.

China primarily pursues the goal of space autonomy. Still it maintains a long history of space cooperation with Europe, like in the Dragon programme or the Double Star mission as stated before. It also reaches out to emerging space actors like Brazil, for example via the well established CBERS programme. CBERS 3 is foreseen to be launched in 2010. Multilateral efforts of China are undertaken in APSCO that can be seen as a rival to APRSAF.

India disposes of a wide international space cooperation portfolio as well. Its Chandrayaan 1 mission does not only carry three instruments from Europe, but also two from the U.S. and one from Bulgaria. The successor Chandrayaan 2 will be developed jointly with Russia. Beyond that and the cooperation on GLONASS stated above, India plans to devise an experimental satellite for university student education with Russia. Furthermore, India is set to contribute a scientific instrument for the upcoming Russian satellite CORONAS-Photon. India also prepares common efforts with France, Italy and Israel.

2. Space transportation

As one the critical enabler of independent access to space, launch vehicles remained at the core of the space policy of all the space faring nations in

2008/2009. All the main actors engaged in ambitious modernisation programmes of their launchers, while for certain emerging space powers, such as Iran, North Korea or South Korea, the mastering of launch technologies as a prerequisite to their nascent space ambitions is an absolute priority. A general trend for new designs is to focus on modular launchers, which allows developing an entire launcher family based on a single design.

2.1. Europe

Launchers remain a policy priority in Europe, as testified by the political and financial support given by the ESA Ministerial Council to launch systems. ESA Member States granted 1.5 billion euros for the 5 launcher-related programmes proposed by ESA.[326]

The European launcher strategy continued to take shape in 2008/2009, with the aim of providing flexible and cost-effective launches for all the mass classes and both for institutional and commercial customers. In this respect, the European launcher family will be composed of three complementary launch systems: Ariane 5 for heavy payloads, Soyuz launched from the GSC for intermediate payloads and the new Vega launcher for smaller missions. In this regard, ESA and Arianespace signed a frame contract for the procurement of launch services in June 2009 as part of a new legal framework for launcher exploitation. Its main purpose was to maximise the use of the three launch systems and to ensure competitive launch prices for ESA missions.[327]

In 2008, Ariane 5 conducted five commercial launches and one institutional launch and was again dominating the market of commercial launches, both in terms of launch revenues and mass placed in orbit. Arianespace earned around 700 million U.S. dollars in 2008, representing 35% of the total annual commercial launch revenues, and Ariane 5 placed a total mass of 37 metric-tons into GTO, a 40% share of the total commercial mass launched into GTO.[328] Development and improvement programmes of Ariane 5 are under way, as the post-ECA development programme was adopted at the ESA Ministerial Council with a focus on the development of the Vinci engine and the conception of a new upper stage. Similarly, an enhanced version of Ariane 5, Ariane 5ME, should be used starting from 2016 to 2017. It will have 20% more payload capacity but the same price than an Ariane 5ECA, and could be used for manned flights around 2025. A decision concerning Ariane 5ME will be taken at the 2011 ESA Ministerial Council.[329]

The Vega programme also progressed in 2008/2009, despite certain difficulties. The first successful test for Vega's Zefiro 9-A third stage solid fuel rocket engine took place in October 2008. With its qualification in April 2009, all the three stages of the Vega launcher are qualified for flight. The announcement of opportunity for the second flight of Vega, scheduled for 2010, took place in July 2008. It will be one of the five VERTA flights aiming at demonstrating the flexibility of the new launcher.[330] However, problems encountered during the qualification of certain subsystems lead to the postponement of the maiden flight, which will take place only in mid-2010.

In parallel, the construction of the Soyuz launch pad in GSC continued. While the launch table and the gas deflectors were completed in spring 2008, Russian hardware arrived in Kourou in August 2008 and the whole facility should be ready by the end of 2009. Some delays in the construction of the mobile tower postponed the first flight to the first week of 2010. In spite of this, Soyuz already has a strong backlog for launches from GSC: Pleiades 1 in 2010, two IOV Galileo satellites in 2010, Pleiades 2 in 2011, Sentinel 1 in 2011 and further Galileo spacecrafts.[331] In parallel, EADS Astrium is already willing to prepare the future, as it wants to develop an intermediary launcher to replace Soyuz, which would enter in service by 2020–2025.[332]

To take into account the delays which accumulated, ESA agreed to pay additional 100 million euros for the Soyuz (60 million euros) and Vega (40 million euros) projects in April 2009.[333]

Besides the implementation of these three launcher programmes, a strategic reflection was also launched in Europe on the future of launch systems. In particular, a report on the future of ESP in the area of launcher was prepared in France by the director of the Atomic Energy Commissariat (CEA), the director of CNES and the General Delegate for Armament. As Ariane 5 already reached its mid-life, the report suggested defining the development of a new Ariane 6 launcher that could fly for the first time in 2020–2025. The decision concerning this programme could be submitted to the next ESA Ministerial Council in 2011. The objective would be to develop a flexible launcher that could replace Soyuz and Ariane 5 at the same time. Another important feature of the report was the fact that it didn't include manned flights in the future missions of Ariane 6.[334] At the technical level, the work on the European Next Generation Launcher already started, as ESA signed a 20 million euros contract with the Joint Propulsion Team to develop the future liquid engine demonstrator in June 2009.[335]

2.2. United States

In 2008/2009, the main focus of the space transportation sector in the U.S. was the development of the next generation launch systems in the frame of the Constellation programme. As the orientation of the U.S. exploration programme is not clear yet, the Constellation programme was marked by a series of hesitations and setbacks in 2008/2009. In parallel, private companies made progresses in the development of alternative launch systems, encouraged by the COTS (Commercial Orbital Transportation Services) contracts awarded by NASA.

Developments in the Constellation programme were strongly conditioned and influenced by the announced five-year gap between the Shuttle retirement in 2010 and the first flight of the replacement system Ares 1/Orion. Indeed, various options were studied to reduce this gap. A NASA study released in October 2008 made recommendations on how to accelerate the development of Ares 1 and Orion in order to complete the first crewed flight in September 2014 instead of March 2015. Similarly, an alternative launch system to Ares, Direct 2.0, was proposed to NASA and its proponents claimed that it would send the U.S. back to the Moon faster and cheaper than the Ares system. The NASA however, refused the proposal and decided to stick with Ares 1. As a whole, the results of the Augustine panel are awaited before taking strategic decisions about the final orientations to give to the Constellation programme.

Despite these uncertainties, which were coupled with budget constraints, 2009 marked the next big rounds of contracts for Constellation systems. In spring 2009, NASA's Exploration Systems Mission Directorate awarded multiple design study contracts for the Ares 5 launch vehicle and the Altair lunar lander. Ares 5 will be the biggest rocket ever built, with a height of 116 m and the ability to carry 180 tons in LEO. It will launch the lunar lander and cargo to the Moon, but will also be used by the scientific community to launch heavy space science missions, such as massive telescopes. In parallel, the procurement strategy for Constellation ground operations and processing support is also ongoing. As for the Ares 1 launch vehicle, NASA is considering cancelling a second Ares 1 test flight (Ares 1Y scheduled in September 2013) to meet its schedule.

NASA launched the COTS programme as it came to the conclusion that firms in a free market could develop and operate an orbital transportation system more efficiently and affordably than the government. After two selection rounds, NASA announced in December 2008 that it had selected Orbital Sciences Corp. and SpaceX to provide cargo resupply services to the ISS. NASA ordered 8 flights for 1.9 billion U.S. dollars from Orbital and 12 flights for 1.6 billion U.S. dollars from SpaceX.[336] Both companies developed their own spacecrafts – Cygnus for Orbital and Dragon for SpaceX – and their own launch systems.

After three failures, SpaceX's Falcon 1 conducted its first successful launch on 29 September 2008 and had its first commercial success on 14 July 2009, with the launch of the Malaysian microsatellite Razkasat. SpaceX is already working on an improved version, Falcon 1e, which will fly in 2010. While Falcon 1 can launch 420 kg in LEO, Falcon 1e will be able to carry 900 kg to LEO. In parallel, work continued on the heavy Falcon 9, with the first nine-engine firing test being successfully conducted in August 2008 and a firing test for a full mission length having taken place in November 2008. The maiden flight of Falcon 9 with a dummy payload simulating the Dragon capsule is scheduled in summer 2009. SpaceX also develops a heavier version of the launcher, Falcon 9 Heavy, with enhanced performances. With its launcher family, SpaceX will offer the lowest prices on the market.[337]

Orbital Sciences on its side, continued to develop the Taurus 2 launcher, based on the technology of its existing Taurus 1 launcher. The first flight of this launcher especially developed for COTS missions, is expected by 2010. In addition, Orbital Sciences is developing the Minotaur 4 launch vehicle, using three Peacekeeper solid rocket stages. The launcher is developed under a 10-year contract with the U.S. Air Force Space and Missile Systems Center to provide low-cost launches for U.S.-government satellites in LEO.[338]

2.3. Russia

In Russia, the main developments in the area of launch systems in 2008/2009 were centred on the Proton replacement, the Angara launcher, and on the next-generation Rus M launcher. In parallel, important milestones were reached in the development of the new Vostochny cosmodrome.

The Angara launcher family first, will be Russia's next generation launch vehicles. It is intended to complement and eventually replace the Proton and Rockot launchers to deliver military, governmental and commercial satellites from a new launch pad in Plesetsk. The rocket is currently under development by Khrunichev, and 2008 and 2009 were mainly devoted to static firing test of the engines. A first flight test will occur in 2011 in a light-class configuration while a heavy-class Angara rocket will be launched at the end of 2011. The first commercial launch is expected to take place around 2013–2014. The rocket will be designed in a modular way, which will allow for a wide range of configurations capable of lifting between 2 and 45 metric-tons into LEO.[339] However, Khrunichev faced financial difficulties, as 10 billion U.S. dollars were still needed to complete the programme by 2011. As a consequence, the Russian government injected 8 billion U.S. dollars in the Khrunichev capital.[340]

Russia is also in the course of developing a next-generation manned spacecraft. In this respects, a new rocket is needed to carry six instead of three crew members and to replace the obsolete Soyuz launcher. Throughout 2008, several rocket manufacturers competed and by September 2008, the system evaluation for a possible architecture of the future launch vehicle was completed and proposals were submitted to Roscosmos. In February 2009, Roscosmos published a set of requirements for the new launcher dubbed Rus M. It should be capable to carry 23 metric-tons to LEO, with the possibility to develop it towards a heavy launcher (50 tons) and a super-heavy launcher (100 tons). These Rus M derived rockets could be used to support Russia's plans for Moon missions. The preliminary design of the rocket was then established in April 2009 and Roscosmos chose a team composed of TsSKB Progress – Energya – Makeev to develop the future launcher.[341]

The new Russian launcher for human rated missions will be starting from the Vostochny cosmodrome currently under development in the Russian Far East. In November 2008, Roscosmos had announced that nearly 50 facilities will be built in Vostochny, including a cosmonaut training centre, R&D areas and residential areas. Preliminary design work will take place in 2010, construction will start in 2012 and the first orbital launch is expected by 2015–2016. The construction of this new launch complex will serve two strategic interests for Russia: it will reduce its dependency on Baikonur and it will enable new investments in a remote area under increasing economic influence from China.[342]

2.4. Japan

In 2008/2009, Japan focused on two interrelated projects in the field of launchers: the development of the HTV on the one hand, and the development of the new H 2B rocket which will carry the spacecraft on the other hand.

The HTV is currently under development by JAXA, and will be able to carry 6 tons of cargo supply to the ISS, both pressurised and unpressurised. Its first flight should take place in the Japanese FY 2009, meaning between April 2009 and March 2010.[343]

JAXA and Mitsubishi Heavy Industries (MHI) unveiled the design of the new H 2B launcher in the beginning of 2009. It is based on technologies used for H 2A, and will be able to offer similar commercial services as Ariane 5ECA, as it will carry 16.5 tons on LEO or 8 tons on GTO. JAXA conducted two static firing tests of its engines in April 2009. A particular feature of this project is that it had been developed as a PPP between JAXA and MHI, allowing for an effective utilisation of the small allocated budget. This approach also facilitates transfer of publicly developed technologies to the private sector, which is one of the goals of the new Japanese Space Law.[344] Japan also hopes to enter the commercial market with the new launcher, even though it is primarily designated to launch institutional payloads.

2.5. China

China's ambition is to enter in the commercial launch market. Even though ITAR regulations are still limiting its ability to catch western payloads, it could already register some commercial successes with the export of its DFH 4 satellite bus. The production of ITAR-free satellites by Thales Alenia Space will also facilitate Chinese commercial launches in the future. Furthermore, China is likely to implement an aggressive pricing policy, made possible by its low production costs.[345] A final factor which will contribute to its rising position on commercial markets is the removal of CGWIC from the blacklist of the U.S. Office of Foreign Assets Control in June 2008. CGWIC is the company selling telecommunications transponders and marketing commercial launches on LM 3.[346]

In order to support its ambitions, both as a global space power and as an actor on commercial markets, China is currently developing a new launcher, the Long March 5 launch vehicle (LM 5). It will have a new design, and be the first pure space launch vehicle in China, as all the previous versions of Long March were derived from the Dong Feng (DF) ballistic missiles. The China Academy of Launch Vehicle Technology (CALT) is working on the launcher, which will have performances similar to those of Ariane 5 and will feature a modular design, allowing to add easily boosters to the core stage. The definition phase of the programme ended in 2008, and the production started in late 2008 while the first launch is expected for 2014. LM 5 is central for the Chinese space strategy, as it will carry heavy GEO satcoms and will be used to implement China's space exploration policy, in particular for Moon missions and the future manned space station.[347]

China will also start the construction of a new space launch centre in the southern most Hainan province. The proximity of the equator and the fact that the centre will be surrounded by the ocean make it an ideal location for a launch complex. The decision to implement this project was taken at the end of 2008 and the centre will have three missions: the transfer, test and launch of rockets. It will be specially designed to launch the LM 5 rocket.[348]

2.6. India

India is currently developing an enhanced version of its GSLV launcher, the GSLV Mk.2. GSLV was developed to lessen India's dependence on foreign launchers, in particular for GEO payloads. It made its first flight in 2001, and is capable of placing 2.2 tons in GEO. Concretely, India is finalising a new cryogenic stage for GSLV and the first flight in this configuration is planned for 2009–2010. In parallel, India is also developing an enhanced version of the launcher, GSLV

Mk.3, capable of orbiting 4.5 tons in GTO. The first flight of this new launch vehicle is scheduled in 2013.[349]

Given the low production costs in India, the GSLV Mk.3 launcher could become a serious competitor on the commercial market as well. Indeed, India has rising commercial ambitions as well, and started to commercialise its PSLV (Polar Satellite Launch Vehicle) recently, having already launched the Italian Agile satellite and the Israeli TecSAR reconnaissance satellite.

Finally, India has also plans to develop reusable launch vehicles in the nest 15 to 20 years. In this respect, a Reusable Launch Vehicle Technology Demonstrator (RLV-TD) will be tested next year as a first step. It will allow testing the aerodynamics and the controllability of the spacecraft.[350]

2.7. Emerging actors

As the critical tool to ensure an independent access to space, launch technologies were at the core of many emerging space faring nation's space programmes. 2008/2009 was marked by the launch attempts by North Korea and Iran, as well as by the developments in the South Korean launcher programme.

On 5 April 2009, North Korea launched a three-stage Taepodong 2 rocket, but it was considered a failure by international observers. According to the North Korean officials, it was an attempt to place the Kwangmyongsong satellite into orbit, but the international community suspected it to be a test for a long-range ballistic missile. As such, it was considered a violation of the 2006 UNSC Resolution 1718, which prohibited North Korea to conduct any further nuclear test or ballistic missile launch. In parallel, intelligence sources suggest that there is a second ballistic missile launch site under construction on the West coast of the country. However, it is not estimated to be usable for launches before 2009 or 2010.[351]

Similarly, Iran launched a Safir-Omid launch vehicle loaded with a dummy satellite on 17 August 2008. The launcher consisted of a Shahab rocket as a first stage and of a liquid-fuelled second stage. Despite claims of success by the Iranian officials, the launch was considered a failure by international observers, since the vehicle didn't reach orbit. In a second attempt in February 2009, Iran succeeded in placing its Omid satellite into orbit. As it is the case with North Korea, international observers continue to believe the Iranian rocket launches to be tests for ballistic missiles.[352]

South Korea finally, rescheduled the first launch of its KSLV 1 in the summer of 2009, although it was initially planned in December 2008. KSLV 1 was developed with the help of Russia, which designed the first stage while South

Korea built the upper stage. Korea has already spent 339 million U.S. dollars since 2002 for the development of KSLV 1, as a key programme of its strategy to become a space power. The development of a fully national rocket to be launched in 2017 and able to carry 1.5 tons in LEO, is already planned. The Korean government will spend 15 million U.S. dollars in 2009 for the development of this KSLV 2, even though this new programme depends heavily on the success of KSLV 1. To launch the new rocket, South Korea built a new space centre in Goheung, the Naro Space Center, which became operational in the summer of 2009.[353]

2.8. Industrial comparison

The possession of launch vehicles and spaceports is a central element enabling independence in space activities. Moreover, the number of launches and the level of activity on the space bases give an indication on the dynamism of a country in the space sector.

In 2008, six countries, plus Europe and a multinational consortium, Sea Launch, conducted 68 launches[354] (Figure 10). This is a similar figure than in the last years, as there were 68 launches in 2007 and 66 in 2006. The hierarchy in this domain is very stable as well, as Russia was again on the first position with 24 launches, followed by the U.S. (15 launches) and China (11 launches). These three countries represent 73.5% of the total launches. Europe performed 6 launches, one more than previous year. A notable evolution is the strong presence of Sea Launch in 2008, as the consortium launched 6 rockets.

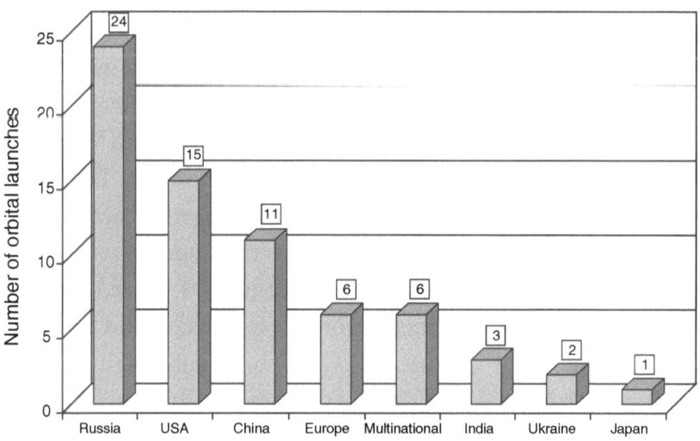

Fig. 10: *Total worldwide orbital launches per entity in 2008.*

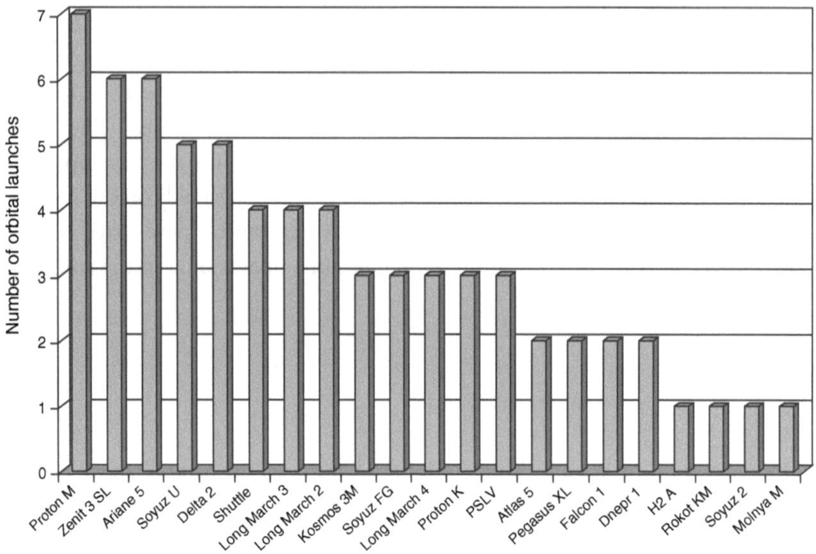

Fig. 11: *Worldwide orbital launches per launch system in 2008.*

The launches were distributed over 21 launch systems, with Russia having used 8 different ones, the U.S. 5 different launchers, China 3 different launch systems and Europe, India, Sea Launch, Japan and Ukraine having used a single launch system (Figure 11). Proton M was the mostly used launch system, with 7 launches, followed by Zenit SL and Ariane 5 with 6 launches each, Soyuz U and Delta 2 completing the podium with 5 launches each. The distribution is more even than in the previous year, with only 5 launch systems having served just once. The five mostly used launch systems represented only 41% of all launches performed in 2008.

Estimated 289 metric-tons were launched in space in 2008, about 35 metric-tons more than in 2007. Russia was the world leader again, as it launched about 95.5 metric-tons into orbit. It was followed by the United States, which launched around 64 metric-tons and by Europe which launched around 58 metric-tons into orbit. China launched about 35 metric-tons into orbit, Sea Launch about 27 metric-tons, and Japan, Ukraine and India together accounted for about 8.5 metric-tons launched in space. Europe was the leader in terms of commercial mass launched into orbit (more than 37 metric-tons), followed by Russia and Sea Launch (around 25.5 and 27 metric-tons, respectively) and the United States (around 11 metric-tons) (Figure 12). The unusually high amount of non-commercial mass launched by Europe compared to the two last years is due to the launch of the ATV Jules Verne. An Ariane 5ES rocket launched the 20 metric-tons spacecraft in March 2008.

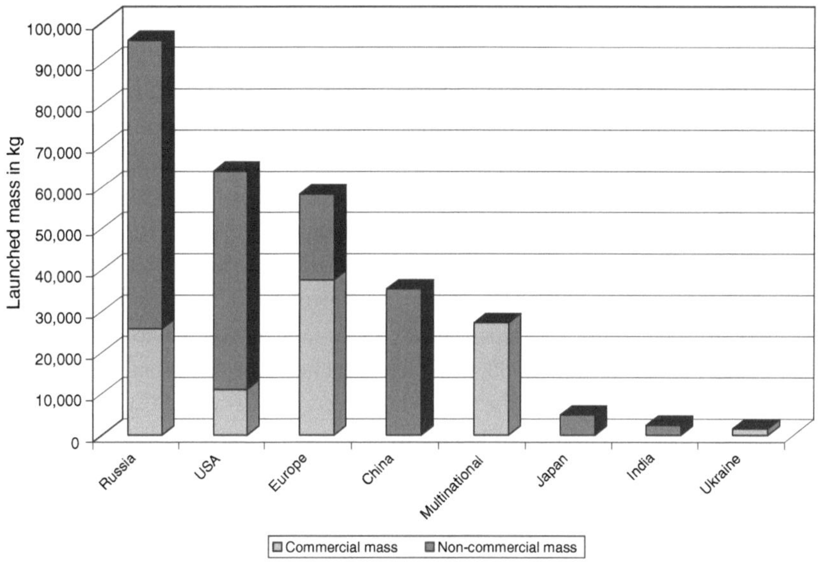

Fig. 12: *Estimate of the mass launched in 2008 by country/entity and by commercial status.*

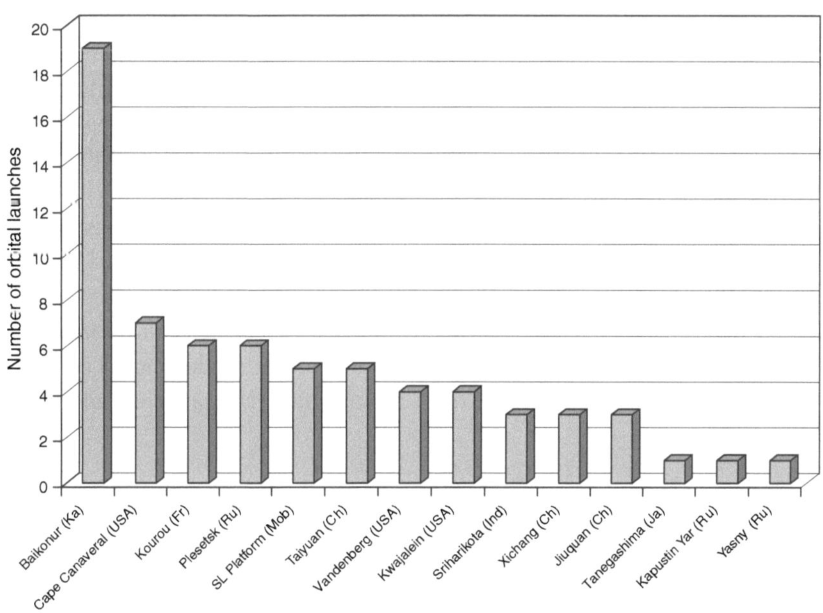

Fig. 13: *Worldwide orbital launches per launch sites in 2008.*

The use of space transportation infrastructure is another indicator that helps assessing the "space hierarchy", as space bases are core assets for an independent access to space. The number of space bases used by a country, as well as the frequency of launches conducted from its different spaceports, are important indicators of the dynamism of a country's space activities. In 2008, 15 launch sites were used, including one mobile platform (Sea Launch Odyssey Platform) (Figure 13).

Russia used 4 launch sites (Baikonur, Plesetsk, Kapustin Yar and Yasny), followed by the USA and China which both used 3 launch sites (Cape Canaveral, Vandenberg and Kwajalein for the U.S., Taiyuan, Xichang and Jiuquan for China). Europe, India and Japan all used one launch site. Baikonur was again the busiest space base in 2008, with 19 launches, one less than in 2007, but representing one third of the total launches in the year. Cape Canaveral (7 launches) and Kourou and Plesetsk (6 launches each) completed the podium. Sea Launch used two launch sites: its Odyssey Platform for 5 launches, and Baikonur for the first commercial launch of Land Launch.

Fifty-eight percent of the total launches conducted in 2008 were non-competed, representing 40 launches, and 42% of them were competed launches, representing 28 launches. Only four actors performed competed launches, whereas seven actors performed non-competed launches. As a whole, there was a significant increase of competed launches in 2008, compared to 2007. This was mainly due to the return to service of the Sea Launch vehicle Zenit 3SL, which performed 6 launches in 2008 and only one in 2007. After a hard year, GEO launches were again the top commercial activity, and the whole space transportation market was largely driven by the demand for GEO satcoms. This is likely to remain the same in the near term. Competed launches were particularly important for Europe and Sea Launch. U.S. launch services in contrast, continued to rely heavily on the governmental market, with only 6 out of 15 U.S. American launches being competed. In Russia as well, the relatively important domestic institutional demand continued to support the launch sector, as 15 out of 24 payloads launched by Russia were of domestic institutional nature. India and Japan focused on non-competed launches. Russian launchers conducted 9 competed launches, followed by U.S. American vehicles and the Sea Launch rocket conducting six competed flights each, and the European Ariane 5 performing five competed launches.

3. Space sciences and exploration

Space sciences and exploration rank highly on political agendas and in public debate worldwide. This is flanked by discussions about the justification of space activities in times of economic crisis as well as by relevant reviews like the report of the so-called Augustine committee that will be addressed below.

3.1. Human spaceflight activities

On board the ISS, the crew doubled to six persons in May 2009, when a Soyuz spacecraft brought ESA Astronaut Frank de Winne along with a Russian and a Canadian astronaut to the ISS.[355] For the first time then, all ISS partners (USA, Russia, ESA, Japan and Canada) had an astronaut on board. The docking also marked the beginning of regular six crew member operations.

In the U.S., a blue-ribbon panel headed by former Lockheed Martin chief executive officer Norman Augustine was charged to assess U.S. human space flight plans and to generate appropriate options, taking into account input from NASA, the Congress, the White House, industry, international space partners and the public. This initiative can be seen as a political effort to give new credibility to plans stemming from the era of the former U.S. administration. The committee had its inaugural meeting in June 2009.[356]

In April 2009, NASA announced plans to spend 150 million U.S. dollars for giving an impetus to commercial efforts to bring astronauts to the ISS. The money would come from the 400 million U.S. dollars that NASA has to spend on human exploration within the 1 billion U.S. dollars that it received in the framework of the American Investment and Recovery Act.[357]

SpaceX had announced in April 2009 that its cargo capsule Dragon could be rated for human spaceflight within 24 months provided NASA grants 200 million U.S. dollars of financial assistance. SpaceX pursues the goal of making human spaceflight cheaper and hopes to play a role in human exploration, possibly being involved in Mars missions.[358]

Russia plays a crucial role in human spaceflight, especially with the upcoming gap in U.S. human space transport capacities after the Shuttle retirement. A series of new launchers is under development in Russia, among others to replace the ageing Soyuz (see also Sect. 2.2.3). One project is the capsule PTK NP that can be re-used at least ten times (except for the motor part) and that can carry 500 kg of payload.[359]

Russia has also completed the design of its segment on the ISS, including the small research modules MRM 1-SGM and MRM 2-S02 and other modules. By 2015, the Russian segment of the ISS will consist of 8 modules, representing a total mass of 122 tons and a volume of 400 cubic meters.[360] The ISS is one of the principal objectives o the Russian Federal Space Programme for 2006–1013, and Roscosmos is in favour of extending ISS operation. Roscosmos also plans to propose a low-orbit space station for facilitating future exploration of the Moon and beyond.

China's human spaceflight activities can be considered as being the sector's highlight in 2008, with the Shenzhou 7 mission having been executed successfully. It was launched using a LM 2F rocket in September 2008 with three taikonauts on board, one of which performed the first Chinese space walk. The whole mission was broadcasted live on TV, which marks an important step towards transparency.[361]

China is also pursuing plans for its own space station that is scheduled to be completed around 2020. As a preparation activity, China will send the unmanned 8.5 ton module Tiangong 1 to orbit around 2010. China hopes to be able to involve international partners in this effort. In parallel, it is developing Shenzhou 8 as an unmanned version of Shenzhou 7. Tiangong 1 is foreseen to perform a docking manoeuvre with Shenzhou 8 in 2011.[362] The actual build-up of the space station could start in 2013 when Long March 5 enters into service.

The European ATV "Jules Verne" completed its successful logistic mission to the ISS in September 2008. There are four more ATV missions scheduled. The second ATV will be called "Johannes Kepler" and is foreseen for launch in 2010. EADS Astrium has suggested a human rated version of ATV through the project "ATV Evolution".[363] The ARV (Advanced Re-entry Vehicle) that could ultimately carry humans following some modifications has been suggested to the ESA Ministerial Council of November 2008 and received subscriptions of 20.93 million euros for phase A extending from 2009 to 2011.[364]

In November 2008, ESA astronaut Frank de Winne was nominated by the ISS Multilateral Crew Operations Panel to serve as the first European ISS commander from October 2009 on. Six new European astronauts were presented to the public in May 2009, following the second ESA astronaut selection process that lasted for one year. More than 8000 candidates had applied. Among the six new astronauts, there is one woman.

In March 2009 a crew of six persons, including two from ESA, entered an isolation facility in Moscow to spend a time of 105 days cut off from the outside world as part of the Mars500 programme. This project is executed between the ESA Directorate of Human Spaceflight and the Russian Institute for Biomedical Problems. It aims at simulating all the elements of a Mars mission. A full-fledged 500 day simulation is due to start in early 2010.[365]

Japanese efforts in the domain of human spaceflight were centred on completion of the Kibo module on the ISS. Kibo's main facility, the pressurised module, had been installed in June 2008 together with the Remote Manipulator System. In July 2008, the Japanese Experiment Module was added with other segments to follow later on. In April 2009, the first experiment began, dealing with crystal growth.[366]

In parallel, JAXA develops the HTV for the ISS, which could be man-rated in the long run.

JAXA's approach to human spaceflight is shaped by its 2005 "Vision" which is not an official government statement. It foresees that Japan uses the knowledge and insights gained on the ISS to develop its own astronaut launching capacity from 2015 on, potentially resulting in a contribution to a lunar human outpost by 2020. The Vision also states that a decision on Japanese human spaceflight involvement is due by the middle of next decade. At the moment, Japan disposes of seven astronauts.[367]

In India, ISRO is pushing to conduct human spaceflight by around 2015. A corresponding five year programme has been adopted, foreseeing a first automated flight in 2013 or 2014 and a manned low Earth-orbit mission with a capsule for two astronauts by 2014 or 2015. This mission is planned to be launched by a modified GSLV. Russia will provide assistance for these endeavours, and an Indian astronaut will fly on a Russian Soyuz in 2013. In the course of preparation, ISRO has acquired land near the Bengaluru International Airport to set up an astronaut training centre.[368]

3.2. Lunar exploration

The U.S. has launched two unmanned missions to the Moon to prepare the return of humans to the Moon. The LRO has the aim to find suitable landing sites, to find potential resources, to measure the radiation conditions and to test novel technologies. LCROSS serves to confirm the absence or presence of water ice in the polar regions of the Moon. Originally foreseen for May 2009, these missions have been launched in June 2009.

A prototype of a future Moon truck had been tested successfully in the Arizona desert in October 2008. This is part of the Constellation Programme that foresees to bring astronauts back to the Moon by 2020. The new truck is meant to move humans on the surface of the Moon and it shows significantly better performance than the ones used in former Apollo missions.[369]

In parallel, activities around the Google Lunar X Prize are soaring. As the first company, Astrobotic Technology has publicly announced to compete for the prize. It plans to send a robotic rover to the historic landing site of Apollo 11 on the Moon.[370] The X Prize Foundation has 21 teams from 11 countries registered to compete for the prize worth 30 million U.S. dollars. The deadline for winning the prize is the end of 2014.[371]

Russia has also been active in the area of lunar exploration and has been setting up relevant cooperation schemes. The company Lavochkin has designed a robotic lunar exploration programme featuring four phases. Phase 1 will consist of the Luna-Glob mission to be launched in 2012, where a Moon orbiter will launch 4 Japanese penetrators on the surface of the Moon. Phase 2 is foreseen for 2012, featuring a Russian rover for the polar regions of the Moon and an Indian orbiter (in the framework of the Chandraayan 2 mission). Phase 3 employing a more capable rover is planned from 2014 on, and Phase 4 is the construction of a lunar base.[372]

In Europe, various projects pertaining to lunar exploration are going on, both at national and European level. The U.K. pursues its MoonLITE mission, which aims at placing a satellite in Moon orbit and deploying four penetrators with geophysical instruments at the surface of the Moon. The launch is foreseen in 2014. Approval for the conduct of a study on the technical feasibility of the project was given in December 2008. NASA will contribute to the study.[373] In Germany, DLR plans for a probe circling the Moon have been abandoned in 2008. In June 2009, Astrium announced that it has been charged by DLR with a project on future Moon landings. Funds for the project are 1 million euros. In a first step, technical requirements are to be defined. In a second phase, then, a lunar lander is to be developed, constructed and tested on Earth. The project is considered as demanding in terms of the complex propulsion system and the autonomous navigation capability it demands.[374]

ESA is pursuing lunar exploration in a number of ways. Beside some international cooperation (see below), there are also projects being realised with industrial partners. For example, the EUROBOT Ground Prototype (EGP) was awarded to Thales Alenia Space with various subcontractors. EGP will involve a robot mounted on a rover type platform that can be used to navigate over a Moon- or Mars-like surface.[375] ESA also partially funds the MELiSSA (Micro-Ecological Life Support System Alternative) project at the Autonomous University of Barcelona. MELiSSA is a life support pilot plant recovering food, water and oxygen from the waste, carbon dioxide and minerals. The technologies tested with the project can contribute to facilitate human space exploration missions of long duration.[376]

India has launched its high-profile lunar mission Chandrayaan 1 on 22 October 2008. The spacecraft reached its operational orbit around the Moon in November 2008 and released a Moon Impact Probe from there. The spacecraft's purpose is to map the lunar surface in chemical, mineralogical and photo-geological manner. ISRO leads the project, with three of the eleven instruments onboard having been supplied by ESA. ESA also assists ISRO in flight dynamics and data handling. In return, data from the scientific instruments are shared. The U.S. and Bulgaria also participate in the project.[377] India's planned follow-up mission Chandrayaan 2 will be conducted together with Russia.

The Japanese Kaguya lunar mission ended successfully on 11 June 2009 by manoeuvring the main orbiter to crash onto the Moon as planned. The mission is also known by the name of SELENE (SELenological and ENgineering Explorer). It had been launched in September 2007. Kaguya was the largest lunar exploration endeavour since the Apollo programme and has enhanced Japan's reputation as a space player, both regionally and globally. The mission objective was to clarify the Moon's origin and evolution by investigating physical key features like elemental composition, surface structure, its gravitation and the remnants of its magnetic field.[378]

China's lunar probe Chang'e 1 already impacted on the Moon in a controlled manner on 1 March 2009. The mission had been launched in October 2007 – one month after the Japanese. It served to map the lunar surface. After the impact of Chang'e 1, further Chinese Moon plans were revealed. They foresee a stepwise approach – subsequent orbiting, landing and returning. Ultimately, China wants to send humans to the Moon around 2020.[379]

3.3. Mars exploration

Various space actors also show strong interest in Mars exploration, with Mars sample return missions being envisaged from 2020 on. NASA has a number of missions that are currently being executed or planned. The Mars Reconnaissance Orbiter (MRO) is still active. It was launched in 2005 with the aim of looking for evidence that water has persisted long enough on Mars to provide a habitat for life. It carries instruments that allow for detailed photography of the Martian surface, for analysing minerals, for searching sub-surface water and for monitoring weather conditions. MRO also served as a relay for the Phoenix lander.[380]

The Phoenix lander of NASA, part of the scout programme featuring small and cheap spacecraft, was shut down in November 2008 due to an anticipated lack of sufficient sunlight to produce power. It had operated on Mars for more than five months after having been launched in August 2007. As an important scientific result, the lander was able to verify the existence of water-ice below the planet surface.[381] The second mission within the scout programme will be Mars Atmosphere and Volatile Evolution (MAVEN), which is due to be launched in 2013. The MAVEN orbiter will investigate the upper atmosphere of Mars and its interaction with the sun. It will also serve as communication relay for future Mars landers.[382]

The launch of the Mars Science Laboratory (MSL), originally planned for 2009, has been postponed to 2011 with a touch down on Mars in 2012. MSL will investigate Mars sites where images from other spacecraft hint to wet conditions in the past. It will check whether former Mars environments offered conditions facilitating microbial life. The shifted launch date will cause a conflict with the

Juno launch that is planned for 2011 as well. This might result in a changed launch trajectory for MSL.[383]

Furthermore, NASA has selected the Lander Radio-Science (LaRa) on ESA's ExoMars as a project within the SALMON (Stand Alone Mission of Opportunity) programme, where planetary science research is conducted on non-agency missions at a cost of less than 35 million U.S. dollars. LaRa will use NASA's Deep Space Network to track the ExoMars mission. This will help to gain new insights about the structure of the interior of Mars.[384]

Europe has extended its Mars Express mission in February 2009 by a decision of the ESA Science Programme Committee. The mission had been launched in 2003 as the first European Mars mission and has already been extended once. Mars Express is orbiting Mars with the aim of studying its surface, observing certain atmospheric layers and solar conditions and mapping of the planet's surface. Thanks to Mars Express and its sub-surface measurements, underground water-ice deposits on Mars have been detected.[385]

Europe is also pursuing its ExoMars mission as part of the Aurora Exploration Programme. This mission will feature landing and manoeuvring a mobile scientific platform on Mars as well as sub-surface sample processing. ExoMars might face financial difficulties, given the cost ceiling of 1 billion euros set by the ESA Member States. The current project proposals amount to 1.2 billion euros, including a launch with Europe's Ariane 5 rocket.

ESA has been seeking international cooperation in Mars exploration. There have been deliberations to launch ExoMars by a Russian Proton rocket as part of a no-exchange-of-funds agreement with the Russian Space Agency.[386] Under the Mars Exploration Joint Initiative between ESA and NASA presented in June 2009, ExoMars might now be launched aboard an Atlas 5 rocket in 2016 – together with the Mars Science Orbiter of NASA.[387]

Russia is pursuing its Phobos-Grunt mission that is foreseen to collect samples from the surface of the Martian moon Phobos and to return them to Earth. Originally it was planned to be launched in 2009. Meanwhile it has been shifted to 2013, due to a tight schedule and testing issues.[388] China prepares the Mars probe Yinghuo 1 that will be launched piggy-back on Russian Phobos-Grunt. As China's first mission to explore another planet, Yinghuo 1 will investigate environmental changes in Martian climate from orbit without landing.[389]

3.4. Saturn exploration

The Cassini mission is a joint project of NASA, ESA and ASI. It was launched from Earth in 1997. It reached Saturn in 2004 to start exploring the planet and its environment, among others dropping the Huygens probe on Saturnian moon Titan. Operations within the Cassini mission came to an end in 2008, but since the spacecraft remains healthy, it is now dedicated to the Cassini equinox mission. This mission extension takes advantage of the fact that Saturn witnesses an equinox in August 2009 and its northern hemisphere along with the north face of its ring will be illuminated from then on. This will results in changing atmosphere conditions.[390]

NASA and ESA also eye a common mission to the Saturnian moons Titan and Enceladus, the Titan Saturn System Mission (TSSM). The mission concept is rooted in ESA's Tandem project that had been considered as part of the Cosmic Vision 2015–2025. It foresees an orbiter built by NASA and a lander and research balloon supplied by ESA. The mission, posing significant technical challenges, will create synergy with the planned Europa Jupiter System Mission (see Sect. 2.3.7).[391]

3.5. Venus exploration

Venus Express, a mission of ESA, was started in 2005 and reached Venus in 2006. It has been mapping the planet's atmosphere in 3D for the first time. This allowed setting up meteorological maps. Among others, Venus Express found evidence for water molecules escaping into space, revealed details about cloud systems and detected indications of lightning in the atmosphere. Its operation time had already been extended once. In February 2009 the ESA Science Programme Committee extended the mission a second time, until December 2009.[392]

Other States also have plans for exploring Venus. The Russian Federal Space Programme for 2006–2015 foresees the Venera D mission to land on the surface of Venus and to explore the environment with an especially long duration of about 30 days. The launch is scheduled for around 2016.[393]

3.6. Mercury exploration

The NASA mission Messenger was launched in 2004 and had several Mercury fly-by's, for example in October 2008. In March 2011, it will go on a science orbit for one year. Messenger takes stereo images of Mercury to investigate its global topography and landscape. This will help to figure out why Mercury is denser and richer in metal than its neighbour planets.[394]

ESA's BepiColombo mission to Mercury is planned to be launched in 2013. It will involve various contributions by Member States and it will feature two spacecrafts. The first one will examine the magnetic field of Mercury, and it will be supplied by Japan. The second one, built by ESA, will feature an orbiter for direct examination of Mercury.[395] In the framework of the SALMON programme described above, NASA also picked a science investigation mission by the name of Strofio. Strofio will feature a mass spectrometer to determine the composition of the Mercury atmosphere. Strofio will be part of the science instruments that ASI flies on BepiColombo.

3.7. Jupiter exploration

In November 2008, NASA approved plans for the Juno mission to Jupiter. Aimed at studying the formation and the composition of the planet, it will be launched in 2011 and will reach Jupiter in 2014. The mission results are expected to enhance the knowledge about the formation of the complete solar system. The spacecraft will employ solar power, which requires very efficient solar panels given the low exposure to sunlight and the high level of radiation. In fact, Juno will follow a special orbit around Jupiter that minimises radiation. In April 2009, the mission successfully passed a critical design review.[396]

NASA and ESA pursue the Europa Jupiter System Mission (EJSM) together. This mission is called "Laplace" in Europe. It goes back to a flagship mission that ESA had selected in a competition initiated in 2007 in the framework of the Cosmic Vision 2015–2025. EJSM will employ two robotic orbiters to undertake studies of Jupiter and its moons Io, Europa, Ganymede and Callisto. One spacecraft will be built by NASA, the other one by ESA. Launch is foreseen for 2020.[397]

3.8. Solar observation

Besides ongoing solar exploration missions like STEREO (Solar TErrestrial RElations Observatory), Hinode (Solar B) and SOHO (Solar and Heliospheric Observatory),[398] some new missions are also being planned and prepared. NASA's Solar Dynamics Observatory (SDO) will image the Sun in a resolution that exceeds the one of HDTV by ten times. By having a look at solar activities, such as sunspot, solar flares and coronal mass ejections, SDO will help to understand space weather effects. SDO is part of NASA's Living With a Star (LWS) programme.[399] It is foreseen to be launched in 2010.

In June 2009, NASA competitively selected the Interface Region Imaging Spectrograph (IRIS) project as part of the Small Explorer (SMEX) programme. SMEX aims at providing cheap access opportunities for missions in heliophysics or astrophysics. Although foreseen to be started in 2015, there is a chance that IRIS could launch as early as 2012, possibly by an air-launched Pegasus rocket. IRIS captures images in the chromosphere and in the transition region by using a solar telescope and a spectrograph. This will contribute to new numerical simulation models.[400]

On 30 June 2009, mission lifetime of the Ulysses Sun probe came to an end when NASA and ESA switched it off due to potential freezing of fuel lines as a consequence of degraded power supply. This freezing might have led to a loss of control in managing the probe. While only foreseen for a lifetime of five years Ulysses had studied the solar wind and the Sun's polar regions for more than 18 years following its launch by a Space Shuttle in 1990.[401] As its own solar mission, ESA assesses the Solar Orbiter mission. It is planned to perform studies of the Sun and the inner heliosphere and it could be launched in 2015.[402]

3.9. Outer solar system exploration and observation

On 6 March 2009, NASA launched Kepler as its tenth Discovery mission. Kepler will look into the structure and diversity of planetary systems, with a focus on Earth-sized planets within the so-called habitable zone that is defined by a suitable distance to stars. The mission employs a photometer measuring the brightness of light for detecting planets that transit stars. This can be used to determine planet sizes and orbital periods. Kepler's monitoring activities will last for four years.[403]

The Spitzer telescope of NASA experienced a loss of cooling liquid in May 2009. At that point of time it had been operational for five and a half years, following its launch in 2003. In this time span, Spitzer discovered planet-forming disks around stars, hidden black holes and unknown galaxies. The exhaustion of coolant did not seriously degrade the performance of two short wavelength channels of the infrared camera on board of Spitzer. Accordingly, the mission will be continued under the new name of "warm Spitzer".[404]

NASA has also selected the Gravity and Extreme Magnetism Small Explorer (GEMS) mission within the SMEX programme (see Sect. 2.3.8). Like IRIS, GEMS could be launched aboard a Pegasus rocket around 2015. It carries a specialised X-ray telescope to detect X-rays emitted from cosmic objects like neutron stars and black holes. Regular operation is planned for nine months. However, the spacecraft is foreseen to be operational for at least two years, so there will be a possibility for researchers to use the spacecraft in a Guest Observer programme.[405]

Apart from that, NASA successfully serviced Hubble in the framework of Shuttle mission STS 125 in May 2009. This will allow continuing Hubble operation until 2014, when Hubble successor New Generation Space Telescope (NGST), also known as James Webb telescope, is foreseen to be launched in cooperation with ESA and CSA. Throughout the 19 years of its operation, Hubble has compiled a wealth of scientific achievements, including evidence of the age of the Universe, that is now estimated to be 13.7 billion years old.[406]

ESA's Herschel and Planck missions were launched in May 2009 aboard an Ariane 5 rocket. Both orbiting around the second Lagrange point of the Sun-Earth system, the two missions will perform specific tasks that exceed it. Herschel is the largest and most powerful infrared telescope to study the birth and the development of stars and galaxies to explain the present state of the Universe. Planck aims at investigating the relic radiation of the Big Bang to enhance knowledge about the formation of galaxies.[407]

The multinational mission of COROT (Convection, Rotation et Transits planétaires), led by France, keeps on operating smoothly. Having been launched in 2006, it senses the vibration performance of stars to figure out their inner structure and age, advancing fundamental research. So far, COROT has discovered 7 planets that orbit stars outside the solar system. Its mission time has been extended for another three years.[408]

3.10. International cooperation in space exploration

Besides cooperation in concrete projects, multilateral high-level endeavours in international space exploration cooperation have been pursued in parallel. Following the Global Exploration Strategy (GES) that was drafted in 2007, the International Space Exploration Coordination Group (ISECG) held its third meeting in March 2009. The meeting was held in Japan and chaired by JAXA. Among other things, the ISECG members published their 2008 annual report and they advocated further study of internationally coordinated lunar exploration.[409]

This shows the significance of lunar activities. In July 2008, a multilateral agreement on lunar exploration was signed by the United States, U.K. Canada, France, Germany, India, Italy, Japan and South Korea at the Ames Research Center of NASA. The agreement foresees a fleet of unmanned spacecraft representing the lunar exploration of the next decade.[410] The emphasis on the Moon as a goal of exploration also shows in other initiatives such as the envisaged International Lunar Network (ILN).[411]

Several important high-level cooperation schemes have also been initiated in a bilateral manner. NASA and ESA agreed to embark on a Mars Exploration Joint Initiative in June 2009 (see Sect. 2.3.3). This initiative is meant to frame the agencies' plans and efforts in regard to Mars. As a flanking measure, a joint architecture review team will help in setting up an appropriate mission portfolio.[412]

In February, NASA and ESA announced that they will explore Jupiter and its four largest moons as well as Saturn's moons Titan and Enceladus. The corresponding missions EJSM, called "Laplace" in Europe, and TSSM are seen to set the stage for science research on planets.[413] NASA and ESA have also discussed a potential mission on dark energy that could be launched in 2017 or 2018.[414] Besides these missions and cooperation on ExoMars and the James Webb Space Telescope, NASA and ESA also consider an International X-Ray Observatory (IXO) with the participation of JAXA.[415]

4. Satellite applications

Two aspects can be identified in the field of satellite applications. On the one hand, space-based assets for communication, navigation and Earth observation are seen as crucial assets by spacefaring nations. As a consequence, established space powers continued to aim at enhancing their satellite fleets in these three areas in 2008/2009, while an increasingly number of emerging space powers developed programmes to acquire such assets, in particular in space-based communications and Earth observation. On the other hand, satellite applications represent a growing market for private initiatives. The most important one is represented by satcoms, but there are also growing market opportunities in the fields of navigation and positioning as well as remote sensing.

4.1. Space-based communications

In 2008/2009, the commercial market for satellite communications remained healthy both for satellite operators and satellite manufacturers, while governments continued to launch communication satellites both for civilian and military purposes.

The crisis didn't affect the activity of the biggest satellite operators in 2008, although the situation is a bit more difficult for regional operators.[416] This was true for the three types of commercial satellite services, DBS, FSS and MSS.

DBS was again the most dynamic commercial market, focusing on satellite TV and satellite radio. The global growth in DTH TV will keep the high demand for C-band and Ku-band satellite capacity. Indeed, the HDTV market is set to take off in Europe, while witnessing a steady growth in India and a booming in the

Middle East and in North Africa. Transponder demand has risen at a rate of 12% per year in the last 5 years in the region and the capacity demand is still higher than the supply.[417] Another trend is the development of Ka-band satellite internet. Indeed, ViaSat and Eutelsat are jointly going to offer such services in North America in Europe, while the operators HuguesNet and O3b Networks began to develop similar services.[418]

FSS also witnessed a steady growth over the last three years, sustained by digital TV. The leading providers of FSS, including Intelsat and SES, reported record revenues in 2008 and this positive trend is expected to continue with increasing demand for satellite transponder capacity in the Middle East, North Africa and the Asia-Pacific region.[419]

MSS revenues remained relatively stable in the last five year, but due to the crisis the MSS market is facing a liquidity shortage, as there is no public funding in the MSS sector.[420] Consequently, the MSS industry may be supported by hedging funds from the financial community, as it already happened in the past for the FSS industry. Despite these difficulties, MSS is an innovative sector, mixing wireless and terrestrial solutions to offer new products that integrate Global System for Mobile communications, GNSS and multiband roaming capabilities for voice, data and video products.[421]

The satellite manufacturing market remained healthy as well, as 80 GEO satcoms and 48 LEO satcoms were on the order lists of the industry by the end of 2008. Replacement orders for aging satellites will likely continue to drive the demand. However, while large satellite manufacturers resist the crisis fairly well, start-ups or companies owned by private equity firms will face increasing credit shortage and a volatile stock market.[422] A striking feature in 2008 was that despite a very unfavourable Euro-dollar exchange rate, Europe almost competed with the U.S. on the commercial satcom market, thanks to the good performances of Astrium and Thales Alenia Space.[423]

In Europe, satellite communication systems played a central role in the ESA Ministerial Council of November 2008. Indeed, 816 million euros were granted for the various ARTES (Advanced Research in Telecommunication Systems) programmes. In particular, the EDRS programme (ARTES 7) was granted 154.2 million euros. It aims at establishing PPPs for geostationary operational capabilities. The new system will replace Artemis, which was launched in 2003 and will be retired between 2010 and 2015. The EDRS will respond to needs in the frame of GMES in the first place, then to those of the commercial sector and finally to security needs. The Iris programme (ARTES 10), on its side, will be a telecommunications system allowing data transmission through satellite for ATM (Air Traffic Management). The preoperational services will start in 2015. Other programmes include the Alphabus platform (ARTES 8) and the small GEO platform (ARTES 11) as well as the satcom applications in the frame of the IAP (Integrated Applications Promotion) programme (ARTES 20).[424]

In 2008/2009, the EC also completed the process towards the attribution of licences to operate S-band MSS in Europe. The goal is to provide video, audio and data transmission services to the most remote areas while ensuring their viability. Indeed, MSS can cover a large territory and reach areas where such services were not economically viable before. The EC decision setting up the selection procedure entered into force on 5 July 2008.[425] In the first step, interested companies could submit their applications until October 2008. The second step encompassed an assessment of the technical and commercial ability of the candidates to reach the stated objectives.[426] On 14 May, the EC announced that Inmarsat and Solaris Mobile were selected, and that each operator will receive 2×15MHz of S-band spectrum and was granted an 18 years-licence. However, an antenna problem onboard the Eutelsat W2A satellite launched on 4 April 2009 may reduce Solaris's ability to provide the service requested by the regulatory licence. Furthermore, two of the losing bidders, ICO Global and TerreStar Network, initiated legal proceedings in the European Court of First Instance to annul the EC decision.[427]

The growing strength of the commercial market for satcom applications in Europe was testified by intense launch activities in 2008/2009. Inmarsat 4 was launched in August 2008 to provide services in the framework of the Broadband Global Area Network (B-GAN). SES launched Astra 1 M in November 2008 to provide Europe, the Middle East and North Africa with HDTV. A similar mission will be performed by the four Eutelsat satellites Hot Bird 9 and W2M launched in December 2008, Hot Bird 100 launched in February 2009 and W2A launched in April 2009.[428]

France and Italy were also active in military telecommunication in 2008/2009. In April 2009, Italy launched its second military telecommunication satellite, SICRAL 1B, while it is developing the SICRAL 2 satellite as well as the Athena-Fidus dual-use broadband system together with France. France on its side ordered the two first ground stations for its military satcom Syracuse 3, while it announced that it would not sell the Syracuse 3 system to private industry and then lease capacity from the company. However, France is almost certain to use a commercial procurement procedure for its next-generation military satcom system.[429]

In the North American market, 2008/2009 was a busy year in terms of commercial launches. Orbcomm first launched its constellation of 6 small LEO satcoms in June 2008, capable of performing classical telecommunication tasks but also able to receive the Automatic Identification System (AIS) for global maritime monitoring. However, after the loss of one of these satellites in February 2009, Orbcomm announced in May 2009 that it will file insurance claim for all six satellites, as the system is operational only as a constellation.[430] Dish Network further launched its EchoStar 11 on 16 July 2008, which will be a backup satellite, while SES Americom launched the AMC 21 satcom on 14 August 2008 to provide services to the U.S. Nimiq 4 launched in September 2008 by Telesat Canada will provide digital TV services in Canada and Intelsat launched Galaxy 19 to cover North America in September 2008 as well. The Canadian operator Ciel Satellite also launched its Ciel 2 satellite to cover North America in December 2008, as well as Loral Skynet with its Telstar 11 N satellite launched in February 2009. Finally, Sirius FM launched the Sirius FM 5 satellite for use in its satellite radio services in June 2009.[431]

The U.S. also focused on military satcom in 2008/2009. Among the main programmes in the FY 2009 defence budget were three communication systems: the fleet satellite communication follow-on/Mobile User Objective System (MUOS) which was granted 507.5 million U.S. dollars for procurement and 516.8 million for R&D, the Advanced Extremely High Frequency Satellite (AEHS) (116 million U.S. dollars for procurement and 388 million for R&D) and the Transformational Communications Satellite (TSAT) (785 million U.S. dollars for R&D).[432] The latter will provide the DoD with next-generation data communications via laser, but had to face several delays. After the launch of the second WGS in April 2009 and the scheduled launched of WGS 3, 4, 5 and 6 between 2009 and 2013, the DoD will move forward on this programme in 2009 with the procurement of WGS 7, 8 and 9.[433] An increasing trend is also the military demand for commercial bandwidth. The DoD relies on satcom for 80% of its needs today, and despite the launch of the WGS series, the demand for commercial bandwidth is likely to grow by 56% between 2010 and 2019.[434]

In February 2009, Russia launched two satcoms to provide the country with advanced communications and broadcasting services, Express AM 44 and Express MD 1, both operated by the Russian Satellite Communications Company (RSCC).[435] There is an increasing integration of Russia's satcom industry in the global market. On the one hand, Russian manufacturers, such as ISS Reshetnev, offered common bids with other manufacturers to build satcoms (with EADS Astrium for the Yamal 400 satellite, and with Thales Alenia Space for the Amos 5 satellite). On the other hand, the first commercial order for a Russian satcom took place in 2008, with RSCC ordering the AM4 satellite from EADS Astrium. In parallel, Russia launched two military communications satellites: Raduga 1 8, launched in February 2009 and Meridian 2, which was launched in May 2009 but didn't reach the intended orbit due to a failure during the end of the second burn.[436]

Japan concluded the restructuration of its satcom sector by the official creation of a single satellite operator on 1 October 2008, Sky Perfect Jsat Group. It gathers the operators Japan Communications Satellite Company, Satellite Japan Corp., NTT Communications Corp., Sky Perfect TV and Space Communications Corp. and operates a fleet of twelve satellites covering the Asia-Pacific area.[437]

China confirmed its increased ambitions on the commercial satcom market in 2008/2009. It started producing its DFH 4 bus in series: three satellites have been ordered for the domestic market (Sinosat 4, 5 and 6) and two export contracts have been signed with Pakistan (Paksat 1 R) and Nigeria (Nigcomsat 1 R). In October 2008, it launched the first Venezuelan satellite Venesat 1. In addition, two ITAR-free satellites built by Thales Alenia Space will be launched by the Long March launch vehicle in the summer of 2009 (Palapa D) and in 2010 (W3B).[438] However, the major in-orbit failure of the Nigerian Nigcomsat 1 satellite in November 2008 casted some uncertainty about the sustainability of the Chinese market presence.

The Indian space agency ISRO is also intending to play a major role in satellite manufacturing in the future. In 2006, its commercial arm ANTRIX entered a partnership with Astrium to enter the market. As a result, EADS and ISRO built the W2M satellite for Eutelsat, which was launched in December 2008. However, the satellite suffered a major anomaly in orbit, which undermines the credibility of the ANTRIX bus on the market.

The most buoyant market for satellite communications is the Middle East, which witnessed a booming growth in 2008/2009. As a consequence, all the operators active in the region developed aggressive capital spending plans. Eutelsat, Intelsat, SES and Russian operators plan to direct more of their current and planned satellite beams over the Middle East and certain regions of Africa. Other regional operators, such as Nilesat from Egypt and Arabsat, a 21-nation consortium based in Saudi Arabia, have launched ambitious expansion programmes. New entrants in the market also appeared, such as Thuraya and Yahsat from Abu Dhabi or S2M and Smartsat from Dubai. As a whole, this strong growth transformed the Middle East TV landscape from a government-dominated one into one dominated by commercial broadcasters.[439]

Emerging countries finally, being increasingly aware of the benefits of satellite communications, started several projects in this field. Venezuela launched its first satellite in October 2008, Venesat 1, also dubbed Simon Bolivar 1. The satellite was launched aboard a Chinese launcher and intends to provide rural communities with an access to educational and medical services. In the same region, Columbia should start a regional satcom project in 2009 while Argentina will launch a domestically produced satcom in 2012. In Africa, apart from Nigeria which will launch a Chinese-manufactured replacement satellite for the lost Nigcomsat by 2011, Angola ordered the Angosat satellite to the Russian manufacturer RKK Energya, Algeria will launch its Alcomsat in 2012 and Egypt ordered the Nilesat

201 to Thales Alenia Space. In Asia Laos started a project with China and Vietnam plans to widely apply space-based communications technologies after the launch of its Vinasat 1 in April 2008. After the loss of its first satcom Kazsat 1, Kazakhstan also has two communication satellites currently under construction by Khrunichev, Kazsats 2 and 3. Similarly, Azerbaijan plans to launch a project for the construction of a satcom covering central Asia, to be launched a 2010. A second satellite is also foreseen for 2012.[440]

4.2. Space-based positioning, navigation and timing systems

2008/2009 was marked by a high level of activity in the field of space-based navigation systems. The U.S. continued to develop its next generation GPS system, while Russia gave a new impetus to its GLONASS programme. China is also rapidly implementing its Beidou system, while India and Japan develop regional GPS augmentation systems. Finally, the Galileo and GNSS programmes in Europe were marked by an important re-profiling in 2008/2009. As a whole, spending in this field could reach 10 billion U.S. dollars in 2010 and 144 satellites are scheduled to be launched between 2008 and 2017.[441] Two important trends regarding PNT systems in 2008/2009 were the growing question of compatibility between the systems, leading to bilateral and multilateral negotiations for frequency allocations, and the development of the economic potential of PNT applications.

In Europe, EU documents recalled that the implementation of Galileo was a priority within the ESP. The ESP Progress report[442], released by the EC on 11 September 2008, as well as the Resolution adopted at the 5th Space Council on 26 September 2008[443], laid down the next steps in this regards. Both documents also emphasised the need to stimulate downstream markets based on satellite navigation applications.

Following the failure of the negotiations for concession contracts with the private sector and the Council's conclusions on the need to restructure the European Global Navigation System Programmes in November 2007, the Council and the EP adopted a regulation on the further implementation of the European satellite navigation programmes on 9 July 2008.[444] The regulation adopted a legal and financial framework for the European GNSS programme for the 2008–2013 period. The main decision of the text was the new role of programme manager allocated to the EC. In this position, the EC took a series of actions in 2008/2009 to ensure that the programmes will be implemented according to the objectives. For EGNOS, the EC took over the programme from ESA on behalf of the

Community on 1 April 2009. It further agreed with ESA that it will take on the design and procurement of EGNOS equipment and software renewals. For Galileo, the EC delegated the infrastructure procurement to ESA, launched the competitive tendering procedure for the procurement of the Galileo infrastructure and provided ESA with the necessary funds to cover the cost overruns of the IOV phase.[445]

The directive tackled two main issues: the legal framework and a new governance structure. As for the legal framework, the text introduced a strict division of responsibilities between the EC, ESA and GSA with the EC having the overall responsibility for the programme management. Furthermore, the Full Operational Capability (FOC) phase from 2008 to 2013 will be financed by the Community budget, amounting to 3.4 billion euros in total. Finally, the regulation sets rules on the governance of security matters and the application of security regulations. For the period starting in 2014, the EC will make a financing proposal by 2010. As for the governance structure, several new bodies were created. The European GNSS Programme Committee was set up to assist the EC in the management of the programme, the Galileo Inter-institutional Panel was created to facilitate each Community institution exercising its respective responsibility and the GNSS Security Board was formed to deal with the security issues of the GNSS programmes. Furthermore, the roles and responsibilities of ESA and the EC were clarified and a proposal was adopted to revise the mandate of the GSA (GNSS Supervisory Authority).[446]

Following the important step of the GNSS regulation, the procurement for the deployment of the full Galileo constellation was launched on 1 July 2008. The IOV which is ongoing will result in the launch of four satellites, while the procurement process concerns the FOC phase. Six work packages were identified for the procurement (system support, ground mission system, ground control system, satellites, launchers and operations). As the procurement procedure is particularly complex due to technical and highly sensitive issues, the EC opted for the procedure of "competitive dialogue". In the first preselection phase, a Tender Information Package (TIP) was released and interested companies could submit their preliminary proposals in November 2008. A shortlist of bidders was then established for the second phase of the procurement, the actual competitive dialogue. The process will be finalised in the course of 2009.[447] The first contracts were signed in June 2009 with EADS Astrium and OHB/SSTL for the hardware of the satellite constellation. The selection of two competitors could indicate a double source for the procurement of the satellites. However, ESA sees this as an insurance against further delays. Another debate focused on the launch vehicle chosen. ESA is currently planning to launch 28 satellites with Soyuz, but Astrium Space Transportation is pledging for launches aboard Ariane 5ES to reduce dependence on Russian hardware and increase European strategic autonomy.[448]

The overall difficulties encountered by the Galileo programme were analysed in detail in a report released by the European Court of Auditors in June 2009. Following an ESA audit, that had concluded that the cost overruns incurred during the IOV phase were justified, the EC conducted its own audit. The report assessed the period of the GJU (Galileo Joint Undertaking), a management tool set up by ESA and the EC between 2003 and 2006. It concluded that the management of Galileo during this time has been inadequate, in particular concerning the negotiations with the private sector. After the release of the report, the EC decided to pay for the cost overruns.[449]

Finally, an important turn in the Galileo programme was marked by the increased emphasis put on downstream markets in the recent months. On the background of the crisis, and given the new priority of space as a contribution to the Lisbon strategy identified at the 5th Space Council, the creation of new markets for downstream services, mainly for SMEs, was pushed forward. Several EU documents highlighted the economic potential of the Galileo programme, including the ESP Progress report and the resolution from the 5th Space Council.

In the U.S., the last satellite from the GPS 2R generation was launched in August 2009, while the development of the enhanced GPS 3 version continued. 2008/2009 was also marked by the publication of a report by the Government Accountability Office (GAO) warning about a possible degradation of GPS services in the coming years and by debates on the eLoran backup system for GPS. GPS as a vital strategic system remained a U.S. priority in space, as testified by the Presidential proposal for the FY 2010 budget. More than 1 billion U.S. dollars were provided in it for GPS and related positioning, navigation and timing (PNT) programmes. As GPS is a military-operated system, the major share of the funding (927.8 million U.S. dollars) would go to the DoD.[450]

The Air Force launched the seventh GPS 2R-M satellite on 24 March 2009 after several delays. The spacecraft is carrying a demonstration payload for the new L-5 civil signal, but signal anomalies were detected after the launch, and the Air Force expects an operational entry into service of the satellite by October 2009.[451] The last launch of a GPS 2R satellite is scheduled for August 2009, which will also mark the final launch of the Delta 2 vehicle for the Air Force. The medium launcher entered into service in 1989 after the Challenger disaster, and launched 48 GPS spacecraft since then.[452] The first launch of 12 Block 2F GPS satellites, the upgraded version, will take place in the beginning of FY 2010. The spacecraft will include new hardware for the civil user community.[453]

In parallel, the development of the next generation GPS 3 continued. Lockheed Martin was awarded a 3 billion U.S. dollars development and production contract in May 2008. The Preliminary Design Review (PDR) was successfully completed in May 2009, validating the design of the spacecraft that would allow it meeting its military and civilian requirements. The programme then entered in the Critical Design Review (CDR) phase and the first launch is scheduled for 2014. The new spacecraft will provide enhanced performances both to civilian users, with a new international civil signal (L1 C), and to military users, with increased anti-jam power and global coverage.[454]

In May however, the GAO report expressed concerns about the future of the GPS programme. It warned that delays in the production and launch of the next GPS satellites could lead to a decline of the constellation. It further criticised the lack of synchronisation between the acquisition and development of the spacecraft and of the ground segment. The Air Force however, downplayed the risks, stating that the GPS constellation is not going to stop functioning, but rather that there is a risk for slight degradation in performance over small portions of the world and over short periods of time.[455]

On the background of these developments, another issue at stake in 2008/2009 was the future of the eLORAN (enhanced LOng RAnge Navigation) system. The funding for this terrestrial radio navigation system, long time considered as a backup for GPS, was cut in President Obama's FY 2010 budget proposal. The Congress however, favours a continuation of the programme, given the potential problems faced in the future by the GPS constellation. The issue was then passed to the House-Senate appropriations conference committee.[456]

On the technical level, a new GPS Standard Positioning Service Performance Specification (SPS PS) has been issued in the fall of 2008. It defines the new technical standards for the GPS system, and represents the first new version of the SPS PS since 2001.[457]

As for compatibility issues with other GNSS systems finally, representatives form the U.S. and from the EC met in Washington D.C. on 23 October 2008 to discuss interoperability and cooperation issues. The meeting took place in the frame of the 2004 cooperation agreement on Galileo and GPS.[458]

In Russia, 2008/2009 was marked by the new impetus given to the GLONASS programme. After the collapse of the Soviet Union, the economic decline hurting Russia coupled with the short lifespan of the first-generation GLONASS satellites had lead to a rapid dwindling of the operational GNSS constellation. On 12 September 2008 however, President Putin took the decision to restore the Russian GNSS system by the end of 2009. 67 billion roubles (2.62 billion U.S. dollars) were consequently approved for the period 2008–2011 to modernise and rebuild the GLONASS fleet. 24 new GLONASS M spacecraft are scheduled for launch, which will extend the constellation to 30 satellites by 2011. Already in 2008, a total of 6 GLONASS M spacecraft were launched, 3 in September and 3 in

December.[459] In May 2009, the Russian government announced that despite the economic crisis, Russia will not cut its funding for the GLONASS system,[460] therefore testifying the strategic importance of a GNSS for Russia.

In parallel, Russia continued to develop its next-generation GLONASS K spacecraft which will be launched in December 2010 for the first time. The smaller design of the satellite will enable double launches aboard Soyuz 2 rockets, instead of using the Proton launcher, thus reducing launch costs significantly. In addition, Russia is developing a GNSS augmentation system similar to EGNOS, the GLONASS System of Differential Corrections and Monitoring (SDCM).[461] Researchers at the Institute of Space Studies from the Russian Academy of Science are also working on a fundamentally new navigation system that would use pulsar X as reference points. This would make the system totally independent from terrestrial stations and would make it suitable for interplanetary journeys.[462]

The legal framework related to GNSS was also precised through the adoption of a Federal Law on navigation activities in February 2009. It is intended to develop and support the GLONASS system and its applications for contributing to the economic growth. Among others, the law establishes a legal framework for navigation activities and defines the powers of federal, regional and municipal bodies.[463] Russian authorities also intend to unleash the economic potential of GLONASS applications. The government indeed, was considering a proposal to stimulate domestic manufacturing of high volume GLONASS and GLONASS-GPS user equipment. In Parallel, the transport ministry is working on a regulation to impose the use of GLONASS receivers on governmental vehicles, vessels and aircrafts.[464]

China is moving fast towards the development of its Compass PNT system. The first Beidou 1 MEO spacecraft, launched in 2007, provided substantially better atomic-clock timing precision than expected. On 14 April 2009, China launched a second GEO PNT spacecraft, the modernised Beidou 2. China plans to launch 10 more spacecraft in 2009/2010, in order to reach a regional PNT capability in the Asia-Pacific region by 2010. The completion of this regional capability will cost around 1.46 billion U.S. dollars. By 2015, the constellation will consist of 30+ satellites. The Compass system will offer 10 services, five free "open" services and five restricted "authorised" services.[465]

A growing concern is the overlay of Beidou signals with the planned Galileo PRS (Public Regulated Service) frequencies, intended to be used for public safety, security and military applications. Europe is concerned that if China completes its regional system by 2010, it will be up to Galileo to adapt its frequency, as the European PNT constellation could enter in service only by 2013 or later. Although negotiations are ongoing, Chinese intentions are still unclear.[466]

Apart from GNSS systems, India and Japan are also developing regional navigation systems. India is working towards the deployment of 3 GEO and 4 SSO satellites for its GPS-Aided Geosynchronous Augmented Navigation System (GAGAN). The first satellite could be launched in 2010, and the system could become fully functional in 2013. GAGAN will bridge the gap between the European EGNOS and the Japanese MSAS (MTSAT Satellite-Based Augmentation System) to provide satellite-based navigation for civil aviation over Indian airspace.[467]

In Japan, finally, the new Basic Plan for Space Policy released in June 2009 put a particular emphasis on the completion of the local GPS augmentation system, the Quasi Zenith Satellite System.[468]

4.3. Space-based Earth observation

EO satellite applications continued to play a central role in 2008/2009, both for established space powers and emerging space actors. In particular, EO satellites often represent a priority for developing countries, as they can help tackling pressing developmental issues. Growing trends in the EO sector include the increasing focus put on the fight against climate change, the development of downstream markets and the military use of EO satellites.

In Europe, a series of official documents paved the way for the emergence of a real GMES programme in 2008/2009 (see Sect. 2.4.2). The Forum GMES 2008, which took place in Lille in September 2008, marked the launch of the first GMES services in pre-operational mode. The pre-operational services include marine and atmosphere services, aiming at monitoring and understanding climate change and the Earth's subsystems, and land, emergency and security services, used for geo-information services, security applications and emergency response.[469] Despite the remaining of several open questions, on financing issues, the decision making process or the governance architecture, GMES will move into full operational phase in 2013.

While Segment 1 of the GMES Space Component covers the period 2009–2018, Segment 2 covering the period 2009–2018 was approved by the ESA Ministerial Council on 25–26 November 2008. Subscriptions from the participating States amounted to 831.4 million euros, adding to the 205 million euros granted by the EC for Segment 2. The Segment 2 of the GMES Space Component comprises the five Sentinel satellites, as well as so-called Contributing Missions. During this second phase, the development of the initial series of Sentinel satellites will be completed, and the access to EO data from Contributing Missions will be ensured. Specifically, the spacecraft Sentinels 1B, 2B and 3B are scheduled for launch respectively in 2015, 2016 and 2017. The two Sentinel 4 instruments will be carried onboard the MTG satellites, while a Sentinel 5 instrument will be carried on a post-EUMETSAT Polar System (EPS).[470] On 28 January 2009, an amendment to the February 2008 ESA-EC agreement on GMES was signed. It will extend the scope of the 2008 agreement to Segment 2 activities, thus paving the way to the ordering of the second units of the Sentinel 1, 2 and 3 satellites and of the Sentinel 4 and 5 instruments.[471]

An important aspect of the GMES programme is the stated will to create new business opportunities for the private sector, and to support the emergence of a downstream sector for EO applications. These aspects were highlighted in all the official EU documents on GMES that were issued in 2008/2009. Finally, high-ranked European officials also made clear that GMES should be used by the military as well, and should be considered a dual-use resource from the beginning.

The Earth Observation Programme of ESA on its side is composed of principal missions and opportunity missions. The current principal missions include GOCE (launched in March 2009), ADM Aeolus (launch in 2011) and Earth-CARE (launch in 2013), while the selection process for the fourth mission is ongoing. The opportunity missions on the other hand, are SMOS, Cryosat 2 (both to be launched in 2009) and Swarm (launch in 2011). All missions have a strong focus on investigating climate change.

GOCE was the first satellite to be launched in the series of EO satellites called Earth Explorers, on 19 March 2009. It will measure the Earth gravity field and map the gravity variations. As the satellite is flying at the altitude of 265 km, an electronic propulsion system keeps it completely free from drag, which represents a main technological achievement.[472] The next Earth Exploration mission will be the Atmospheric-Dynamics Mission (ADM)-Aeolus, which will be launched around 2010 and will measure wind profiles on a global scale. Around 2013, the EarthCARE (Earth Clouds, Aerosol and Radiation Explorer) mission is planned for launch to investigate the interaction between clouds, radiations and aerosols in climate regulation.[473] The selection process for the 7th Earth Explorer mission (and the 4th principal mission) is ongoing. In January 2009, 6 concepts were pre-selected after a 2 years assessment cycle. ESA's Programme Board for Earth Exploration then selected three missions for feasibility study in February 2009: BIOMASS, PREMIER and CoReH2O. The winning mission will be launched around 2016.

As for the opportunity missions, SMOS (Soil Moisture & Ocean Salinity), also called ESA's water mission, is scheduled to be launched in November 2009. Its task will be to measure the moisture in salt and soil in the surface waters of the ocean, in order to contribute to the understanding of the water cycle and therefore of climate change. Cryosat 2, or ESA's ice mission, will be launched in December

2009 to investigate whether climate change is causing the polar ice caps to shrink. The magnetic mission Swarm, finally, will be launched in 2010 and will provide high-resolution and high-precision measurements of the strength and direction of the Earth's magnetic field.[474]

Two current ESA EO missions were extended in 2009: the Cluster mission, investigating the Earth's magnetosphere, in February 2009 and the environmental satellite Envisat in June 2009.

In the U.S., NASA experienced a failed launch for its OCO (Orbiting Carbon Observatory) mission on 24 February 2009. This important mission was supposed to study the entire carbon cycle and it was the first spacecraft dedicated to studying atmospheric carbon dioxide.[475] However, NASA and NOAA put emphasis on climate change through other meteorological missions as well. The Jason 2 spacecraft, launched in June 2008 and resulting from a cooperation endeavour between NASA, NOAA, CNES and EUMETSAT, started to disseminate operational data in December 2008. In February 2009, NOAA 19 was launched, carrying a EUMETSAT instrument. Furthermore, the GOES O (Geostationary Operational Environmental Satellite) spacecraft was launched on 27 June 2009. NASA is also working on the improved GOES R version to replace the current weather satellites. The contracts for the instruments were awarded in 2008, while Lockheed Martin has been selected in May 2009 to build the two satellites.[476]

The development of the NPOESS (National Polar-orbiting Operational Environmental Satellite System) on its side, continued to face financial and technical problems in 2008. The first launch, initially planned in January 2013, will not take place before May 2014, and cost overruns amounted around 1 billion U.S. dollars above the 12.5 billion U.S. dollars originally foreseen. The programme, managed by NASA, the DoD and the Department of Commerce (DOC), is intended to develop the next generation of LEO environmental satellites. It will pool into one programme the operational requirements from the military and the civilian users, to replace both the DoD Defense Meteorological Satellite Programme (DMSP) and the NOAA Polar-orbiting Operational Environmental Satellite (POES) series.[477]

EO is also one of the main priorities of China, in particular to monitor natural disasters. China is moving towards the deployment of its own satellite fleet in this area, after having been a leader in helping to coordinate international EO initiatives such as GEO. It launched the small environmental optical EO satellites Huang-Jing 1A and 1B in July 2008, and the Huang-Jing 1C radar EO satellite is scheduled for launch in 2009. These spacecraft will be part of an 8-satellites optical and radar-imaging constellation, which will fly in formation to provide more coordinated information. The entire constellation will be operational by 2010.[478] In addition, China launched three remote sensing satellites in 2008/2009, Yaogan

4 and 5 in December 2008 and Yaogan 6 in April 2009. It also launched its third experimental Earth mapping satellite Shiyan 3 in November 2008, which was China's first digital mapping system. Although all these satellites are for civilian use, the possibility of dual-use is not excluded.[479]

Finally, China is also building up its meteorological satellite fleet. It launched its 5th GEO meteorological satellite Feng-Yung 2 (FY 2) in December 2008 and the last FY 2 should be launched in 2013. By then, the new generation of GEO satellites FY 4 should be launched. In parallel, China is replacing its first generation polar weather satellites FY 1 by FY 3 satellites: FY 3A was launched in May 2008 and FY-B is scheduled for launch in 2010.[480] The launch of an ocean survey satellite Haiyang 2A is also planned for 2010, to monitor ocean wind fields, sea levels and temperatures.

Japan also contributed to the fight against climate change through satellite applications with the launch of its Greenhouse Gases Observing Satellites (GOSAT) on 23 January 2009. The project, jointly developed by JAXA and the Japanese Environment Ministry, aims at monitoring the distribution of the density of carbon dioxide.[481]

Russia on is side, mainly launched military EO satellites, all optical systems: Kosmos 2441 in July 2008, Kosmos 2445 in November 2008 and Kosmos 2450 in April 2009. It was imitated by Germany, which launched its radar reconnaissance spacecraft SAR-Lupe 5 in July 2008, Italy, which launched its radar EO satellite COSMO-Skymed 3 in October 2008 and India which developed its SAR imaging Radar Imaging Satellite (RISAT) 2 based on the Israeli TecSar 1 satellite and launched it in April 2009.[482]

A growing trend is also represented by the rising number of satellite programmes from emerging space-faring nations. Over the next 10 years, 30% of all EO satellites will be developed within government space programmes, in particular in developing countries. A number of factors is explaining this trend, among which the low cost of EO satellites, their high-value contribution to the solving of concrete problems, such as disaster management, natural resources monitoring or cartography, the possible commercialisation of data and the technology transfer.[483] Thailand launched its THEOS in October 2008 and will provide the country with worldwide geo-referenced image products. Similarly, Nigeria continued to develop its Nigeriasat 2 EO satellite, which is being built by SSTL and will be launched in 2010 and Turkey made progresses in the development of its Göktürk military EO satellite, as Telespazio was chosen as the prime contractor in December 2008.

Finally, a growing trend is the maturation of the global remote sensing market, which was soaring in 2008. The market consists of three segments: Very High Resolution (VHR) optical imagery, Medium Resolution (MR) optical imagery and SAR imagery. In the last years, more imagery of better quality was made available, and the data distribution became faster. In 2008/2009, 6 commercial EO satellites were launched: the 5 RapidEye spacecraft in July 2008 and GeoEye 1 in September 2008.

5. Technology developments

Several areas of technology are very relevant for discussion of space policy, since they constitute important boundary conditions for decision making. In the following, selected topics will be shortly sketched to provide a high-level overview and to convey a sense of what is going on and what will be important for the future.

5.1. Propulsion

Europe pursues its work on the "cryogenic high thrust Vinci rocket" engine within the Future Launcher Preparatory Programme (FLPP) of ESA since 2005. The efforts are coordinated by Snecma, involving partners like Astrium GmbH, Avio, Volvo Aero and Techspace Aero. Vinci is primarily aimed at the next upgrade of the Ariane 5 launcher, but it might be as well adapted to suit other upper stages or space vehicles. As of mid 2008, 37 test firings have been undertaken, and the engine had spent a total of 2200 seconds under simulated flight conditions.[484]

On the low thrust side, Europe for the first time is using a regulated ion engine with permanent thrust control aboard the GOCE satellite that was launched in March 2009. EADS Astrium was in charge of the satellite platform. The ion power unit serves to compensate the drag that is caused by residual atmospheric molecules. This attitude control keeps the satellite on an exact orbit to be able to perform its task of measuring the Earth's gravitational field in a uniform way.[485]

Outside Europe, propulsion science and technology is also carried out in the high and low thrust domain. In the high thrust segment, the first segment of the five-segment U.S. development motor DM-1 with new solid boosters for the Ares 1 rocket as part of the Constellation programme was moved from the ATK production site to the nearby test stand in April 2009. Later in 2009, a firing test was completed successfully, and the Ares 1X rocket passed a flight test.[486]

As for the low thrust domain, NASA is using ion thrusters on its interplanetary Dawn mission aimed at a target asteroid. Before, ion propulsion had been tested and proved in the Deep Space 1 mission. Dawn is equipped with three ion thrust units that can be moved in two directions to compensate the spacecraft's changing centre of mass and to support attitude control. The three thrusters provide enough lifetime and reserve for the required mission time. Only one thruster will be operating at a time. The challenging mission objective of Dawn could not be met without ion engines.[487]

5.2. Information technology

Laser communication continues to be a field of intense research and of several testing endeavours. As a high frequency carrier, laser allows to transmit more information per time unit than conventional radio signals. Besides, it features vastly reduced beam divergence, which relieves the burden on power budget calculations and antenna dimensions. However, laser-based space communication is challenging due to the high velocities involved and to potential disturbance by atmospheric conditions.

In 2008, there had already been successful attempts to set up a laser communication link between the German TerraSAR-X and the U.S. American NFIRE (Near Field Infrared Experiment) satellite, bridging a distance of 5000 km and realising a bidirectional data transfer rate of 5.5 Gbit per second. This is about 20 times more than conventional data links employing microwaves.

In October 2008, the German Aerospace Center succeeded to contact TerraSAR-X from the ground station by laser.[488] The experiments included signal acquisition, mutual terminal tracking and signal quality assessment. Within the trials, no coherent signal was received, but measurements of received power and wave front quality were performed, along with an assessment of the associated adaptive optics.

One possible application of laser communication would be real-time data links from Earth to high performance Earth observation satellites that produce large amounts of data. These links could also be established via geostationary satellites to make up for limited visibility of Earth observation satellites in low orbit. In fact, the European Data Relay System is planned to employ laser communication.[489]

5.3. Spacecraft operations and design

In the area of spacecraft operations (and its ground segment complement), there are two major trends to be observed – automation and outsourcing. While the latter is not a technology in itself it influences technology and its utilisation. Both developments follow the overarching aim of enhancing quality, improving reliability and reducing costs. The 8th workshop of a series called "Reducing Cost on Ground Segment and Operations" was held in Tsukuba, Japan, in May 2009. It gathered representatives from many space agencies and industry.

Generally, automation supports operation by taking over repetitive or labour intensive tasks and by decreasing reaction times. This allows the staff to concentrate on high level tasks and to perform overview. It is obvious that only well understood tasks can be automated and that less well understood tasks have to be performed manually. For unique science missions, the cost of automation would be too high. Moreover, automation demands a high degree of reliability for subsystems involved, and it will have implications for work profiles and required competences.

As for outsourcing, the classical mode is to transfer operation of a complete business segment to an external service provider. Space agencies can transfer ground station operation or IT support to commercial entities. However, they must keep a minimum degree of knowledge in-house, and they must avoid the monopolisation of service provision. Outsourcing of ground station operation is only possible within commercially viable space segments like Earth observation. At the launch of ESA's GOCE mission, three of the four ground stations involved were commercially operated. The lessons learnt in the process are being reviewed and are used for future optimisation.

Regarding the purely technological aspects, various successes have been achieved. The Delta-DOR (Differential One-way Range) method employed by ESA for determination of angular positions of spacecraft has been enhanced to facilitate utilisation of non-European ground stations like those from NASA or JAXA as well. Interoperability allows for working together in suitable missions.

Shortly after launching the Herschel spacecraft in May 2009, a GMSK-modulated high data rate downlink of 1.5 Mbps was established. GMSK (Gaussian Minimum Shift Keying) is very efficient in terms of bandwidth and transmission power. It is the first time this modulation was used in space. GMSK has been used by ground mobile telephone networks before, but at a lower data rate.

Apart from this progress, there are growing concerns of spacecraft operators about having to compete for frequency bands with terrestrial systems. In some domains, S-band frequencies used for communication with spacecraft are also used by UMTS or other standards. This has already led to conflicts of interest.

5.4. Suborbital activities

As an important part of human spaceflight, various commercial entities develop reusable launch vehicles for suborbital flights for space tourists.[490] Virgin Galactic officially presented its White Knight 2 (WK2) vehicle in July 2008. WK2 will carry the rocket plane Space Ship 2 (SS2), and launch it at an altitude of 15 km with 6 passengers. The entry into service is forecasted for 2010. The first flight of WK2 took place in December 2008, while SS2 should fly into space by the end of 2009 or the beginning of 2010.

Armadillo Aerospace and Rocket Racing Inc. created a joint venture in October 2008 to fly paying customers to the edge of space as soon as 2010. The company's plan is to build and exploit a fleet of reusable vertical take-off and landing vehicles avoiding heavy wings. The vehicles are foreseen to feature deep throttling, regeneratively cooled, alcohol burning motors, which are not commercially available. The vehicles could also carry scientific equipment or conduct meteorological measurements. First flight tests were conducted in 2009.

XCOR Aerospace is developing the Lynx spacecraft, a rocket-powered vehicle which takes-off and lands like planes and is capable of several flights per day. Featuring liquid fuel engines, it will be much smaller than other vehicles for suborbital flight, and it will only climb to an altitude of 60 km. Flying opportunities for tourists will be considerably cheaper than those planned by space travel competitors, and the passengers will seat besides the pilot, which ensures an optimum view. Commercial flights are expected to start in 2010. Moreover, Lynx can also be used to test hardware in weightlessness.

Bigelow Aerospace, a start-up company founded in 1999, has plans for expandable space station modules. It developed and launched two prototypes in 2006 and 2007, while it cancelled the launch of its third prototype due to rising launch costs. There have been plans to directly develop and launch Sundancer, a spacecraft capable of supporting human crew, by 2010. The concept would serve industrial and scientific purposes, but it would also hold a potential for space tourism.

The activities above are being run in the United States. To promote the emerging industry and to create a clear legal, regulatory and safety regime, the

Tab. 4: *FAA permitted suborbital flight events in 2008 (source: FAA).*

Date	Vehicle	Company	Site
25 October 2008	Pixel	Armadillo Aerospace	Las Cruces, USA
24 October 2008	MOD-1	Armadillo Aerospace	Las Cruces, USA
24 October 2008	MOD-1	Armadillo Aerospace	Las Cruces, USA
24 October 2008	Ignignokt	Scott Zeeb d/b/a TrueZer0	Las Cruces, USA
24 October 2008	MOD-1	Armadillo Aerospace	Las Cruces, USA

U.S. Congress passed the Commercial Space Launch Amendment Acts in 2004. These acts established an experimental permit regime, and made the FAA responsible for regulating human spaceflights. In 2008, the FAA authorised 5 launches as listed in Table 4.

5.5. Other technologies

The new computerised Russian space suite Orlan-MK was tested in June 2009 at the International Space Station during a five hour extravehicular activity undertaken by commander Gennady Padalka and flight engineer Michael Barrat. The new Orlan-MK spacesuit features a mini-computer in the portable life support system backpack. It monitors the state of the spacesuit and issues a warning in the case of problems. To facilitate adequate response to the malfunction, a LCD display attached to the breast outlines a contingency plan.[491]

The concept of a space elevator receives considerable attention. Japan drew up an international conference on this topic in November 2008. A space elevator would consist of a stretched cable between a satellite docking station in geostationary Earth orbit and a ground station. Carbon nanotubes are considered to be candidates for these cables that need to be extremely light, mechanically stable and resistive to impacts originating from within and outside the atmosphere. Carbon nanotubes files might also be employed to provide electrical power to the elevator.[492]

5.6. Innovation policy

A series of prizes is running in the space sector, with the aim of supporting and speeding private endeavours and commercial technological developments. The Northrop Grumman Lunar Lander X Prize Challenge and the Google Lunar X Prize support an effort to establish international partnerships between space agencies and private entrepreneurs.[493]

The Northrop Lunar Lander Challenge will have direct application for NASA's space exploration as well as for the personal spaceflight industry. The aim of the competition, which comprises two levels, is to simulate lunar lift-offs and landings. For level 1, a rocket has to take off from a designated launch area, to fly up to an altitude of 50 m, and then hover for 90 seconds while landing precisely on a landing pad 50 m away. The flight must then be repeated in reverse. For level 2, the rocket has to hover for twice as long before landing precisely on a simulated lunar surface. The prize for the winner is worth 2 million U.S. dollars. Nine teams engaged in the competition in 2008, and Armadillo Aerospace won level 1 in October 2008, earning 350,000 U.S. dollars. However, Armadillo Aerospace failed to win level 2 later on.[494]

The Google Lunar X-Prize is a 30 million U.S. dollars international competition to safely land a robot on the surface of the moon, travel 500 meters and send images and data back to the Earth. The participating teams must be at least 90% privately funded, and registered by 31 December 2010. The first team to succeed will win 20 million U.S. dollars, the second team will get 5 million U.S. dollars, and the additional 5 million U.S. dollars are for bonuses. Astrobotic Technology was the first team to officially register for the prize. Meanwhile, around 20 teams have signed up.[495]

[324] "International Cooperation with the USA." 12 Dec. 2009. European Commission. 13 Jan. 2010. http://ec.europa.eu/enterprise/policies/space/esp/international-cooperation/usa/index_en.htm.
[325] "International Cooperation with Russia." 12 Dec. 2009. European Commission. 13 Jan. 2010. http://ec.europa.eu/enterprise/policies/space/esp/international-cooperation/russia/index_en.htm.
[326] Lardier, Christian. "L'Europe Mise sur les Applications." Air & Cosmos 2150. 5 Dec. 2008: 38.
[327] ESA and Arianespace Sign a Frame Contract For the Procurement of Launch Services. 15 June 2009. ESA. 19 Aug. 2009. http://www.esa.int/esaCP/SEMAP73XTVF_index_0.html.
[328] Rathgeber, Wolfgang. "Space Policies Issues and Trends in 2008/2009." 18 May 2009. ESPI Report 18. 19 Aug. 2009. http://www.espi.or.at/images/stories/dokumente/studies/espi%20report%2018.pdf.
[329] Lardier, Christian. "EADS Relève les Défis Spatiaux." Air & Cosmos 2178. 26 June 2009: 22+.
[330] ESA website. 19 Aug. 2009. http://www.esa.int/.
[331] Taverna, Michel A. "Race for Soyuz." Aviation Week & Space Technology 8 Sept. 2008: 64; Lardier, Christian. "Les Lanceurs Spatiaux du XXIe Siècle." Air & Cosmos 2176 12 June 2009: 110+; Lardier, Christian. "ESA: Premiers Contrats Galileo." Air & Cosmos 2177 19 June 2009: 38+.
[332] Lardier, Christian. "Des Projets Pour l'Avenir d'Astrium." Air & Cosmos 2167 10 Apr. 2009: 16+.
[333] de Selding, Peter B. ESA Invests 100 Million Euros in Delayed Soyuz, Vega Programs. Space News 20 Apr. 2009: 6.
[334] Bigot, Bernard/d'Escatha, Yannick and Laurent Collet-Billon. "L'Enjeu d'une Politique Européenne de Lanceurs: Assurer Durablement à l'Europe un Accès Autonome à l'Espace. " 18 May 2008. 11 Aug. 2009. http://lesrapports.ladocumentationfrancaise.fr/BRP/094000223/0000.pdf.
[335] "ESA Signs High Thrust Engine Demonstrator Contract for Next Generation Launcher." 17 June 2009. ESA. 19 Aug. 2009. http://www.esa.int/esaCP/SEMHBUZO0WF_index_0.html.
[336] "NASA Awards Space Station Commercial Resupply Contracts." 23 Dec. 2008. NASA Press Release C08-069. 20 Aug. 2009. http://www.nasa.gov/home/hqnews/2008/dec/HQ_C08-069_ISS_Resupply.html.
[337] SpaceX website. 20 Aug. 2009. http://www.spacex.com/.
[338] Orbital Sciences Corporation website. 20 Aug. 2009. http://www.orbital.com/.
[339] "Russia to Test Launch Angara Carrier Rocket in 2011." 19 Mar. 2009. Space Daily. 20 Aug. 2009. http://www.spacedaily.com/reports/Russia_To_Test_Launch_Angara_Carrier_Rocket_In_2011_999.html; "Rockets: Angara Family." Russianspaceweb. 20 Aug. 2009. http://www.russianspaceweb.com/angara.html.
[340] Lardier, Christian. "Les Lanceurs Spatiaux du XXIe Siècle." Air & Cosmos 2176 12 June 2009: 110+.

[341] Russianspaceweb website. 20 Aug. 2009. http://www.russianspaceweb.com/ppts_lv.html.
[342] "Centers: Svobodny." Russianspaceweb. 20 Aug. 2009. http://www.russianspaceweb.com/svobodny.html.
[343] JAXA website. 20 Aug. 2009. http://iss.jaxa.jp/en/htv/index.html.
[344] Ibid.
[345] Bigot, Bernard/d'Escatha, Yannick and Laurent Collet-Billon. "L'Enjeu d'une Politique Européenne de Lanceurs: Assurer Durablement à l'Europe un Accès Autonome à l'Espace. " 18 May 2008. 11 Aug. 2009. http://lesrapports.ladocumentationfrancaise.fr/BRP/094000223/0000.pdf: 17.
[346] Couvault, Craig. "Great Wall Sanctions Lifted." Aviation Week & Space Technology 8 Sept. 2008: 31.
[347] "Long March 5 Will Have World's Second Largest Carrying Capacity." 4 Mar. 2009. Space Daily. 20 Aug. 2009. http://www.spacedaily.com/reports/Long_March_5_Will_Have_World_Second_Largest_Carrying_Capacity_999.html.
[348] "New Space Launch Center to Be Built in China's Southernmost Hainan." 3 Mar. 2009. Space Daily. 18 Aug. 2009. http://www.spacedaily.com/reports/New_Space_Launch_Center_To_Be_Built_In_China_Southernmost_Hainan_999.html.
[349] Lardier, Christian. "L'Inde Augmente son Budget Spatial de 27%." Air & Cosmos 2161 27 Feb. 2009: 42+; Jayaraman, K.S. "ISRO Gets Boost in Annual Budget." Space News 23 Feb. 2009: 16.
[350] Subramanian, T.S. "An Indian Space Shuttle Takes Shape." The Hindu 1 Apr. 2009. 20 Aug. 2009. http://www.hindu.com/2009/04/01/stories/2009040161821100.htm.
[351] Shin, Jenny. "Chronology of North Korea's Missile Flight Tests." 13 July 2009. Center for Defense Information. 21 Aug. 2009. http://www.cdi.org/pdfs/NKMissileTestTimeline7.16.09.pdf.
[352] Shine, Jenny. "A Chronology of Iran's Space Activities." 2009. Center for Defense Information. 21 Aug. 2009. http://www.cdi.org/pdfs/IranSpaceTimeline09.pdf.
[353] Kim, Tong-hyung. "Local Space Program Hits Snag." 8 Dec. 2008. The Korea Times. 21 Aug. 2009. http://www.koreatimes.co.kr/www/news/tech/2009/04/133_35754.html; Kim, Tong-hyung. "Rocket Launch Delayed Until July." 12 Mar. 2009. The Korea Times. 21 Aug. 2009. http://www.koreatimes.co.kr/www/news/nation/2009/03/123_41199.html.
[354] This figure includes 62 successful ELV launches, 2 ELV failures and 4 Shuttle flights. The Iranian attempt on 17 August 2008 is not included, as it might not have targeted anything than a sub-orbital trajectory.
[355] "NASA Assigns Space Station Crews, Updates Expedition Numbering." 21 Nov. 2008. NASA. 9 Dec. 2009. http://www.nasa.gov/home/hqnews/2008/nov/HQ_08-306_Expedition_crews.html.
[356] "Europe's Reaction to the Augustine Report." Oct. 2009. ESPI, DGAP, FRS, IFRI, SWP, London Institute of Space Policy and Law. 10 Dec. 2009. http://www.espi.or.at/images/stories/dokumente/studies/joint%20memorandum%20_%20europe%5C%27s%20reaction%20to%20the%20augustine%20report.pdf.
[357] Iannotta, Becky. "NASA puts $150 Million into Commercial Crew Transport." Space News 4 May 2009: 6.
[358] Iannotta, Becky. "SpaceX Continues Push for Role in Human Exploration." Space News 6 April 2009.
[359] Lardier, Christian. "Satellites: Entre Exploration et Télécommunications." Air & Cosmos 2176 12 June 2009: 116.
[360] Lardier, Christian. "Nouveau Segment Russe Pour L'ISS." Air & Cosmos 2136 29 Aug. 2009: 35.
[361] Moskowitz, Clara. "China Again Steps into the Space Limelight With Shenzhou 7 Launch." Space News 29 Sept. 2008: 24.
[362] Iannotta, Becky. "China Developing Unmanned Module as Space Station Precursor." Space News 6 April 2009 : 7.
[363] Musquere, Anne. "ATV: la succession du cargo de l'espace." Air & Cosmos 2137 5 Sept. 2009: 38.
[364] Lardier, Christian. "Satellites: Entre Exploration et Télécommunications." Air & Cosmos 2176 12 June 2009: 116.

[365] "Mission Accomplished." 14 July 2009. ESA. 11 Dec. 2009. http://www.esa.int/esaMI/Mars500/SEMIS47CTWF_0.html.
[366] "KIBO Japanese Experiment Module." October 2009. JAXA. 11 Dec. 2009. http://kibo.jaxa.jp/en/.
[367] Kallender-Umezu, Paul. "JAXA Focused on Kibo for Short Term; Long-Term Policy Uncertain." Space News 2 Feb. 2009: 12.
[368] Jayarman, K.S. "India Plans First Manned Mission with Assistance From Russian Space Agency." Space News 2 Feb. 2009: 11.
[369] Malik, Tariq. "NASA Astronauts Ride in Style While Testing New Moon Truck." Space News 3 Nov. 2008: 12.
[370] Berger, Brian "Astrobotic Plans Robotic Lunar Mission to Apollo 11 Site." Space News 3 Nov. 2008: 12.
[371] "About the Google Lunar X Prize." 2009. X Prize Foundation. 14 Dec. 2009. http://www.googlelunarxprize.org/lunar/about-the-prize.
[372] Lardier, Christian. "Le Retour de Lavotchkine sur la Lune." Air & Cosmos 2151 12 Dec. 2008: 38–39.
[373] "MoonLITE Mission Gets Green Light for Next Step." 5 Dec. 2008. BNSC. 11 Dec. 2009. http://www.bnsc.gov.uk/10278.aspx.
[374] Lardier, Christian. "Satellites: Entre Exploration et Télécommunications." Air & Cosmos 2176 12 June 2009: 116.
[375] "EUROBOT Ground Prototype." July 2008. Joanneum Graz. 11 Dec. 2009. http://dibcms.joanneum.at/projects/eurobot.
[376] "Life support Pilot Plant Paves the Way to Moon and Beyond." 5 June 2009. ESA. 11 Dec. 2009. http://www.esa.int/esaCP/SEMKNIVTGVF_index_2.html.
[377] "Lunar Spacecraft Reaches its Final Orbit." 13 Nov. 2009. European Space Agency. 10 Dec. 2009. http://www.esa.int/esaCP/SEM9X74DHNF_index_2.html.
[378] "SELenological and ENgineering Explorer KAGUYA." Nov. 2009. JAXA. 14 Dec. 2009. http://www.jaxa.jp/projects/sat/selene/index_e.html.
[379] "China Able To Send Man to Moon Around 2020." 12 Mar. 2009. Space Daily. 14 Dec. 2009. http://www.spacedaily.com/reports/China_Able_To_Send_Man_To_Moon_Around_2020_999.html.
[380] "Mars Orbiter Completes First Phase of Science Mission." 11 Dec. 2008. NASA. 14 Dec. 2009. http://www.nasa.gov/mission_pages/MRO/news/mro-20081211.html.
[381] "Mars Phoenix Lander Finishes Successful Work on Red Planet." 10 Nov. 2008. NASA. 14 Dec. 2009. http://www.nasa.gov/mission_pages/phoenix/news/phoenix-20081110.html.
[382] David, Leonard. "NASA Selects Colorado's MAVEN for Mars Scout Mission." Space News 22 Sept. 2008: 15.
[383] Iannotta, Becky."Conflict with 2011 Launch of Juno Prompts NASA To Rework MSL Plan." Space News 26 Jan. 2009: 1+.
[384] "NASA Selects Future Projects to Study Mars and Mercury." 4 May 2009. NASA. 14 Dec. 2009. http://www.nasa.gov/home/hqnews/2009/may/HQ_C09_020_NASA_Salmon_Awards.html.
[385] "ESA Extends Missions Studying Mars, Venus and Earth's Magnetosphere." 10 Feb. 2009. ESA. 11 Dec. 2009. http://www.esa.int/esaCP/SEMACI05VQF_index_2.html.
[386] de Selding, Peter. "ESA's Exomars Likely to Launch on Russian Proton." Space News 19 Jan. 2009: 11.
[387] de Selding, Peter. "NASA Poised To Take Key Role in European ExoMars Rover Mission." Space News 22 June 2009: 1+; "NASA and ESA Establish a Mars Exploration Joint Initiative." 8 July 2009. NASA. 15 Dec. 2009. http://www.nasa.gov/mission_pages/mars/news/mars-20090708.html.
[388] "Russian Sample Mission to Martian Moon Delayed Until 2010." 22 Sep. 2009. RIA Novosti. 14 Dec. 2009. http://en.rian.ru/science/20090922/156209097.html.
[389] "China to Launch Mars Space Probe." 29 May 2009. Space Daily. 14 Dec. 2009. http://www.spacedaily.com/reports/China_to_launch_Mars_space_probe_999.html.

[390] "Cassini Equinox Mission: Introduction." 2009. NASA. 15 Dec. 2009. http://saturn.jpl.nasa.gov/mission/introduction/.
[391] "NASA and ESA Prioritise Outer Planet Missions." 18 Feb. 2009. ESA. 20 Dec. 2009. http://www.esa.int/esaCP/SEMPHGWX3RF_index_0.html.
[392] "ESA Extends Missions Studying Mars, Venus and Earth's Magnetosphere." 10 Feb. 2009. ESA. 11 Dec. 2009. http://www.esa.int/esaCP/SEMACI05VQF_index_2.html.
[393] "Venera-D." 2 Dec. 2009. Russian Space Web. 14 Dec. 2009. http://www.russianspaceweb.com/venera_d.html.
[394] "Messenger: The Extreme Machine." 20 July 2004. NASA. 15 Dec. 2009. http://www.nasa.gov/missions/solarsystem/f_messenger-extreme.html.
[395] "NASA Selects Future Projects to Study Mars and Mercury." 4 May 2009. NASA. 14 Dec. 2009. http://www.nasa.gov/home/hqnews/2009/may/HQ_C09_020_NASA_Salmon_Awards.html.
[396] Werner, Debra "NASA Moving Ahead with 2011 Juno Mission to Jupiter." Space News 8 Dec. 2008: 12.
[397] "NASA and ESA Prioritise Outer Planet Missions." 18 Feb. 2009. ESA. 20 Dec. 2009. http://www.esa.int/esaCP/SEMPHGWX3RF_index_0.html.
[398] Schrogl, Kai-Uwe, Charlotte Mathieu and Nicolas Peter, eds. Yearbook on Space Policy 2007/2008: From Policies to Programme. Vienna: SpringerWienNewYork, 2008: 88.
[399] "Solar Dynamics Observatory – Exploring the Sun in High Definition." 2008. NASA. 16 Dec. 2009. http://sdo.gsfc.nasa.gov/assets/docs/sdo_factsheet.pdf.
[400] Berger, Brian. "NASA Selects Winners of Small Explorer Competition." Space News 29 June 2009: 13.
[401] "Ulysses overview." 21 Aug. 2009. ESA. 16 Dec. 2009. http://www.esa.int/esaSC/120395_index_2_m.html.
[402] "Solar Orbiter." 19 August 2009. ESA. 20 Dec. 2009. http://www.esa.int/esaSC/120384_index_0_m.html.
[403] "Discovery Mission: Kepler." 2009. NASA. 17 Dec. 2009. http://discovery.nasa.gov/kepler.html.
[404] "NASA's Spitzer Sees the Cosmos Through 'Warm' Infrared Eyes." 5 Aug. 2009. NASA. 17 Dec. 2009. http://www.nasa.gov/mission_pages/spitzer/news/spitzer-20090805.html.
[405] Berger, Brian. "NASA Selects Winners of Small Explorer Competition." Space News 29 June 2009: 13.
[406] Lardier, Christian. "Succès pour Atlantis et Hubble." Air & Cosmos 2174 29 May 2009: 12–13.
[407] "Hershel and Planck Commissioning Has Begun." 21 May 2009. ESA. 16 Dec. 2009. http://www.esa.int/SPECIALS/Operations/SEME990OWUF_0.html.
[408] "Discovering the Stars." Oct. 2009. CNES. 17 Dec. 2009. http://www.cnes.fr/web/CNES-en/1401-corot.php.
[409] "3rd Meeting of International Space Exploration Coordination Group." 16 Mar. 2009. ESA. 20 Dec. 2009. http://www.esa.int/esaHS/SEMLU2JTYRF_index_0.html.
[410] "Multinational Agreement Signed to Carry Out Lunar Exploration." 28 July 2008. Moon Daily. 11 Dec. 2009. http://www.moondaily.com/reports/Multinational_Agreement_Signed_To_Carry_Out_Lunar_Exploration_999.html.
[411] "International Lunar Network." 2009. NASA. 20 Dec. 2009. http://nasascience.nasa.gov/missions/iln.
[412] "NASA and ESA Establish a Mars Exploration Joint Initiative." 8 July 2009. NASA. 20 Dec. 2009. http://www.nasa.gov/mission_pages/mars/news/mars-20090708.html.
[413] "NASA and ESA Prioritise Outer Planet Missions." 18 Feb. 2009. ESA. 20 Dec. 2009. http://www.esa.int/esaCP/SEMPHGWX3RF_index_0.html.
[414] Lardier, Christian. "Le Cnes Sélectionne de Nouveaux Projets." Air & Cosmos 2168 17 Apr. 2009: 36.
[415] Lardier, Christian. "Un Budget Nasa en Hausse de 5,1%." Air & Cosmos 2172 15 May 2009: 39.
[416] Lardier, Christian and Theo Pirard. "Satcoms: Ralentissement des Systèmes Régionaux." Air & Cosmos 2167 10 Apr. 2009: 36+.

[417] Lardier, Christian and Theo Pirard. "Satcoms: Ralentissement des Systèmes Régionaux." Air & Cosmos 2167 10 Apr. 2009: 36+; de Selding, Peter B. "Capacity Demand Outpaces Supply in Middle East and Africa." Space News 13 Apr. 2009: 17.
[418] "Space Business Review." December 2008. Milbank. 14 Jan. 2010. http://www.milbank.com/NR/rdonlyres/B34A51C3-5450-463C-B8C1-008D261CD07A/0/December_2008_Space_Business_Review.pdf.
[419] Ibid.
[420] Lardier, Christian. "L'Europe Rivalise avec les Etats-Unis." Air & Cosmos 2137 5 Sept. 2008: 16+.
[421] Rousseau, Claude. "MSS Industry Sees XM-Sirius Syndrome." Space News 8 Sept. 2008: 19+.
[422] Iannotta, Becky. "Sat Demand Considered Strong Enough to Weather Downturn." Space News 20 Oct. 2008: 4.
[423] Lardier, Christian. "L'Europe Rivalise avec les Etats-Unis." Air & Cosmos 2137 5 Sept. 2008: 16+.
[424] Lardier, Christian. "L'Europe Mise sur les Applications." Air & Cosmos 2150 5 Dec. 2008: 36+.
[425] European Parliament and Council of the European Union. Decision of the European Parliament and of the Council on the Selection and Authorisation of Systems Providing Mobile Satellite Services (MSS). Decision 626/2008 EC of 30 June 2008. Brussels: European Union.
[426] Lesser, Hartley. "Insight: the European Union + MSS." Sept. 2008. SatMagazine. 27 Aug. 2009. http://www.satmagazine.com/cgi-bin/display_article.cgi?number=318671572.
[427] de Selding, Peter B. "Antenna Problem Clouds Prospects for Solaris Mobile S-Band Venture." Space News 18 May 2009: 1+; de Selding, Peter B. "TerreStar Protests European Commission's Licensing Decision." Space News 1 June 2009: 14.
[428] Skyrocket website. 27 Aug. 2009. http://space.skyrocket.de/.
[429] de Selding, Peter B. "France Hedges on Commercialization of Syracuse 3." Space News 11 May 2009: A3.
[430] Skyrocket website. 27 Aug. 2009. http://space.skyrocket.de/.
[431] Ibid.
[432] Congressional Research Service. "Defense: FY 2009 Authorization and Appropriations." 25 Feb. 2009. 27 Aug. 2009. http://www.fas.org/sgp/crs/natsec/RL34473.pdf.
[433] Center for Defense Information website. 27 Aug. 2009. http://www.cdi.org/.
[434] Brinton, Turner. "Military Demand for Commercial Satcom Projected to Increase." Space News 26 Jan. 2009: 12+.
[435] Skyrocket website. 27 Aug. 2009. http://space.skyrocket.de/.
[436] Ibid.
[437] Lardier, Christian and Theo Pirard. "Satcoms: Ralentissement des Systèmes Régionaux." Air & Cosmos 2167 10 Apr. 2009: 36+; de Selding, Peter B. "Capacity Demand Outpaces Supply in Middle East and Africa." Space News 13 Apr. 2009: 17.
[438] Ibid.
[439] Ibid.
[440] Lardier, Christian. "L'Europe Rivalise avec les Etats-Unis." Air & Cosmos 2137 5 Sept. 2008: 16+.
[441] Toussaint, Marcel. "World Space Programmes and Budgets." European Space Directory 2009. Endorsed by Eurospace. Paris: Eurospace, 2009: 72–75.
[442] Commission of the European Communities. Commission Working Document. European Space Policy Progress Report. COM (2008)561 final of 11 September 2008. Brussels: European Union.
[443] Council of the European Union. Council Resolution Taking Forward the European Space Policy. Doc. 13569/08 of 29 September 2008. Brussels: European Union.
[444] European Parliament and Council of the European Union. Regulation of the European Parliament and of the Council on the Further Implementation of the European Satellite Navigation Programmes (EGNOS and Galileo). Doc. (EC) No. 683/2008 of 9 July 2008. Brussels: European Union.
[445] Commission of the European Communities. Report From the Commission to the European Parliament and the Council on the Implementation of the GNSS Programmes and on Future

Challenges Pursuant to Article 22 of Regulation (EC) No. 683/2008. COM(2009) 302 final of 26 June 2009. Brussels: European Union.

[446] European Parliament and Council of the European Union. Regulation of the European Parliament and of the Council on the Further Implementation of the European Satellite Navigation Programmes (EGNOS and Galileo). Doc. (EC) No. 683/2008 of 9 July 2008. Brussels: European Union.

[447] Commission of the European Communities. Report From the Commission to the European Parliament and the Council on the Implementation of the GNSS Programmes and on Future Challenges Pursuant to Article 22 of Regulation (EC) No. 683/2008. COM(2009) 302 final of 26 June 2009. Brussels: European Union.

[448] Lardier, Christian. "ESA: Premiers Contrats Galileo." Air & Cosmos 2177 19 June 2009: 38; de Selding, Peter B. "With Galileo Contracts at Stake, Disagreement Over Way Forward." Space News 22 June 2009: 4.

[449] de Selding, Peter B. "Galileo Cost Overruns Passes European Commission Audit." Space News 29 June 2009: 15; "Galileo Global Nav System Examined by European Court of Auditors." 1 July 2009. Satnews. 13 Oct. 2009. http://www.satnews.com/cgi-bin/story.cgi?number=1653440132.

[450] "Obama 2010 GPS Budget Calls for $1 Billion+." 21 May 2009. Inside GNSS. 20 Oct. 2009. http://www.insidegnss.com/node/1512#Baseband_Technologies_Inc_.

[451] Gibbons, Glen. "What Race? What Competition? The Four GNSS Systems." March/April 2009. Inside GNSS. 19 Oct. 2009. http://www.insidegnss.com/node/1425#Baseband_Technologies_Inc_; "Air Force May Be Able to Fix Problem With Latest GPS Satellite." 16 June 2009. Inside GNSS. 19 Oct. 2009. http://www.insidegnss.com/node/1573#Baseband_Technologies_Inc_.

[452] "Upcoming Block II-R (21)M Launch Wraps Up Generation of GPS Satellites." 6 Aug. 2009. Inside GNSS. 19 Oct. 2009. http://www.insidegnss.com/node/1622#Baseband_Technologies_Inc_.

[453] "AFSPC Exceeds Global Positioning System (GPS) Standard." 26 May 2009. GPS Daily. 19 Oct. 2009. http://www.gpsdaily.com/reports/AFSPC_Exceeds_Global_Positioning_System_Standard_999.html.

[454] "Lockheed Martin GPS III Team Completes Major Design Review." 25 May 2009. GPS Daily. 20 Oct. 2009. http://www.gpsdaily.com/reports/Lockheed_Martin_GPS_III_Team_Completes_Major_Design_Review_999.html.

[455] Gibbons, Glen. "GAO Report on GPS Satellite Constellation Status: The Pushback." 25 May 2009. Inside GNSS. 20 Oct. 2009. http://www.insidegnss.com/node/1527#Baseband_Technologies_Inc_.

[456] Gibbons, Glen. "U.S. Congress Working on Appropriations for Loran-C and eLoran as GPS Backup." 29 July 2009. Inside GNSS. 20 Oct. 2009. http://www.insidegnss.com/node/1615#Baseband_Technologies_Inc_.

[457] "U.S. Publishes New GPS Standard Positioning Service (SPS) Performance Standard." 6 Oct. 2008. Inside GNSS. 20 Oct. 2009. http://www.insidegnss.com/node/847#Baseband_Technologies_Inc_.

[458] "U.S. European Meeting Reaffirms GPS/Galileo Cooperation." 25 Oct. 2008. Inside GNSS. 23 Oct. 2009. http://www.insidegnss.com/node/894#Baseband_Technologies_Inc_.

[459] "Putin Backs $2.62-Billion Addition to Glonass Budget." 16 Sept. 2008. Inside GNSS. 22 Oct. 2009. http://www.insidegnss.com/node/790#Baseband_Technologies_Inc_; "Glonass Constellation Grows to 17 as Latest Satellites Come On-line." 14 Nov. 2008. Inside GNSS. 22 Oct. 2009. http://www.insidegnss.com/node/916#Baseband_Technologies_Inc_.

[460] "Russia Glonass System to Get Full State Support." 15 May 2009. GPS Daily. 22 Oct. 2009. http://www.gpsdaily.com/reports/Russia_Glonass_System_To_Get_Full_State_Support_999.html.

[461] "Russia Launches Three More GLONASS-M Space Vehicles." 26 Dec. 2008. Inside GNSS. 22 Oct. 2009. http://www.insidegnss.com/node/982#Baseband_Technologies_Inc_; Gibbons, Glen. "Russia Building Out GLONASS Monitoring Network, Augmentation System." 19 Aug. 2009. Inside GNSS. 22 Oct. 2009. http://www.insidegnss.com/node/1631#Baseband_Technologies_Inc_.

[462] "Le Bulletin de Spheris N.37." April–May 2009. Spheris. 22 Oct. 2009. http://www.spheris-fr.com/Info/index.php?RID=5&ACT=telechargerDoc&ID=44.

[463] Information-Analytical Center website. 22 Oct. 2009. http://www.glonass-ianc.rsa.ru/pls/htmldb/f?p=202:1:1244260643211211752.
[464] Gibbons, Glen. "Russia Building Out GLONASS Monitoring Network, Augmentation System." 19 Aug. 2009. Inside GNSS. 22 Oct. http://www.insidegnss.com/node/1631#Baseband_Technologies_Inc_.
[465] "China to Launch 10 Compass Satellites over Next Two Years." 8 Dec. 2008. Inside GNSS. 23 Oct. 2009. http://www.insidegnss.com/node/924#Baseband_Technologies_Inc_; "China Aerospace Official Say Compass/Beidou Will Be Complete by 2015." 19 Jan. 2009. Inside GNSS. 23 Oct. 2009. http://www.insidegnss.com/node/1152#Baseband_Technologies_Inc_; "China Launches Second Compass (Beidou-2) Satellite in $1.46 Billion First Phase." 18 Apr. 2009. Inside GNSS. 23 Oct. http://www.insidegnss.com/node/1435#Baseband_Technologies_Inc_.
[466] de Selding, Peter B. "China's Satnav Progress Stokes Concerns About Galileo Overlap." Space News 12 Jan. 2009: 1+.
[467] "ISRO Extends Raytheon Contract for GAGAN GPS Augmentation System." 20 July 2009. Inside GNSS. 23 Oct. 2009. http://www.insidegnss.com/node/1604#Baseband_Technologies_Inc_.
[468] Kellender-Umezu, Paul. "Missile Warning System at Forefront of Japan's New Space Policy." Space News 8 June 2009: 8.
[469] Commission of the European Communities. Communication from the Commission to the European Parliament, the Council, the European Economic and Social Committee and the Committee of the Regions. Global Monitoring for Environment and Security (GMES): We Care for a Safer Planet. COM (2008)748 final of 12 Nov. 2008. Brussels: European Union.
[470] "Second Phase of GMES Gets Go-ahead at Ministerial Council." 9 Dec. 2008. ESA. 27 Oct. 2009. http://www.esa.int/esaLP/SEMHO6STGOF_LPgmes_0.html.
[471] "New Steps in ESA/Commission Cooperation on GMES." 28 Jan. 2009. ESA. 27 Oct. 2009. http://www.esa.int/esaCP/SEM27RWPXPF_index_0.html.
[472] ESA/GOCE website. 27 Oct. 2009. http://www.esa.int/SPECIALS/GOCE/index.html.
[473] "Landmark Year Ahead for Earth Observation Science Missions." 16 Jan. 2009. ESA. 27 Oct. 2009. http://www.esa.int/esaCP/SEMFCIVPXPF_index_0.html.
[474] Ibid.
[475] "OCO Mission Overview." NASA. 28 Oct. 2009. http://www.nasa.gov/mission_pages/oco/mission/index.html.
[476] NOAA website. 28 Oct. 2009. http://www.noaa.gov/.
[477] NOAA Satellite and Information Service. 28 Oct. 2009. http://www.ipo.noaa.gov/index.php; Brinton, Turner. "Continuing Instrument Troubles Lead to More Delays for NPOESS." Space News 20 Apr. 2009: 1+.
[478] Covault, Craig. "Reopening China." Aviation Week & Space Technology 8 Sept. 2008: 28+.
[479] Skyrocket website. 28 Oct. 2009. http://space.skyrocket.de/.
[480] Lardier, Christian. "La Chine Lance son Dernier Satellite de l'Année." Air & Cosmos 2153 2 Jan. 2009: 40+.
[481] Skyrocket website. 28 Oct. 2009. http://space.skyrocket.de/.
[482] Ibid.
[483] Keith, Adam. "Emerging Satellite Programs – Threat or Opportunity." Space News 1 Sept. 2008: 19+.
[484] "Vinci Space Propulsion." June 2009. Snecma. 5 Jan. 2010. http://www.snecma.com/IMG/pdf/VINCI_ang-4.pdf.
[485] "More about GOCE and Astrium." 24 Feb. 2009. Astrium. 5 Jan. 2010. http://www.astrium.eads.net/en/news/more-about-goce-and-astrium.html.
[486] "Ares I Five Segment Development Motor on the Move." 17 Apr. 2009. NASA. 5 Jan. 2010. http://www.nasa.gov/mission_pages/constellation/ares/aresI_dm1.html.
[487] "Dawn Spacecraft." 2009. NASA. 5 Jan. 2010. http://www.nasa.gov/mission_pages/dawn/spacecraft/index.html.

[488] "DLR Kontaktiert Erderkundungssatelliten TerraSAR-X per Laserstrahl." 10 Oct. 2008. DLR 5 Jan. 2010. http://www.dlr.de/desktopdefault.aspx/tabid-4237/6803_read-14120.
[489] de Selding, Peter B. "ESA and Germany Reach Agreement on Data Relay System." Space News 19 Oct. 2009: 6.
[490] For the following sections, see also: Rathgeber, Wolfgang. "Space Policies Issues and Trends in 2008/2009." 18 May 2009. ESPI Report 18. 2 Jan. 2010. http://www.espi.or.at/images/stories/dokumente/studies/espi%20report%2018.pdf.
[491] "ISS Astronauts Complete Spacewalk, Test New Russian Spacesuits." 5 June 2009. RIA Novosti. 7 Jan. 2010. http://en.rian.ru/science/20090605/155180266.html.
[492] "Japan Hopes to Turn Sci-fi Into Reality With Elevator to the Stars." 22 Sep. 2008. The Times Online. 7 Jan. 2010. http://www.timesonline.co.uk/tol/news/science/article4799369.ece.
[493] "2009 Northrop Grumman Lunar Lander X PRIZE Challenge." 2009. X Prize Foundation. 8 Jan. 2010. http://space.xprize.org/lunar-lander-challenge.
[494] Golden, Leonard David. "Armadillo, Rocket Racing To Build Tourist Space Vehicle." Space News. 24 Nov. 2008: 16. Berger, Brian. "Astrobotic Plans Robotic Lunar Mission to Apollo 11 Site." Space News 3 Nov. 2008: 12.
[495] "About the Google Lunar X PRIZE." 2010. X Prize Foundation. 8 Jan. 2010. http://www.googlelunarxprize.org/lunar/about-the-prize.

PART 2

VIEWS AND INSIGHTS

1. The new geostrategic context for space and the positioning of Europe

Bertrand de Montluc[496]

1.1. A new strategic context for space policies

With the globalisation of the last few years placing the economy at the centre of world affairs, it is tempting to conclude that political entities and the State are outmoded concepts. The driving force behind globalisation is well known: it is the wave, or possibly cycle, as the "wave" may not necessarily progress in a linear fashion, of economic, commercial and financial liberalisation – both within States and internationally – together with technological revolutions and progress in transport, telecommunications, information technologies, etc.[497] At the same time though, another phenomenon can be observed: throughout the world, national political entities are again making their presence felt. Both the re-emergence of independent States since decolonisation and the rise of new economic powers stimulated by globalisation tend to strengthen national pride as well as a claim for international recognition. The experts talk of the "geopolitics of globalisation", of the "global puzzle",[498] or of the return of geopolitics and power struggles.[499] As in previous times the principal factors are natural ones: resources, populations and geography. The international system may well be becoming increasingly abstract or fluid thanks to information and communications technologies, however it reveals a renewed emphasis on geostrategic identities and diplomacy based on spheres of influence (such as the bargaining power of energy-rich countries over greedy energy consumers). Power relationships are indeed subject to countervailing forces resulting from global interdependence. They can swing back and forth, up and down. And, all the players do not play by the same rules... But overall, the emergence of new players, whether major or minor, the growing difficulty of imposing the rule of law over brute force and the intensification of asymmetrical threats all open the way to what is called competitive "multipolarity" that is more likely to lead to tension, violent confrontations and a strategic arms race than the collective security that, it was hoped, would guarantee peace after the end of the Cold War.[500]

So when geopolitics combines with economic globalisation, political entities require more independent "strategic" means to visibly establish and concretely symbolise their status in relationships of power. To be specific, space capabilities are becoming one of these means.

1.2. International security and space capabilities

Although the danger of war between the superpowers has receded, international security is increasingly threatened by violence that is spreading as the world order is taken less and less seriously by newcomer nations and loses influence. It is beset by three main internal problems: the authority of the western institutions on which it is based is in jeopardy; the new state and non-state forces now asserting themselves are not principally in the West, but in the "Rest"; and lastly, antagonisms of a cultural or religious nature are intensifying.[501] Under such conditions, the proliferation of arms and particularly nuclear weapons becomes especially dangerous for both technical and cultural reasons: in some respects these weapons are easier and cheaper to obtain,[502] so that their actual use is less unthinkable.[503] Overall, the central difficulty is that the world is becoming both more asymmetrical or heterogeneous in terms of perceptions and passions (under the influence of religious feelings and pressures), and less asymmetrical or unequal in terms of power (because of the new potency of the "weapons of the weak" and the new difficulty for the powerful in exploiting their greater military strength).[504]

Space is affected by these contradictory trends – by an international environment in which the old order is contested and in which the balance of power is unstable. It is clear that, for some years now, we have been witnessing a tendency for several of the world's space programmes, traditionally aimed at developing "useful" space applications (or at least non aligned relative to the American approach), to take on a distinctly nationalistic character. It seems that we can now expect China, India, and soon South Korea and maybe others, to adopt more political objectives for their programmes, as symbols of national pride. These countries have announced their intentions of putting their own citizens in orbit (even Malaysia), of going to the Moon and of participating in further space exploration missions. At the same time, the same players, or others (Iran), are making no secret of their intention to acquire space capabilities for defence purposes, whether anti-missile (Japan) or anti-satellite (China).[505] The absolute pre-eminence of America, Russia and Europe will soon be a thing of the past. A study by the CSIS shows that although few nations have independent launch capability (eight at present; ten in the not too distant future) or the technology for manned flight, the number of players controlling their own satellite telecommunications systems has doubled since 1980 (Indonesia was one of the first and Vietnam the most recent, but there is also

Arabsat, etc.). There are now 27 countries with satellite-based Earth observation resources, compared with three in 1980, not to mention the increasingly numerous countries that, owing to the reduction in acquisition costs, now have their own image receiving stations for remote sensing systems. The world now has 25 Space Agencies, large or small. States such as Nigeria, Algeria, Argentina, Malaysia and Indonesia, and many others, have established space programmes as symbols of independence, national pride and the desire to inspire their youth. Moreover, governments have authorised public expenditure for R&D or development costs, as required (usually between 10 and 60 million U.S. dollars, or more than 100 million U.S. dollars in the case of countries such as Australia, Brazil or South Korea).[506] Countries with emerging economies really are emerging fast, and they do not merely wish to take their place in the global economy but to be accorded the geopolitical role that they feel is their due.[507]

This is the new backdrop against which we need to consider space policies around the world in order to fully comprehend the external causalities, both political and strategic, and to specify the positioning of Europe, with its strengths and weaknesses, now and in the long term.

1.3. A geostrategic approach to understanding space policies

The main premise behind this approach is simple: the changing international scene post-1989, followed by 9/11 and the war in Iraq, has meant that space capabilities are becoming, at least for the major States possessing nuclear weapons, an integral part of their "strategic posture" - even if the space layer may appear to be of limited or marginal importance. For others, it is an "asymmetrical" capability aiming to discredit certain advantages of the potential enemy (implicitly the United States), advantages which could directly or indirectly diminish the effectiveness of their national deterrent if they could not be countered. For the emerging space powers, on the other hand, space technologies and systems act rather as a symbol (one among many) of a determination to achieve national independence, regional influence and technological maturity, in short: a national identity. Beyond this, the question remains as to whether it is conceivable that the global arms race is likely to develop in the next years in fields other than nuclear,[508] such as space, since it is now clear that satellites are increasingly one of the key elements of any modern strategic posture and of anti-missile defence systems.[509]

We referred in the introduction to the diplomatic context: it is that of an unbalanced multi-polar world where strategic stability has become unsettled and where there is proliferation from Asia to the Middle East, without mentioning the danger of new uses of weapons of mass destruction by irresponsible or suicidal actors.[510] The deployment of anti-missile interceptors, capable of reaching targets in space, and therefore able to destroy a satellite as demonstrated by the Pentagon in an operation in February 2008 involving a failed NRO satellite,[511] can have a dissuasive effect on States intending to proliferate and also, in certain cases, the unintended effect of giving certain States that habitually hide their true intentions a pretext for modernising their arsenals.

We may simply consider three main examples of "continent-states" (the United States, Russia and China) each possessing nuclear weapons, for which space, in different degrees, is part of their strategic posture.

Concerning the United States, the most remarkable geostrategic change, well beyond what is to be decided about NASA's future programmes, is the implementation of a new strategic "triad" as defined by the *Nuclear Posture Review* of 2002.[512] This new triad, whose explicit aim is to provide the President with options other than solely one recourse to nuclear threat, includes: (1) offensive strike systems (nuclear or non-nuclear), (2) defensive systems (active and passive) and (3) a renewed and reactive defence infrastructure. It is clear that space resources ("command and control", "intelligence and surveillance" and "counter-space systems") are an important part of this third "pillar" and act as a common supporting structure for all three pillars.[513] They are considered as strategic and they correspond to national vital interests that have to be protected by all means available. The new administration will be anxious to avoid the rhetoric of space weaponisation but will stick to the point of the protection of space assets.

Russia no longer has a real arsenal of high-level operational space resources.[514] But Moscow's determination since 2000 to confirm that it can rely on its nuclear weaponry implies modernisation that must include its satellite technology.[515] In addition, what is known of Russian perceptions about future war scenarios confirms this orientation. As early as 1993 the Minister of Defence, P. Gratchev, observed that future wars would have to begin with aerospace operations. The Russian military consider space as a potential strategic theatre. There is a feeling in Russia that space resources are necessary to confront the others, to create anti-missile systems and to keep a close watch on what is happening in space.[516]

Lastly China, since it is unable to keep up with the United States in the field of conventional weapons and is confronted, like Russia, with the future Missile Defence systems, attaches great importance to maintaining the credibility of its deterrent potential and to catching up in areas in which it lags behind (C2/C3I, guidance and accuracy, mobility, and miniaturisation).[517] In addition, as shown by its ASAT anti-satellite missile test in January 2007, China is seeking to acquire asymmetric capabilities to compensate for America's advantages elsewhere and to be a credible player in the field of civil space exploration.[518] There is absolutely no doubt now that China will soon include space in its strategic arsenal and in its international political stance.

1.4. The position of Europe

Europe, meanwhile, is in an unusual position. It has only a partially-integrated space policy through the civilian programmes run by the European Space Agency (ESA) and the "European Space Policy" recently adopted by the Council of the European Union.[519]

Programmes considered as strategic for national sovereignty are run by one or more States (e.g. *Helios*). Furthermore, although Europe has taken the first steps

Fig. 1: *18 December 2009, an Ariane 5 GS launcher lifted off from Europe's Spaceport in French Guiana on a journey to place the French military reconnaissance satellite Helios-2B into Sun-synchronous polar orbit (source: ESA/CNES/Arianespace – Optique vidéo du CSG, L. Boyer).*

towards a common defence stance with the European Security and Defence Policy (ESDP), it does not currently possess either an army or a nuclear deterrent.

Consequently, in the foreseeable future, Europe's space-based military capability and that of its Member States will not be properly integrated as a key element of a European strategic posture.[520]

On the other hand, at the national level, one may assume that programmes such as *Syracuse* and *Helios* in France, *Cosmo* in Italy and *SAR-Lupe* in Germany meet certain strategic needs. According to Michèle Alliot-Marie (a former French Minister of Defence), "mastering space technology has become an essential factor of power and sovereignty. It has the same importance today as that of deterrence in the 1960s".[521]

At the European Union level, the definition of space capacities should most likely be standardised by the European Defence Agency (EDA) to meet requirements such as certain C4ISR (Computerised Command, Control, Communications, Intelligence, Surveillance, Reconnaissance) or telecommunications needs that are seen as indispensable for implementing the ESDP (with ESA being responsible for developing the hardware if necessary for dual civil-military use). The relevant EU authorities would be responsible for coordinating their use, at least for some of them, and only for certain uses (the Petersberg tasks).[522]

1.5. Europe's role in the global context

From the very beginning, overall European space policy has been based on, and exists basically for, programmes. This is true for ESA, a technical intergovernmental organisation, the governance of which is structured on a programmatic basis. It is equally true for the main countries of Europe which individually invest in this sector of high technology (although most of them are Member States of the ESA), each of them has in practice its own rationale for action, thematic priorities (science, Earth observation, telecommunications, manned spaceflights) and industrial policy objectives. Sometimes they have, or have recently created (Italy), their own agency, institute or bureau.

These national agencies (when and where they exist), the European Space Agency and now the European Commission (since the year 2000, for the first joint meeting of the ESA Council and the Union Council) publish from time to time long term vision documents, strategic plans or Green and White papers. These documents are, of course, very useful for policy and lawmakers in particular. But their substance remains completely different from the work on strategy produced by nations whose space policy is strongly integrated at the highest political level.

Indeed, this contradiction is neither a surprise nor an aberration. The economic, financial and political context regularly reminds us that Europe, as a major player on the global scene, is not yet, in terms of international law, a sovereign integrated State. Even if the Lisbon Treaty brings about improvements, they would still not be of a decisive nature. Irrespective of this, culturally, politically or even economically speaking Europe does not constitute a uniform block. In many fields – industry, energy, technology, research and economy – the Member States do not really have the same priorities. They invest the taxpayer's money in varied sectors and conduct quite different industrial policies. This is not in itself an obstacle to Europe's performance and dynamism. It is a fact of life – which shows that Europe is an original unique model where integration and diversity combine to fulfil objectives that are most of the time shared but in some cases are different ("The European Interest"). In the space sector, most States support by consensus the launchers, space science and critical applications (meteorology, Earth observation, navigation ...) but their priorities concerning human flights and programmes for security vary according to national considerations.

Another feature specific to the case of Europe in space is that it does not benefit from a strong efficient institutional mechanism of governance covering both civil and defence dimensions of the space programme.[523] At present, the relationship between the ESA and the EU is still organised only through a framework agreement and joint meetings. ESA itself is an inter-governmental structure ruled by its own convention, but its executive in no way is truly comparable to that of an international political organisation. EUMETSAT and the Galileo structure are autonomous. Eutelsat is a private entity. Even Arianespace operates like a company.... Lastly, in spite of all the efforts made by the ESA, several Member States do have a national space policy.

This rather "soft" mode of governance of the European Space Programme – typically visible today in the complex model of organisation of a so called "flag" programme run by the EU and ESA – has not produced weak results in terms of programmes. On the contrary,[524] while its capacity to respond quickly is questionable in some visible fields of activity (space exploration and space for defence and security), it is rather good in others: Earth observation and environmental sciences, meteorology and weather forecasting, GNNS and, recently, the surveillance of space (SSA).

1.6. Conclusion

A final observation could be that when Europe manages to reach a consensus concerning one objective – space situational awareness, radio-navigation, Earth environment, science, launchers – it becomes powerful and meets the target, even if programme governance is sometimes far from being as efficient it should be. In some cases, the ESA is the leading force. In other situations, the action results from the willingness of some Member States. And, for some ambitious programmes such as Galileo recently, the impulse comes from a combination of EU leadership, ESA pragmatic backing and the Member States' desire to capture industrial markets,[525] the whole project being clearly stamped as an initiative of Europe for the "European Interest".

To conclude, one might reflect on the statement proffered by a British philosopher: "If Europe is still capable of teaching us something, it is certainly not power."[526] But it is educating us, and it prepares the future. If the future in the space endeavour appears to be – after many ups and downs and nationalistic crises – a worldwide enterprise aimed at building global infrastructures either for the monitoring of the Earth environment or for the exploration of the outer space environment, or both, there is no doubt that Europe will act as a strategic partner. The role of Europe would be far from insignificant in that it could – thanks to political will and belief in its own values – provide original (GMES-like) solutions to "cross-cutting" or common issues facing mankind (the climate, the environment, communications and transportation, security and safety and new frontiers for science and exploration). So if tomorrow, as unbelievable as it might currently appear, a few powerful States in Europe, to send a signal, would choose to act with more boldness and creativity in the European Interest through new projects, there is no doubt that, in its turn, the civilian and security space policy of our continent could reach the status of a strategic pillar of the overall European geostrategic posture.

[496] This contribution is a development based on the following article: de Montluc, Bertrand. "The New International Political and Strategical Context for Space Policies." Space Policy 25.1 (2009): 20–28.

[497] Juppé, Alain and Louis Schweitzer (sous la présidence de) La France et l'Europe dans le Monde. Livre Blanc sur la Politique Etrangère et Européenne de la France 2008–2020. March 2008. Paris: Ministère des Affaires Etrangères. On the notion of 'power', see Malis, Christian. "Raymond Aron et le Concept de Puissance." 2005. ISC-CFHM-IHCC. 2 Nov. 2009. http://www.stratisc.org/act/Malis_POWERII.html: power, both an end and a means, is the capacity to make, produce or destroy, to influence the behaviour of others ("command and inducement"). Power is not an absolute but rather a potential; its nature changes over the course of history. So we could say that it is 'strategic': what (messages and means) increases a nation's security, independence and global power.

[498] Gnesotto, Nicole and Giovanni Grevi (Directed by). The New Global Puzzle. What World for the EU in 2025? Pars: European Institute for Security Studies, ISS-EU, 2006.

[499] Contribution by P. Levy to the 5th Prospective seminar. "Quel(s) Monde(s) en 2040?" Délégation aux Affaires Stratégiques (DAS), Ministry of Defence, Paris, France. 13 Feb. 2008.

[500] Zecchini, Laurent. "Espace, Missiles et Satellites, Nouvelle Frontière Stratégique" Le Monde 23 Feb. 2007.

[501] Hassner, Pierre. "Who Killed Nuclear Enlightenment?" International Affairs 83:3 (2007). See also the view of an anthropologist: Godelier, Maurice. Au Fondement des Sociétés Humaines. Paris: Albin Michel, 2007.
[502]Tertrais, Bruno. "L'Europe face à la prolifération nucléaire." Le défi de la Prolifération – Perspectives Européennes. Eds. Lindstrom, Gustav and Burkard Schmitt. Cahiers de Chaillot No. 66. December 2003. Paris: ISS-EU, 2003.
[503] Hassner, Pierre. "The Fate of a Century." The American Interest July/August 2007.
[504] Hassner, Pierre. "Le Siècle de la Puissance Relative." Le Monde 3 Oct. 2007.
[505] Malis, Christian. "L'Espace Extra-atmosphérique, Enjeu Stratégique et Conflictualité de Demain". 2005. ISC-CFHM-IHCC. 3 Nov. 2009. http://www.stratisc.org/act/Malis_Astropol.html.
[506] Center for Strategic and International Studies (CSIS). "Briefing of the Working Group on the Health of the U.S. Space Industrial Base and the Impact of Export Controls." February 2008. 3 Nov. 2009. http://csis.org/files/media/csis/pubs/021908_csis_spaceindustryitar_final.pdf; Euroconsult. 2006.
[507] Védrine, Hubert. "Rapport pour le Président de la République sur la France et la Mondialisation." 4 Sept. 2007. 3 Nov. 2009. http://lesrapports.ladocumentationfrancaise.fr/BRP/074000535/0000.pdf.
[508] See Taubman, Philip. "Driving to Abolish Atomic Weapons." International Herald Tribune 11 May 2009: 6.
[509] See the testimony before the *House of Representatives* for the presentation of the 2008 Ballistic Missile Defence (BMD) budget by Lt. Gen. H.A. Obering, director of the *Missile Defence Agency* (27 March 2007): "I believe the performance of the BMD system could be greatly enhanced by an integrated space-based layer." Some U.S. thinktanks and now the new American Administration are advocating progressive nuclear disarmament. The fact that President Obama cancelled the M.D. facilities planned to be deployed in Eastern Europe does not mean that the U.S. is going to give up missile defence.
[510] Hassner, Pierre. "Ethique et Relations Internationales." Presentation. CEREM, Ecole Militaire, Paris, France, 19 Feb. 2008.
[511] See de Montluc, Bertrand and Patrice Brudieu. "Destruction du Satellite Américain NRO." CNES note. 21 Feb. 2008.
[512] Gormley, Dennis M. "Silent Retreat: The Future of U.S. Nuclear Weapons." The Nonproliferation Review 14:2; Woolf, Amy F. "U.S. Strategic Nuclear Forces: Background, Developments, and Issues." CRS Report for Congress, 5 Sept. 2007; Website of armscontrol.org. "Nuclear Posture Review."
[513] Becher, Klaus. "High Noon in Orbit: Will More Satellites be Shot Down?" Presentation. ESPI. Vienna, Austria. 9 Apr. 2008.
[514] Grouard, Serge and Odile Saugues. Rapport d'Information sur les Enjeux Stratégiques et Industriels du Secteur Spatial. Rapport d'Information No.688. 5 Feb. 2008. Paris: Assemblée Nationale; Facon, Isabelle and Isabelle Sourbès-Verger. "La Place du Spatial dans le Projet de Restauration de la Puissance Russe." 19 May 2007. FRS. 3 Nov. 2009. http://www.frstrategie.org/barreFRS/publications/notes/20070519.pdf; de Montluc, Bertrand. "L'Evolution du Complexe Militaro-industriel Russe: Perspectives et Contraintes pour la Coopération Spatiale." 12 Nov. 2007. Note for the French Foreign ministry's policy planning staff, C.A.P.
[515] Sokov, Nikolai. "The Origins of and Prospects for Russian Nuclear Doctrine." The Nonproliferation Review 14:2 (2007): 208–210.
[516] Fitzgerald, Mary, C. "The Impact of the Military-technical Revolution on Russian Military Affairs." 20 August 1993. Report for the Hudson Institute. quoted in: Malis, Christian. "L'Espace Extra-atmosphérique, Enjeu Stratégique et Conflictualité de Demain." 2005. ISC-CFHM-IHCC. 3 Nov. 2009. http://www.stratisc.org/act/Malis_Astropol.html.
[517] Yuan, Jin-dong. "Effective, Reliable, and Credible: China's Nuclear Modernization." Nonproliferation Review 14:2 (2007): 226–301; Pollpeter, Kevin. "Building for the Future: China's Progress in Space Technology During the Tenth 5-year Plan and the U.S. Response." Mar. 2008. Strategic Studies Institute. 3 Nov. 2009. http://www.strategicstudiesinstitute.army.mil/pubs/display.cfm?PubID=852.

[518] See publications by A. Tellis for the Carnegie Foundation or Tellis, Ashley J. "China's Military Space Strategy." Survival 49:3 (2007). Also, several Chinese authors, for example, Shixiu, Bao. "Deterrence Revisited: Outer Space." China Security 3:1 (2007): 2-1; de Montluc, Bertrand. "Chinese Space Policy: Military and Strategic Implications." Presentation. Centre d'Etudes Asie, Paris, France. June 2008.
[519] Council of the European Union. Resolution on the European Space Policy. Doc. 10037/07 of 25 May 2007. Brussels: European Union.
[520] For an outsiders' view of European capabilities, see Hitchens, Theresa and Tomas Valasek. European Military Space Capabilities: A Primer. Washington DC: CDI, 2006.
[521] Groupe d'Orientation Stratégique de la Politique Spatiale de Défense. Orientations d'une Politique Spatiale de Défense pour la France et l'Europe. Paris: French Ministry of Defence, 2007.
[522] See ongoing study by P. Cardot for the French Conseil Général de l'Armement, MoD, D.G.A. Paris, June 2008.
[523] Gaubert, Alain and André Lebeau. "Reforming European Space Governance." Space Policy 25.1 (2009): 37–44.
[524] See European Space Policy Institute. "A New Paradigm For European Space Policy: A Proposal." ESPI Report 1; Nov. 2005; 5 Nov. 2009. http://www.espi.or.at/images/stories/dokumente/studies/espi-report1-nov2005-final.pdf; de Montluc, Bertrand and Florent Perrache "L'Espace, Facteur d'Intégration pour la Gestion de la Sécurité en Europe?" Les Annales des Mines. Réalités Industrielles. Mai 2006: 61–65.
[525] See de Montluc, Bertrand. "Galileo, A European Project?" Europe and Power. CulturesFrance, ed. Penser l'Europe. Paris: La Documentation Française, July 2008: 184–196.
[526] Steiner, George. "Diversity and Culture." CulturesFrance, ed. Penser l'Europe. Paris: La Documentation Française, Aug. 2007.

2. Forecasting the consequences of the "Crash of 2008" on space activities

Walter Peeters

2.1. Introduction

On 20 April 1961, 8 days after the mission of Yuri Gagarin, President J.F. Kennedy issued a memorandum to the Chairman of the Space Council (the U.S. Vice-president) asking him to come up as soon as possible with a spectacular proposal to "beat the Russians" in a space programme.[527] Noteworthy in this memorandum is that the options given were:

- To build a laboratory in space;
- To land on the Moon;
- To make a trip around the Moon;
- To land a rocket with men onboard on the Moon and return them to Earth.

It was partly the entrepreneurial spirit of the U.S. but certainly also the unique opportunity to advance space exploration with a gigantic step that obviously drove the NASA administrator to propose the most challenging of these options.

It seems very evident from this reaction that the main driver for such a programme was geopolitical. Still we cannot ignore that the order was given during a time of healthy economic growth which enabled the U.S., at the peak of expenditure, to allocate not less than 0.8% of its GDP alone to NASA to execute the Apollo programme. If we add to this other (military) space expenditure, we note that at its highest level, in 1967, space expenditure reached more than 1% of U.S. GDP that year, a level never reached again.

This geopolitical framework has been used by the OECD Futures Project group in its study to forecast space activities for the year 2030. In an interim report, three scenarios are developed based upon geopolitical, socio-economic and energy forecasts:[528]

- "Smooth Sailing", based upon the assumption that a global world order will be implemented under the benevolent guidance of international organisations and where free markets and democracy become gradually the acceptable universal model for national institutions.

- "Back to the future", under this scenario three major economic powers will strive to dominate the world (United States, Europe and China).
- "Stormy Weather", where strong disagreements between major powers lead to a gradual erosion of international institutions and increasing conflicts.

It is evident that there will be considerable differences and shifts between civil, commercial and military space expenditure under these three scenarios. For the final predictions in the report it was decided to base further work on the first scenario.[529] Nevertheless the strong diversification and effects of the other scenarios clearly indicate the interrelationships between space activities and geopolitical as well as socio-cultural factors.

One of the elements that was certainly not taken into account was the stock market crash of 2008. Historically, crashes of this nature happen often. The first well-documented crash was in the 17th century and was based upon speculation on ... tulips. Between 1634 and 1637 the price of tulips rose by a factor of 50. The price even tripled in one week at the highest point of the "bubble" and, similar to later crashes, the over-speculative market collapsed in February 1637, with prices dropping 95% resulting in a serious financial crisis.[530]

In general, such crashes happen after periods of excessive economic optimism and continuously rising stock prices and often follow very speculative stock market bubbles. Some more recent major crashes were the *Wall Street Crash* of 1929 (with a cumulative drop of not less than 89%) and the 1987 crash, also known as the *Black Monday Crash*, with a drop of 22.6% in one day.

The *Crash of 2008* clearly had its origins in the (equity) financial and banking sector, symbolised by the failure of Lehman Brothers. It quickly led to the collapse of 14 other banks in the U.S. that year and then expanded internationally.

At this stage it is difficult to predict when the effects of this latest crash will smooth out on economies in general and on the space sector in particular, but some elements are worth noting when comparing the 2008 crash with the *Black Monday Crash* of 1987:

- The total drop in stock prices was not as severe as in 1987.
- The crash of 1987 did not result in a subsequent "bear market" of constantly declining stock values.
- A full recession seems to have been avoided (at the time of writing, this point is strongly debated as many economists do not exclude a second wave).

– There is obviously better knowledge of how to apply economic countermeasures to avoid long-term effects at present (in the light of lessons learnt from previous events).

What is already clear, however, is that there will be a re-evaluation of banking and financial institution procedures, particularly with respect to securities. Commentators have started to suggest how this paradigm shift could be implemented.[531]

The space sector was not hit as hard as other sectors in 2008. In its yearly report, the Satellite Industry Association (SIA) (based upon Futron research) estimated commercial space turnover in 2008 at over 144 billion U.S. dollars, an increase of 19% compared to 2007.[532]

As far as global expenditure is concerned, the 2009 Space Report presents a worldwide space turnover figure for 2008 of 257 billion dollars, which still represents a modest increase (2%) compared to 2007.[533] However, this low increase is also due to a readjustment of the calculation method in order to avoid some overlaps reported in the 2007 report.

In particular, European statistical figures are based upon a stable methodology and less susceptible to some unknown factors in the figures of the U.S., China and Russia such as military and security expenditure. Eurospace figures show a consolidated space sales increase of 12% in 2008 over 2007, now reaching 5.9 billion euro for Europe.[534]

However, a distinction has to be made between sales and new orders. Indeed, one of the major reasons for the relative sales stability is no doubt the long-term expenditure profile of space projects that were commenced a few years ago and cannot be stopped. Typical examples are mega-projects such as Yahsat (UAE), or Shuttle and ISS operations. Also, many major scientific projects have long-lead times and decision cycles covering many years. The effect on space sales figures will therefore likely be seen in the coming years.

In its recent forecast study, Euroconsult foresees that sales, in particular in the field of telecommunication satellites, will continue to grow until 2011.[535] The main reason for this is the replacement of existing satellites. Indeed, following the entry of private equity firms in this market segment, replacement of satellites has been postponed, but this replacement is now becoming urgent in order to ensure continuity of service.

There is, however, no doubt that declining GDP growth will influence space activities in general as well as new projects in the mid-term, thus making it important to concentrate on this economic factor.

2.2. Macroeconomic effects on space activities

2.2.1. Space and GDP

Per country expenditure on public space activities is strongly dependant on the national strategic interest in space. However, expressed in percentage of GDP, there is a relatively constant pattern. In Europe, public space expenditure over the last two decades has fluctuated in the range of 0.6–0.7% of total GDP.

Despite the many studies and econometric analyses that have tried to correlate space expenditure with economic parameters, no such correlation has been found for Europe.[536]

In order to evaluate the relationship between space expenditure and economic conditions, the evolution of the U.S. space sector will be used as an illustration. Indeed, there have been many more fluctuations in the U.S. programme than in European space expenditure mainly because:

- Europe has mostly participated in mega-projects in relation to its capabilities (cf. ISS) and, contrary to the U.S., has not initiated such projects.
- There has been no major coherent pan-European political incentive to embark on such projects.
- Space budgets in Europe are largely based upon mid-term budgetary expenditure profiles and less on short-term fluctuating cycles (cf. NASA's yearly budget approval cycle).

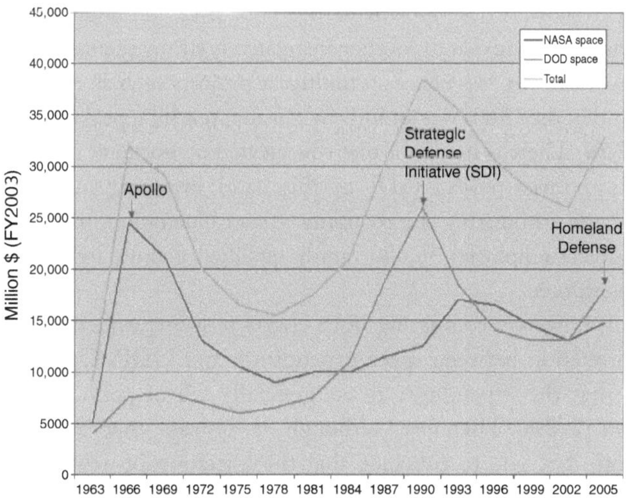

Fig. 2: *Space expenditure in the U.S.*[537]

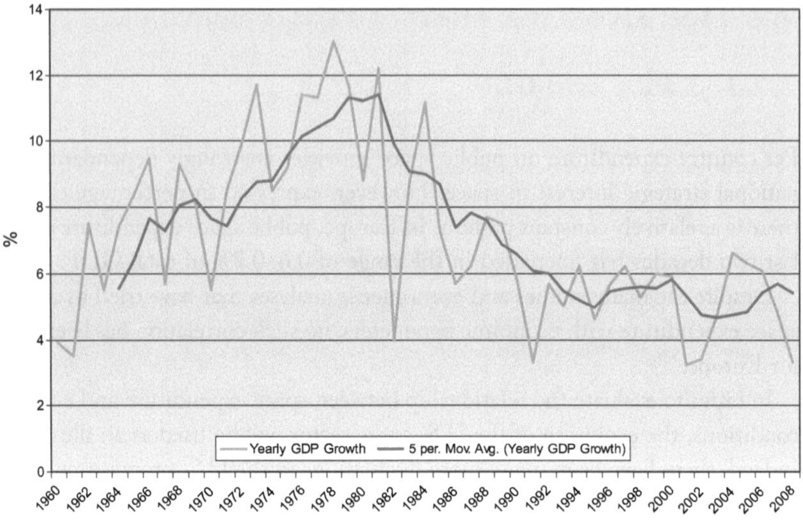

Fig. 3: *Historical GDP evolution in the U.S.*[538]

In Figure 2 some major fluctuations in U.S. space expenditure over the last decades can be noted, particularly two major expenditure peaks in the late 1960s (Apollo programme) and the late 1980s (Strategic Defense Initiative or SDI). There is no doubt that both were driven by strong geopolitical motivations in the Cold War era, but they were evidently only possible in times of sufficient budget availability.

Figure 3 shows GDP growth in the U.S. over the same period. Whereas fluctuations from one year to another are relatively strong (hence the five year trend line added), this has less effect on multiyear projects such as Apollo. Clearly the GDP trend in the 1960s was steadily increasing, reaching peaks in the second half of the 1970s. There is no doubt that this created an excellent financial climate to undertake a mega-project (SDI in this case) involving many thousands of engineers and stimulating the economy. As an illustration, in NASA alone the Apollo project employed 36,000 highly qualified people, twice the number of today's workforce.

Overall, however, comparing both charts it seems evident that there is no direct correlation between space expenditure and GDP. On the other hand, we note that the expenditure peaks generally fall in periods of yearly GDP growth of 7–8%. Therefore, although it appears that there is no direct correlation, it is safe to conclude that the initiation of new programmes can only be expected in times of considerable economic growth; hence clearly there is a relation.

2.2.2. Effects of the present financial crisis

In terms of forecasting, the concept of the BRIC countries was introduced as result of a study by Goldman-Sachs on the next generation of rising countries.[539] BRIC stands for Brazil, Russia, India and China. While it is not clear which country will have the fastest growth, as a group they are particularly fast growing.

Subsequently, the research team identified a new series of countries as the next group of emerging countries, namely (in alphabetic order) Bangladesh, Egypt, Indonesia, Iran, Korea, Mexico, Nigeria, Pakistan, Philippines, Turkey and Vietnam.[540] This group of countries is labelled as the N-11 (Next 11). The selection was done on the basis of present performance and Growth Environment Scores (GES).

Measuring growth and competitiveness originated from the standard work of Porter, who added a number of factors that would influence future growth, such as human resources, knowledge resources and infrastructure, to the pure macroeconomic indicators that had been used until then.[541] The model proposed by Goldman-Sachs has the advantage of being based upon a number of tangible benchmark figures that are published by organisations such as the OECD, IMF and the World Bank.

Measurable parameters can be grouped into a number of categories, namely:

- Macroeconomic stability (inflation, governmental deficit, external debt);
- Macroeconomic conditions (investment rates, trade openness);
- Human Capital (education, demography);
- Political environment (political stability, rule of law and corruption);
- Technology level (penetration of telephones, internet and PCs).

From these 13 parameters a score was calculated which led to the group of aforementioned 11 countries.

It is to be noted that Korea, Mexico and Vietnam lead this group in order of ranking scores.

Figure 4 depicts the forecasts of when BRIC and N-11 countries are expected to overtake the GDP of G7 industrialised countries. Considered initially as too optimistic, the present financial crisis gives greater realism to these projections.

To illustrate this: in the initial simulation, GDP growth in 2009 for Germany was calculated at 1.9% (whereas in reality a negative figure could be expected) while for China the projection was 7.1% (which will be reached, if not exceeded).

There are additional reasons to assume that the present crisis will justify this forecast:

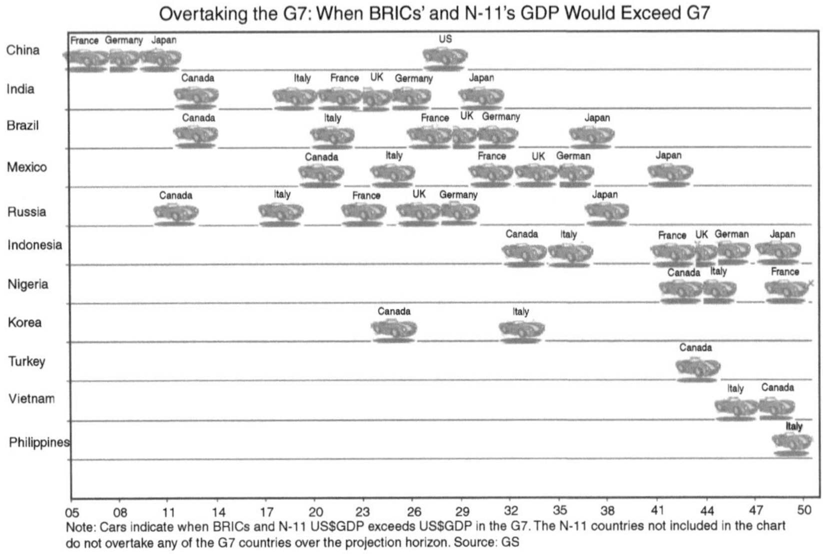

Fig. 4: *Comparative GDP evolution forecast.*[542]

- Industrialised countries are currently forced to subsidise their secondary and tertiary sectors, banking and the automobile industry in particular.
- Emerging countries can still invest in activities that are traditionally known to increase employment, such as infrastructure and road works (e.g. it is reported that China is presently putting 60% of its estimated 450 billion euro support to the economy into infrastructure).
- Even if this would be attractive for e.g. European countries, this would not be possible due to the high infrastructure density that already exists (for example Germany, France, Austria and the U.K. have infrastructure indices which are high above the EU-15 average).
- As a comparison, the U.S. is also presently putting 50% of its present economic support into infrastructure.
- Infrastructure support was encouraged as a stimulus by Noble Prize winner Paul Krugman because it has a GDP economic multiplier of 1.5 (i.e. each euro invested in infrastructure increases the GDP by 1.5 euro).[543]

2.2.3. Forecasting space expenditure on the basis of GDP

To get an impression of how space expenditure will likely evolve, the following approach was taken:

– Countries were categorised into six groups:

- the U.S. and Japan as stand-alones;
- European countries and Canada;
- BRIC countries;
- N-11 countries;
- Rest of the World (ROW).

– The GDP for 2008 for these blocks/countries was calculated.
– The estimated GDP for 2030 was taken from the Goldman-Sachs report (12).
– The expected 2030 relative space budgets per aforementioned block were extrapolated using the ratio GDP2030/GDP2008 for each of the blocks.

The last aspect can be disputed but is based upon the following observations:

– In many countries, particularly in Europe, space expenditure is relatively stable as a percentage of GDP, even over decades.[544]
– In terms of the total annual budget, the Chinese space budget is 0.9% (equivalent to 1% of the U.S. space budget). It is unlikely that either will change considerably in relation to the other over the next few decades.[545]

The result of these calculations is shown in Figure 5, which clearly highlights the following trends by 2030:

– The space budget of the BRIC countries, as expected, will have grown proportionally at the expense of the vested space nations;
– Europe, Canada and Japan risk being the biggest relative losers in this forecast;
– The N-11 portion is still moderate by 2030 but in clear progress.

Moreover, the clear shift depicted in Figure 5, is a continuing trend and estimated GDP forecasts for 2050 show that:[546]

– N-11 aggregate GDP could reach two-thirds of the size of the G7 by 2050;
– The rise of BRIC countries could shift economic trading patterns towards Asia and therefore additionally stimulate some of the N-11 countries in the region;
– Incremental new demand from the N-11 could conceivably be twice the G7 demand in 2050.

In terms of space activities, expected growth in the N-11 after 2040 may lead to their share reaching the 10% range even by 2050.

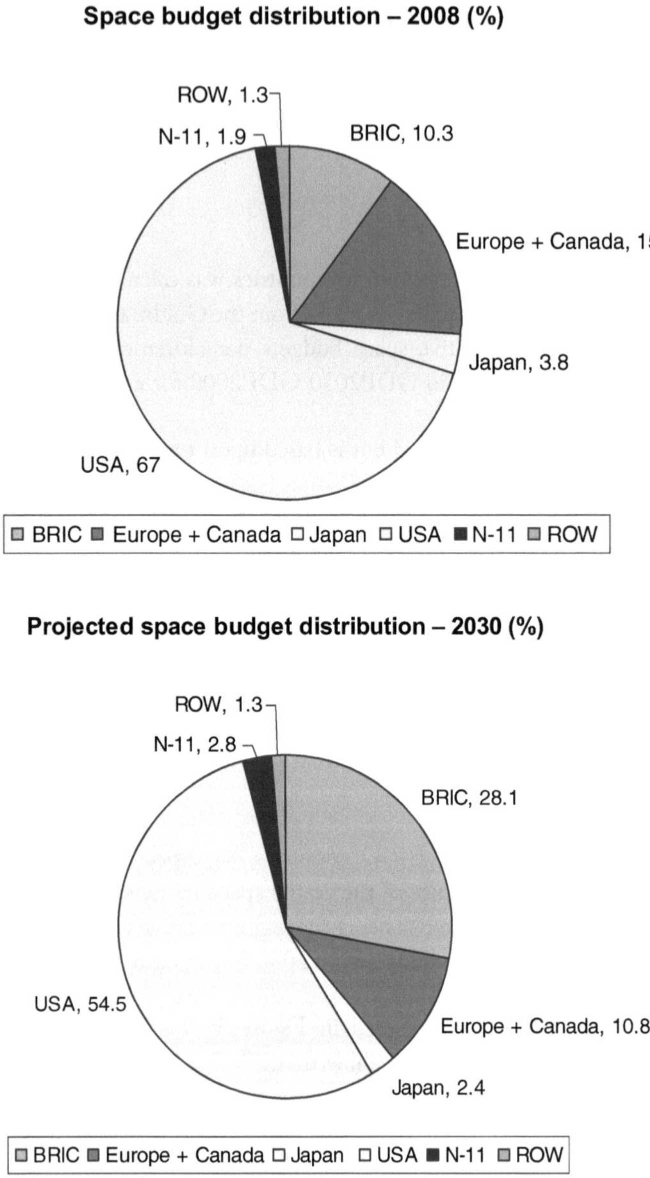

Fig. 5: *Forecast evolution of space activities per region between 2008 and 2030.*

Many analysts have started to work in this direction and early evaluations are being made on how to advance cooperation not only with the BRIC countries but also with these new emerging N-11 partners.[547]

2.3. Microeconomic effects

2.3.1. Workforce effects

In 2008 it was already noted that although there was an overall increase in recruitment in the European space industry, this increase was absorbed by limited duration contracts. It should be also mentioned here that according to recent surveys, the European space sector still has problems to find the right skill set in manpower. Indeed, a survey showed that:[548]

- Nearly 50% of space companies in Europe still have unfilled positions;
- Communication skills are the soft skill quoted most frequently as the reason for candidates failing recruitment interviews;
- Besides engineering skills, there is an increasing demand for a combination of engineering/science backgrounds with business degrees.

The effects of the present crisis in this area are described below.

2.3.1.1. Higher demand for definite term contracts

On the one hand there will be a government desire to increase employment opportunities but, on the other, employers will be reluctant to enter into longer term commitments until prospects have been stabilised. As a compromise, limited contracts will have to be facilitated (2–4 years) with appropriate amendments to labour legislation.

The effect of the prolongation of definite labour contract durations from 18 to 24 months in Germany, for example, has more than doubled the number of short-term contracts in early 2009.

2.3.1.2. Higher demand for multidisciplinary workforce

Higher recruitment levels will be excluded due to cash flow restrictions. Indefinite contracts will be relatively scarce and there will be a careful definition of required skills and selection. Multidisciplinarity will be in higher demand as emphasis will be placed on business development and obtaining new orders.

It should be noted that young graduates have sensed this trend and in general are more inclined to obtain multiple degrees in order to position themselves better in these recruitment cycles. Similarly, within the framework of the BMD system (Bologna Agreements), more students feel the need to advance to Master degrees rather than seeking jobs after obtaining Bachelor degrees.

2.3.2. Financing of space projects

Financing sources of a space project evolve over the lifetime of a company as is depicted as in Figure 6. The present financial crisis has an effect on most phases.

In view of the effects on the banking sector, debt financing is evidently becoming a major obstacle.

Many space companies still have a good turnover, which is the result of the fact that many space projects are planned over and for the long-term and cover an irreversible relatively long time span.

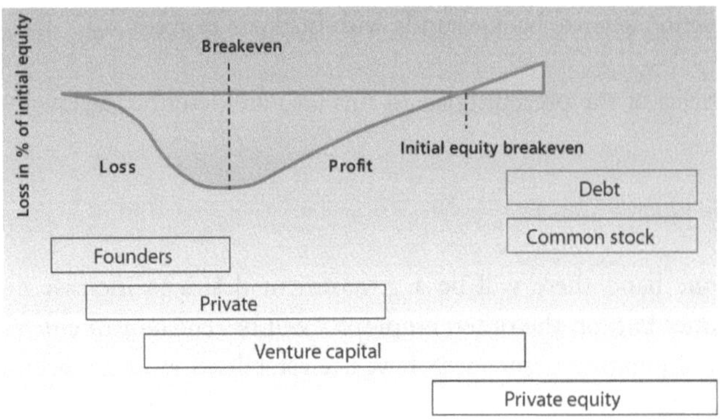

Fig. 6: *Financing sources over the company lifecycle.*

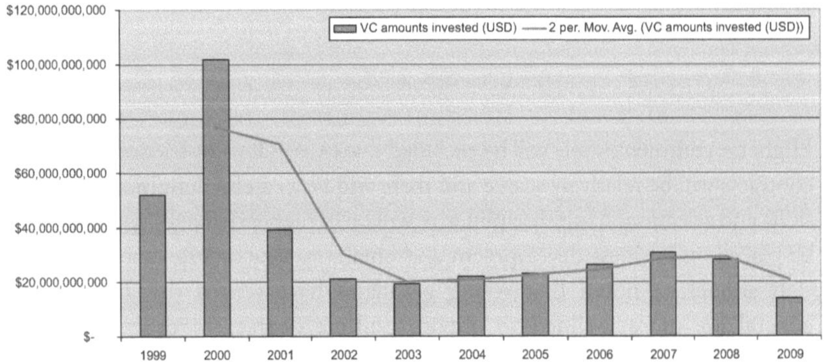

Fig. 7: *Evolution of availability of venture capital.*[549]

The effect of these long-term projects will however slow down in the years to come with an increasing need for cash. While this is less the case for companies belonging to large consortia, it will be even stronger for SME's. Many of these companies will have problems to survive and, irrespective of their know-how and potential, will be forced to look for partners. It is clear in Europe that an increased number of mergers and acquisitions are to be expected, in particular in the SME area, forced by cash-flow problems.

On the other hand, the lack of venture capital (VC) will hurt the start-up sector. Risk avert financers will shy away from such endeavours and the number of start-ups will reduce drastically over the coming years.

Figure 7 clearly shows the strong reduction of available VC in the last decade. The trend line shows the considerable drop after the Internet-boom and the gradual recovery from 2003 onwards. The crisis of 2008 is, however, leading to a second drop resulting in the lowest availability for many years. It is evident that VC investors will be very selective and will probably rather invest in more established and secure proposals than accepting the high risks of start-ups.

2.3.3. Increased vigilance against cost overruns

In times of economic expansion and GDP growth, cost overruns can be absorbed more easily for long duration contracts. Depending on the contract type, escalation can be integrated and will partially level out some cost overruns. This will not be the case for the next few years and companies may count on more stringent vigilance and auditing and cost control measures.

Contractors are aware of this and, especially when forced to submit fixed price proposals, will increase the margins within the costs. However, driven by cash-flow problems, some contractors will have to take risks to survive but may in times of restricted budgetary flexibility have more problems to pass Contract Change Notices or ask for price readjustments. Again this may lead to more pressure on companies with high debt/equity ratios and indirectly lead to mergers and acquisitions.

A method of countermeasure against cost overruns, called the 5C method, has been proposed and is based upon the following steps:[550]

Step 1: Realistic **Cost** Estimation
Due to the uniqueness of the space sector, this is not as evident here as in other sectors. Nevertheless excellent (parametric) costing models that allow accurate own cost estimates before receiving offers are available.

Step 2: Consideration of the Life **Cycle** Cost
The longer lifetime of space projects necessitates giving greater attention to the operational phases, including appropriate budget provisions. In our phased-approach project management philosophy, this aspect tends to be underestimated. Moreover, total costs are already fully committed at the early phases of the project, emphasising the need for good system engineering during the first phases (A/B).

Step 3: An appropriate **Contractual** framework
The type of contract price is an underestimated tool to keep costs under control. It is of paramount importance to determine the contract type to be used for each particular project, especially under tight budget scrutiny conditions.

Step 4: Cost **Control** and Risk Management during the project phase.
It is clear that external cost scrutiny will become more stringent, but many cost overruns can be detected internally at an early stage and partially remedied by using a number of proven tools, such as the Earned Value Method.

Step 5: A **Communication** managed Insurance Approach
Space insurance back-funding will also become more restricted due to the continuing problems in the financial sector. Brokers will therefore be forced to be more vigilant when confronted with claims and timely communication may avoid claims discussions afterwards.

A good illustration of this point is the new measures that have been imposed on NASA, which oblige it to report on projects that risk having cost overruns over 15%.

Alarmed by announced cost overruns on the Mars Science Laboratory (originally estimated at 1.6 billion U.S. dollars by NASA in 2003 and now estimated at 2.3 billion U.S. dollars), the Government Accountability Office (GAO) conducted a survey in early 2009 and reported to Congress on 5 March 2009 that only 5 out of the 40 major NASA programmes are on time and schedule whereas 9 major NASA projects are suffering considerable delays and associated cost overruns. It is highly likely that these reports to Congress will result in new instructions being given to the new NASA management.[551]

Even in the absence of organisations in Europe comparable to the GAO, we can assume that comparable reporting and scrutiny will take place in the European public space sector over the coming years.

It should ne noted that this GAO report also reopened the debate on the accuracy of the initial cost estimates, as early phase O and phase A estimates were used as unreliable benchmarks. This reinforces the point that a good initial cost estimate is the ultimate tool to control cost overruns.

2.4. Conclusion

It will still take a long time before it will be possible to evaluate the effects of the *Crash of 2008.* Only after recovery and stabilisation of the economy can the full impact be measured, as was demonstrated by the analyses of the 1987 crash that were made in the mid-1990s. The longer governments have to support the financial sector, the higher the risk of subsequent inflation and the longer the recovery period will take. Nevertheless, we can attempt to forecast the effects in a qualitative way as many of them are both unavoidable and predictable.

On a macroeconomic scale, strongly reduced world trade and financial losses will affect GDP growth, to the extent that the IMF forecasts economic growth of –1.4% in 2009 (status, July 2009).[552] The forecast expects that emerging economies will recover faster in 2010 and 2011 (with e.g. a return to 8.5% growth for China).

If we assume that this GDP growth will reflect on the possibility of funding space activities, we can forecast a trend whereby initially the so-called BRIC countries, around 2030, will become very strong space powers, second to the U.S. but considerably above European space budgets. Following that, a new generation of countries, the N-11, will gain momentum and are expected to start playing a major role from 2050 onwards. Within this group of countries Korea, Mexico and Vietnam seem to have the highest potential.

It is also worth noting that many of these countries will have the necessary purchasing power to acquire services without having the desire to acquire the basic knowledge. They can therefore be rather seen as potential markets and not as future competitors.

This is clearly a strong strategic element for export oriented space companies. As building up relations in some of these countries is a lengthy process, there may be a strategic interest to now refocus efforts onto future target markets.

In the shorter term, on a microeconomic scale, the present crisis will have effects on the workforce. Recruitment selection will be made with much more care and a combination of hard and soft skills will be required for fixed term contracts. In order to stimulate employment, short-term contracts will be made more flexible and will be more frequently used.

The lack of debt as well as equity financing will have considerable consequences. Venture capital availability is at its lowest point for many years and will be a hindrance for start-up companies. But also established companies, with reducing turnover for the next few years, may face cash-flow problems and will have problems to find debt financing. Mergers and acquisitions seem to be a logical consequence of this situation, while governmental organisations in particular will increase cost control procedures to reduce cost overruns.

[527] U.S. Government. White House Memorandum for Vice-President. 20 April 1961.
[528] OECD. The Commercialisation of Space. 17 Oct. 2003. OECD Futures Programme Steering Group paper. Paris: OECD, 2003.
[529] OECD. Space 2030. Tackling Society's Challenges. Paris: OECD, 2005.
[530] Van Suntum, Ulrich and Ewiges Muster. "Finanzinnovationen lösen Spekulationsblasen aus, haben aber auch einen hohen volkswirtschaftlichen Nutzen." Focus-money 10 June 2009: 48–49.
[531] Soros, George. The Crash of 2008 and What it Means: The New Paradigm for Financial Markets. New York: Public Affairs, 2009.
[532] SIA/Futron. State of the Satellite Industry Report. June 2009.
[533] Space Foundation. The Space Report 2009. Colorado Springs, CO: Space Foundation, 2009.
[534] ASD-Eurospace. The European Space Industry in 2008. June 2009. Paris: ASD-Eurospace.
[535] Euroconsult. Satellites to be Built & Launched by 2018. Paris: Euroconsult, 2009.
[536] Zervos, Vasilis. "Demand for Space Expenditure – an Econometric Analysis." Arms Trade, Security and Conflict. Levine, Paul and Ron Smith. London: Routhledge, 2003.
[537] PROSPACE: The American Space Program. 2007.
[538] U.S. Department of Commerce. Bureau of Economic Analysis.
[539] Goldman-Sachs. "Dreaming with BRICs: The Path to 2050." Global Economics Paper No. 99. October 2003. 2 Nov. 2009. http://www2.goldmansachs.com/ideas/brics/book/99-dreaming.pdf.
[540] Goldman-Sachs. BRICs and Beyond. November 2007. 2 Nov. 2009. http://www2.goldmansachs.com/ideas/brics/BRICs-and-Beyond.html.
[541] Porter, Michael E. The Competitive Advantage of Nations. New York: The Free Press, 1990.
[542] Goldman-Sachs. BRICs and Beyond. November 2007. 2 Nov. 2009. http://www2.goldmansachs.com/ideas/brics/BRICs-and-Beyond.html.
[543] Krugman, Paul. The Return of Depression Economics and the Crisis of 2008. New York: Norton, 2009.
[544] Peeters, Walter. Space Marketing. Dordrecht: Kluwer, 2000.
[545] Euroconsult. Satellites to be Built & Launched by 2018. Paris: Euroconsult, 2009: 119.
[546] Goldman-Sachs. "Dreaming with BRICs: The Path to 2050." Global Economics Paper No. 99. October 2003. 2 Nov. 2009. http://www2.goldmansachs.com/ideas/brics/book/99-dreaming.pdf: 134.
[547] Peter, Nicolas and Katharina Stoffl. "Global Space Exploration 2025. Europe's Perspectives for Partnership." Space Policy 25: 1 (2009): 29–36.
[548] Doule, Ondrej and Walter Peeters. "Study on HR Needs in the Space Sector." Presentation. Eurisy Workshop. Ministry of Transport, Prague, Czech Republic. 13–14 March 2009.
[549] PricewaterhouseCoopers/National Venture Capital Association MoneyTree™ Report, Data: Thomson Reuters.
[550] Peeters, Walter and Bernd Madauss. "A Proposed Strategy Against Cost Overruns in the Space Sector: the 5 C Approach." J Space Policy 24:2 (2008): 80–89.
[551] Lawler, Andrew. "Trouble on the Final Frontier." Science, Vol. 324 (2009): 34–35.
[552] IMF. Economic Forecast 2009–2010: IMF Raises GDP Growth Expectations. 9 July 2009. 2 Nov. 2009. www.economywatch.com.

3. Space as a strategic policy area for Europe and the European Union

Jürgen Turek

3.1. Space as a strategic policy for Europe

Space is challenging Europe as well as the rest of the world more than ever before. In the 21st century, in contrast to old industries, this area will determine the future of mankind. Space is becoming increasingly important for the solution of key problems such as security, counter terrorism, climate change, new energy and materials beyond steel and petrol, and orientation and steering through modern navigation. As a strategic area, in combination with other strategic policy areas such as energy policy, security, economic competitiveness and overall social welfare in the European Union as well as in the rest of the world, it determines and defines the future of industrialised knowledge societies. This fact underlines the strategic quality of space and space policy for Europe. In other words: it is becoming vital for the future, particularly for Europe.

In practice, space policy is a self-evident part of Research and Technology Development (RTD) in Europe. This is an area that will play an increasingly important role in determining Europe's competitiveness in the future. It includes not only economic and military factors, but also aims to optimise the political scope of action. The European Commission as well as individual Member States are now doing much more to strengthen their capabilities. Increasingly, space and space policy are becoming a fundamental and strategic aspect of economic and political life. Space policy can no longer be seen as an isolated field of activity. Within a larger perspective, it is part of a greater strategic framework that will strengthen Europe's competitiveness with regard to its economic, military and political abilities and capacities. As such, space policy logically belongs to the larger initiative for economic growth and innovation.

This initiative in Europe currently brings together three integrated strategies: "the European Research Area" (ERA) and the Framework Programme for Research, Technological Development and Demonstration activities (FP7) that includes space for the first time; the Competitiveness and Innovation Framework Programme (CIP); and, last but not least, the Lisbon Strategy for Economic Growth and Employment, which was revitalised in 2005 and represents the

Fig. 8: *The strategic triangle of research and technology development, competitiveness and innovation, employment and economic growth.*

overarching strategy to make the EU the most competitive and dynamic knowledge-based economy in the world.[553] These combined efforts aim to promote pure research and applied research, economic power and – with regard to the industrial and military capacities of other rapidly growing countries – stronger competitiveness of European industries. This last aim precisely demonstrates the great importance of combining science and research (inventions and patents), innovations (marketing and implementation) and economic applications (knowledge, economic growth, employment and social welfare). This relates not only to the great economic and military power of the U.S. but also to the increasing abilities and capacities of the so called BRIC-Countries,[554] which themselves have a strong and increasing interest in space policy and will undoubtedly also compete with Europe's efforts.

As part of this strategic framework, space policy can help to achieve greater competitiveness, prosperity and security in a fast moving, complex and less calculable world. It is part of the developing ERA, which allows researchers, scientific knowledge and technology to circulate more freely than in the past. Financial and organisational support is provided by the 7th Framework Programme of the European Commission. Between 2007 and 2013 it aims to promote productivity, the application of innovations and sustainable growth in Europe. For its part, CIP works to encourage entrepreneurial initiatives and innovations, the use of modern information and communication technologies as well as the use of renewable energy and environmental technologies. Very new and quite forward-looking are the establishment of a European Research Council (ERC) and the foundation of a European Technology Institute, which started its work in 2008 and is now based in Budapest. The 7th Framework Programme has a budget of 53 billion euros and is thus better financially supported than were former Union programmes. This is a result of the goal, agreed upon in 2002, to spend 3% of the EU budget on research and technological development.

FP7 is comprised of four specific parts: "Cooperation", "Ideas", "People" and "Capacities":

- Cooperation – for transnational cooperation within science and research (around 32 billion euros);
- Ideas – for the application of basic research through the European Research Council (ERC), which shall work independently from the European Commission and shall itself define research milestones (around 7 billion euros);
- People – for so called "Marie-Curie-Measures" and other initiatives for the exchange of scientists (around 5 billion euros);
- Capacities – for supporting research infrastructure, knowledge areas and small and medium sized companies (around 4 billion euros).[555]

In addition to information and communication technologies, bio- and genetic technology and new materials, very important research fields now also include nuclear energy technologies, fusion energy and, for the first time, space. This last action field of policy is now a strong part of Research and Technology Development in Europe. Through the 7th Framework Programme, it receives funding of 1.36 billion euros for the period 2007–2013. It will not only benefit navigation and monitoring, but also create synergies for other policy sectors such as agriculture, traffic and telecommunication.

Until 2013 it will support the following sectors:

- GMES-Services for environmental management, security, agriculture, forestry, meteorology, risk management and civic safety;
- General exploration of space;
- Strengthening space foundations.[556]

From the standpoint of the European Union and, surely, the European Space Agency as well, one of the most prominent projects in this context is the project "Global Monitoring for the Environment and Security" (GMES). It complements the complex goals of space policy, which include ensuring global coverage with the Galileo system – which will come online in 2014 – and maintaining access to new strategically important technologies.[557] GMES, for example, operates with a budget of 2.3 billion euros. The first services of this system, which is designed to eventually collect information from 30 satellites, are set to begin working in 2009. In five years (2014) the full satellite network – which is part of a large system of both permanent and mobile stations (e.g. on ships for data collecting and assessing) – will be ready. Europeans and other users will have access to sharp pictures of, and data about, every point on Earth. This will make it possible, for instance, to measure changes in sea levels or observe illegal burning of rainforests.[558] Another example is the "Initiative i2010" with regard to the implementation of the European information society. Furthermore, with the Herschel and Planck missions, the ESA has spectacularly and ambitiously entered a fascinating research area. Herschel carries the largest infrared telescope ever flown in space and represents a pioneering mission to study the origin and evolution of stars and galaxies. It will help to understand how the Universe came to be what it is today. Planck is the first European space observatory whose main goal is the study of cosmic microwave background radiation – a relic of the Big Bang. After a perfect launch sequence by the Ariane 5 rocket on 14 May 2009, the critical Launch and Early Orbit Phase (LEOP) for Herschel and Planck has started to wind down, while the operating of the scientific instruments and subsystems on both spacecraft has begun.[559] This all shows the importance the European Union attaches to space policy and illustrates Europe's strengthening strategic capacities in space.

3.2. Europe's need for a space policy and the new European space politics

In the past few years the EU Commission and many politicians in EU Member States have, with increasing urgency, pointed to the economic and strategic significance of space and the technological dominance of the U.S. in terms of the development and use of corresponding technologies. With the Ariane rocket and its own satellites, Europe has already developed technological capacities in the field of astronautics as well as created an institutional basis through the European Space Agency (ESA). Allegations have remained, however, that previous activities have been too expensive, erratic, and ultimately purposeless. In this context the Commission became active in early 2003 and presented the Green Paper titled "European Space Policy" on 21 January 2003, which contained many proposals towards creating a more coherent space policy. After a broadly applied consultation process, it summed up the concept of such a policy on 11 November 2003 in the White Paper "Space: A new European frontier for an expanding Union – an action plan for implementing the European Space Policy" and presented it to the public. Furthermore, in 2003, the Commission and the European Space Agency reached a consensus regarding a framework agreement that defines the relationship between the two institutions in terms of the new concept. According to this, Europe will establish itself as an important player in the space sector. Although Europe had already commissioned communications and meteorology systems and had tackled ambitious programmes for satellite-based positioning, navigation, and time synchronisation (Galileo), as well as the GMES initiative, broader efforts were now seen as necessary. Accordingly, European space policy was to be implemented through a European space programme spanning several years, during which priorities and goals should be established, tasks and responsibilities should be distributed, and the yearly budget should be determined. The programme's areas of focus were to include research and development, development of infrastructure, services, and technologies, and should be evaluated regularly. The Commission has pointed out that this would require a significant increase of total expenditure for development and commissioning of applications as well as for support of research and development, technology, and infrastructure. The implementation of the European space programme was designed to be carried out over two phases. During Phase One (2004–2007), the guidelines of the common framework agreement between EU and ESA were to be implemented. Essentially, this concerned better coordination of activities. Phase Two (which should have started in 2007) was to begin only after the European Constitutional Treaty entered into force. Chapter Three (Articles III-254 and III-255) provided that space policy would be a sector in which the European Union and its Member States shared competencies. As the European constitutional process faltered in 2007, the Treaty of Nice (without a guiding space policy) was still in force until 1 December 2009, when the Lisbon Treaty came into power.

In view of this tumultuous institutional background, 2007 and the period immediately after were very important months and years for European space politics. In 2007, the European Space Policy (ESP) was published, which underlined the strategic importance of space technology and space applications for Europe. It is an official common document of the European Commission and the European Space Agency. The programme became reality in May 2007 with impressive political support from all parties involved. The European Space Policy will coordinate civilian space programmes in Europe and should leverage synergies between civilian and military activities. In this way, it should ensure the competitiveness of the European space industry and support the scientific and technological underpinnings of space research. The European Space Policy has the political backing of all the 29 Member States of the EU and the ESA. A key feature is the link between civilian and military applications. Projects will become more efficient by avoiding redundant investments. The ESP will preserve Europe's potential for an independent entry into space. This independence – again, with regard to the U.S. and Russia – is certainly one of the strongest motives for European activities in space. That does not mean, however, that cooperation and good relationships (including with India and China) are unimportant. The new space policy since 2007 is not only an instrument for independence but also for defining a coordinating framework for international cooperation and for European involvement with the International Space Station (ISS), where the Europeans play an important strategic role with the Columbus Orbital Facility (COF) and the Automated Transfer Vehicle (ATV) in particular.[560]

It is important to note that, because of its inclusion in the European Constitution Treaty, the European Space Policy would have gained constitutional status if the treaty had been ratified in all Member States in 2007. Although the Constitution failed, the Lisbon Treaty does include space policy as an important part of the European research agenda too. At a time when new world players are emerging with a keen interest in establishing space projects, the Lisbon Treaty now provides a new legal basis for the creation of a coherent space policy: a clear acknowledgement that Europe cannot afford to overlook the economic and strategic benefits of a coherent space policy.[561] The Treaty survived the second Irish vote and institutional or political problems in other countries such as the Czech Republic or the U.K. could be solved. In the spirit of the European Constitution, space policy is now part of the strategic positioning of the EU, and, more importantly, it has become a constitutional part of Europe's legal framework.

3.3. Capabilities to act in space in the 21st century

The EU Commission has warned that Europe will experience a setback as a "space power" without the proposed approach because it would not be in a position to develop new technologies and to sustain projects, which would ultimately compromise Europe's ability to work effectively. Even if some critics of the plan remain doubtful, the enormous civilian and military significance of a coherent space policy is evident. In 2005, Heiko Borchert clearly described the enormous implications of space policy for Europe's role in global security: "There are strong correlations among the demand of the EU to act globally, a higher security networking ability and the use of space. If the EU wants to act globally for security, stability and prosperity, the Union needs a security approach that balances economic, political, societal and military factors systematically [. . .]. From this perspective, space policy becomes a decisive criterion for the political relevance of an actor – being an international organisation, a state group or a separated nation state".[562] Thus, against the background of the European Security and Defence Policy (ESDP) and the European Council's Lisbon strategy of making Europe the most competitive, knowledge-based market in the world by 2010, space-based civilian and military projects will become more important in the future. For example, satellites for navigating changing traffic flows and space-based security components for guiding military capacities within the framework of European security strategy are increasingly in demand. Projects like the GMES initiative and Galileo are now considered flagship projects. According to the European Parliament, space business represents a market of 90 billion euros worldwide and has a growth rate of 7% per year. The European space industry has a market share of 40% and employs around 28,000 people, mainly in Germany, France, Italy, the U.K., Spain and Belgium. In this context, public investments are around 6 billion euros per year. By contrast, the U.S. with its roughly 900,000 employees and budget of around 40 billion euros is by far the leading space power.[563] With this level of commitment, the U.S. demonstrate the sector's importance, which is increasingly also recognised by Russia, China, and India. Thus it is not a surprise that the French EU Presidency and the EU Commission started an initiative to strengthen the European Space Policy in 2008. Both the French Government and Commissioner Günther Verheugen asked for a larger EU financial commitment to space policy. Under their plan, the EU would grant space policy its own budget line in the Union's budget until 2014. Neither the French Government nor the European Commission stated how large the budget should be in the future. Regardless of the possible amounts involved, a stronger financial engagement should be considered openly and seriously.

Space flight explicitly serves European goals in the areas of environmental protection, mobility, security, and the information society. Besides developing suitable technological capacities in the fields of launchers, and satellite manufacturing, as well as new infrastructure, this requires adaptation of the current competency framework. Accordingly, the competencies of the EU Commission will have to be established by contract, and the responsibilities of ESA will have to be further specified. The agency, which is endowed with an annual budget of roughly 3 billion euros, followed civilian goals in the past. These must either be expanded to include military aspects in a transparent manner, or another structure should be pursued. In this context, in 2004 the present Swedish Foreign Minister Carl Bildt and Mike Dillon, at the request of the European Commission, designed a modified institutional concept for the future of Europe in space.[564] The idea behind this initiative was the realisation that "a Europe without a clear space policy is a Europe that lacks ambition [...]. Ultimately, governments must recognise that Europe's success and competitiveness on Earth will depend in part on its success and competitiveness in space".[565] In addition, there must be a transatlantic dialogue to address American concerns regarding the growing military leadership capabilities of the Europeans. Thirdly, international cooperation that covers the issues of security policy as well as legal problems of utilising space and accessing its planets will be necessary. The "Global Expansion Theory", which was drafted in 2007 by NASA (USA), ESA (Europe), Roscosmos (Russia), CNSA (China) and 10 other important space agencies and which outlines the basics of a common space strategy for the first time, could be useful in this regard. While thinking conceptually and strategically on the European space policy, the possibility of an arms race in space should not be discounted. The plausibility and serious of this scenario was illustrated by China in 2007 when, truly remarkably, it successfully shot down a satellite with an anti-satellite weapon.

3.4. Trying something completely new?

With these initiatives in research and science, the Europeans are not trying something completely new. However, by creating the strategic triangle of research area, innovation strategy and Lisbon strategy, the European Commission has recognised the signs of the times and is more than ever becoming an innovative power by itself. This is "just in time" – with regards to progress made by other international competitors, it has been a situation of "five minutes before twelve"! As it faces an increasingly competitive world, research, science and innovations are becoming even more important for the old continent. The strategic goal of this policy is to generate political independence and economic competitiveness. Moreover, awareness of the importance of research and technological development in Europe and most Member States has reached a better balance, as illustrated by the Pillar Strategy and a number of economic measures. Within all of this, the economic, civilian, as well as the military space policies of the European Union play a prominent and increasing role. Even though national interests in space policy are very different,[566] a unified policy enables the EU to play an active role in space and to realise space-based capacities. These capacities facilitate integration within information and decision structures and widen the scope of action – industrially and militarily. An individual State in Europe would not be able to achieve this by itself. Furthermore, investments in space reinforce the necessary technologies (e.g. information and communication technologies, cryptology, data collection and data interpretation) and strengthen economical abilities in areas that will be of increasing importance for the industrialised and knowledge-based European economy.[567] Insofar as innovation, research and technology development will remain important, space policy in both this context and the wider political and socio-economic framework will become even more important in the future. The Europeans were well advised in the first decade of the 21st century to concentrate their efforts in space and space policy and to implement a broader approach that creates a complex and more coherent field of space policy in the context of scientific research, technological innovations and economic incentives!

[553] Commission of the European Communities. Communication to the Spring European Council. Working Together for Growth and Jobs. A new Start for the Lisbon Strategy. COM (2005)24 final of 2 February 2005. Brussels: European Union.
[554] Brazil, Russia, India and China.
[555] In addition, around 4 billion euros are for the Joint Research Center (1751 million euros for direct non-nuclear research under FP7) and for Euratom/Joint Research Center nuclear research under FP7. See Turek, Jürgen. "Forschungs-, Technologie- und Kommunikationspolitik." Jahrbuch der europäischen Integration 2008. Eds. Weidenfeld, Werner and Wolfgang Wessels. Baden-Baden: Nomos, 2009: 150–151.
[556] European Commission CORDIS website. 20 Nov. 2009. http://cordis.europa.eu/fp7/cooperation/space_en.html.
[557] Turek, Jürgen. "Raumfahrtpolitik." Europa von A–Z. 11th edition. Eds. Weidenfeld, Werner and Wolfgang Wessels. Baden-Baden: Nomos, 2009: 363–365.
[558] Frankfurter Allgemeine Zeitung. 6 Jan. 2010. www.faz.net/s/Rub99C3EECA60D84C08AD6B3E60C4EA807F/Doc~E5B6C46ABD57948FBAD765CF3CE0011F9~ATpl~Ecommon~Scontent.html.
[559] "ESA en Route to the Origins of the Universe." 14 may 2009. ESA Science & Technology. 19 Nov. 2009. http://sci.esa.int/sciencee/www/object/index.cfm?fobjectid=44853.
[560] "Ein Neues Politikfeld der EU: Europäische Weltraumpolitik." Eu4journalists. 19 Nov. 2009. www.eu4journalists.eu/index.php/dossiers/german/C79.
[561] "The Treaty at a Glance." Europa.eu. 19 Nov. 2009. http://europa.eu/lisbon_treaty/glance/index_en.htm.
[562] Borchert, Heiko. "Europas Zukunft zwischen Himmel und Erde: Einleitung." Europas Zukunft zwischen Himmel und Erde. Weltraumpolitik für Sicherheit, Stabilität und Prosperität. Ed. Heiko Borchert. Baden-Baden: Nomos, 2005: 11.
[563] "Rüstung im Weltall. Europa-Abgeordnete für zivile und militärische Weltraumpolitik." 8 May 2007. Ngo-online. 19 Nov. 2009. www.ngoonline.de/ganze_nachricht.php?Nr=15897.
[564] Bildt, Carl and Mike Dillon. "Europe's Final Frontier." Europe in Space. CER Report. Ed. Daniel Keohane, London: Centre for European Reform, 2004.
[565] Ibid: 8.
[566] Jäger, Thomas and Mischa Hansel. "Nationale Weltraumpolitiken im Vergleich. Frankreich, Deutschland, Italien und Großbritannen und die Kooperationsoptionen im Rahmen der Europäischen Union." Europas Zukunft zwischen Himmel und Erde. Weltraumpolitik für Sicherheit, Stabilität und Prosperität. Ed. Heiko Borchert. Baden-Baden: Nomos, 2005: 17+.
[567] Borchert, Heiko, ed. Europas Zukunft zwischen Himmel und Erde. Weltraumpolitik für Sicherheit, Stabilität und Prosperität. Baden-Baden: Nomos, 2005: 12.

4. GMES – Status review and policy developments

Josef Aschbacher and Maria P. Milagro Pérez

4.1. Background

Global Monitoring for Environment and Security (GMES) has been established to fulfil the growing need among European policy-makers to access accurate and timely information services to better manage the environment, understand and mitigate the effects of climate change and ensure civil security. The initiative is led by the European Commission, which is responsible also for setting requirements and managing the services, while ESA is responsible for the Space Component.

4.1.1. GMES brief history

Eleven years have passed since the need for a Global Monitoring for Environment and Security initiative was described in the Baveno Manifesto. Since then significant achievements have been made, some of which are summarised in the following table:

1998:	"The Baveno Manifesto", foundation of the GMES initiative, is drawn up at a meeting between the EC, ESA and national space agencies.
2000:	The GMES partnership is formed with participation of Member States, major space agencies, industry representatives and the EC.
2001:	The ESA Ministerial Council in Edinburgh provides major funds for GMES services; supplemented comparable funds through the EC 6th Framework Programme.
2001:	At the Gothenburg EU Summit in Sweden Heads of State and Government request that "*the Community contribute to establishing by 2008 a European capacity for Global Monitoring for Environment and Security*".

2004: The Commission Communication "*GMES: Establishing a GMES capacity by 2008*" introduces an Action Plan to address the Gothenburg challenge.

2004: A Framework Agreement is signed between the EC and ESA providing the basis for cooperation on GMES.

2005: The ESA Ministerial Council in Berlin provides first significant funds for the GMES Space Component.

2007: The European Space Policy Communication, COM(2007)212, recognises GMES as a flagship of the European Space Policy.

2008: The Commission Communication "*GMES: We care for a Safer Planet*" establishes the basis for financing, operational infrastructure and management of GMES.

2008: The ESA Ministerial Council in The Hague, Netherlands, obtains major funding for the GMES Space Component.

2009: The Commission Proposal for a Regulation on "*the European Earth Observation Programme (GMES) and its initial operations (2011–2013)*" proposes a legal basis and EC funding for the GMES programme. The EC launches a Regulation for the GMES Initial Operations (GIO) programme, which is submitted to the European Parliament and EU Council for decision.

Fig. 9: *ESA Council at Ministerial Level, The Hague, November 2008 (source: ESA/A. Le Floc'h).*

4.1.2. GMES components

Autonomous access to information relating to environment, climate change and security is of strategic importance for Europe. In this context, GMES will improve Europe's monitoring and assessment capacity in environment policy and contribute to addressing security needs.

To accomplish this, GMES has been divided into three main components: Space, In-situ and Services.

The Space Component, led by ESA, comprises two main contributions (i) dedicated space infrastructure called Sentinels and (ii) GMES Contributing Missions (GCM) owned by Member States, EUMETSAT and third parties. An integrated Ground Segment infrastructure ensures access to Sentinels and Contributing Missions data.

The In-situ Component, composed of atmosphere-, ocean- and land-based monitoring systems, makes use of established networks and programmes at European and international levels and on ad hoc measurement campaigns enabling the collection of environmental data.

The In situ component is coordinated by the European Environment Agency (EEA).

The European Commission is in charge of implementing the Services Component of GMES and of leading GMES overall. These services cover five main domains: Land, Marine, Atmosphere, Emergency Response and Security. Climate Change services are also planned to be developed as they cross all five service domains.

4.1.3. Funding strategy

The GMES programme is co-funded by ESA and the EC. In addition, some funding has been complemented at national level.

So far, most of the funds have been allocated to the Space Component.

The total funding available for the build-up phase of the Space Component is 2.3 billion euros, which includes the development of the first generation of Sentinel satellites, data access from Contributing Missions and the development of the ground segment. 1.6 billion euros have been made available by ESA Participating States.

The EC has contributed 1.2 billion euros for the whole programme, out of which 700 million euros are allocated to the Space Component, while the remaining 500 million euros are used for the Services and In-situ Components.

Fig. 10: *The amendment to the EC-ESA GMES agreement was signed on 28 January 2009 by Mr. Jean-Jacques Dordain, ESA Director General (left), and Mr Heinz Zourek, Director General of the European Commission's Directorate General for Enterprise and Industry (source: EC).*

Additional funds will be required in 2011/2012 to complete the build-up phase of the Space Component and ensure a smooth transition towards the operational programme.

The Space Component operational programme, which will be composed of the procurement of recurrent Sentinel satellites, operational access to Contributing Missions, routine operations and evolution of the Space Component, will require an investment of approximately 600 million euros per year.

4.2. GMES services serving EU policies

4.2.1. GMES services

EU priorities on environmental policies are combating climate change, protecting biodiversity, reducing the impact of pollution on health and fostering a better use of natural resources. Concerning foreign and security policy, the EU has created formal instruments for both diplomacy and intervention to cope with regional conflicts and to fight against terrorism. The GMES services aim at developing a range of reliable, affordable and cost efficient European services that support the realisation of these policies and their national and regional implementation.

Fig. 11: *Overview of Earth sub-systems (ocean, land and atmosphere) (source: ESA).*

GMES services are divided into five main categories:

- Marine services;
- Land monitoring services;
- Atmosphere services;
- Emergency response services; and
- Security services.

Climate change services are also included as a cross-cutting domain.

4.2.1.1. Marine services

These services perform analysis, predictions and scenario simulations for applications that improve the safety and efficiency of maritime transport and naval operations. They thereby facilitate decisions on how to exploit and sustainably manage ocean resources, such as offshore oil reserves and fisheries, or lessen the impact of environmental hazards, such as oil spills and harmful algal blooms. These services will also contribute to ongoing climate variability studies and forecasts by providing a better understanding of the ocean and its ecosystems.

MyOcean is the implementation project of the GMES Marine Core Service. Maritime security, oil spill prevention, marine resources management, climate change, seasonal forecasting, coastal activities, ice sheet surveys, water quality and pollution, among others, are the targeted applications.

4.2.1.2. Land services

These services produce basic geo-information on land cover and land use and their annual and seasonal changes. They also include a variety of additional biophysical parameters describing the continental vegetation state, the radiation budget at the surface and the water cycle, and also address a wide variety of thematic fields of application such as water quality, forest management, spatial planning, agri-environmental management, carbon cycle and food security.

They are directed towards the support of policies addressing Global Change and Sustainable Development such as the UN Framework Convention on Climate Change, the Forest and Development Communication of the Commission and EU Council Regulations on Food Aid Policy.

The operational capabilities of the GMES Land Core Service are provided by the GeoLand project. Regional land services are focused on the implementation of newly established European directives such as the Convention on Biological Diversity, the Thematic Strategy for Soil Protection and the Water Framework Directive.

4.2.1.3. Atmospheric services

MACC, which stands for Monitoring Atmospheric Composition and Climate, is the pilot for the core global and regional atmospheric services to be delivered under GMES. MACC has started its work of monitoring the global distribution and long-range transport of greenhouse gases such as carbon dioxide and methane, aerosols that result from both natural processes and human activities, and reactive gases such as tropospheric ozone and nitrogen dioxide. It evaluates how these constituents influence climate and estimates their sources and sinks.

MACC also provides specific products covering Europe:

- Maps and data for regional air-quality forecasts;
- Retrospective assessments of air quality;
- Identification of sources of pollution episodes;
- Toolbox for evaluating possible emergency emission control measures and
- Inputs to local air-quality forecasts, health information and warnings.

4.2.1.4. Emergency response services

The pre-operational emergency service of GMES is currently provided through the SAFER project which addresses the following domains: civil protection, humanitarian aid and security crises. SAFER addresses potentially all types of disasters or crises: natural disasters (floods, fires, landslides, storms, earthquakes, etc.), technological accidents, humanitarian crises (for instance after a severe drought period) and civilian-military crises.

4.2.1.5. Security services

Peacekeeping, nuclear proliferation, piracy at sea, illegal immigration, drug trafficking, protection of vital infrastructure such as pipelines, and assistance to European residents in crisis areas are some of the areas where GMES can provide Europe with an autonomous source of information and with products and services that will deliver timely and reliable information to European decision-makers.

The pre-operational security service of GMES is currently provided through the G-MOSAIC and LIMES projects. These two projects will develop services to support intelligence and early warning through the development of crisis indicators and support crisis management operations for the planning process for EU intervention during crises, the EU intervention itself and citizen evacuation during crises, crisis consequences management, reconstruction and resilience.

4.2.1.6. Climate change services

Climate change services focus on mitigation and assessment of the impacts of climate change on the human living environment. This service is of a horizontal nature, benefiting from measurements taken for land, ocean and atmosphere services that have relevance to climate change parameters and their impact on the natural environment. Typically, this includes issues such as desertification, biomass fires, wetlands, land-use change, trace gas dynamics, sea level rises, ice and polar cap melting, among others. This service is being consolidated and funding will be allocated as from 2010.

4.2.2. Governance of GMES

GMES is funded primarily by the EU and by ESA Member States. The governance of GMES has largely been aligned according to the main funding streams that are channelled via the EC and ESA.

It was agreed in 2005 that the overall lead of GMES is with the EC, which determines policy domains that shall be served with GMES infrastructure. In addition, the EC leads the Services Component which is tightly linked to the policy drivers. ESA, the other major funding route, has been requested to lead the Space Component where the majority of GMES funds are allocated today. More recently, the EEA was tasked to coordinate the In-situ Component, although with significantly smaller funding compared to the other two components.

Accordingly, GMES is governed by the policy and decision bodies of the EU and ESA respectively, some of which are bodies involving both organisations while others are specific to each organisations' decision mechanism.

4.2.2.1. Space Council

The highest joint EU-ESA governance body is the Space Council which unites the ESA Council and the EU Competitiveness Council with representatives at Ministerial level from both organisations' Member States. The Space Council allows all ESA and EU Member States, including Cooperating States, to jointly discuss the development of a coherent overall European space programme, leading to the issuance of joint orientations and resolutions regarding space.

Several recommendations have been formulated at this level concerning GMES, which are expressed, for example, in the resolution adopted at the 5th Space Council ("Taking forward the European Space Policy"), held on

Fig. 12: *Participants in the first "Space Council" held on 25 November 2004 where the development of a coherent overall European space programme was discussed (source: ESA/S. Corvaja).*

26 September 2008, where issues such as GMES Initial Operations, and the long-term arrangements for the GMES Space Component are addressed.

4.2.2.2. GMES Advisory Council

The GMES Advisory Council (GAC) unites EU and ESA Member States in a forum dedicated to GMES, while the Space Council deals with all issues related to space. However, the GAC has an advisory function only, as decisions impacting on budget spending are taken in the respective decision bodies of ESA and the EC. During 2010 it is envisaged that the GAC is replaced by the GMES Partners Board with a similar mandate and composition as the GAC.

4.2.2.3. ESA Council and Sub-ordinate Bodies

Decisions related to the implementation of the GMES Space Component (GSC) programme are made through the ESA decision bodies, most notably the Programme Board for Earth Observation (PBEO) and, for higher level decisions, the ESA Council. Other horizontal bodies are involved, in particular the Industrial Policy Committee for procurement issues and the Administrative and Finance Committee. The contribution by the EC to the GSC Programme is administered through a bilateral agreement (see below).

4.2.2.4. EU FP7 Programme Committee

The EC contribution to the ESA GSC Programme is provided through the "Space" budget line within FP7. A corresponding FP7 Space Programme Committee (PC) is responsible for budgetary decisions including the approval of the EC-ESA agreement and its amendments as well as the release of annual instalments according to the agreement. It also assesses the regular reports provided by ESA in order to support decisions related to the release of the annual budgets.

4.2.2.5. The EC-ESA Agreement on the GMES Space Component

The GMES Space Component (GSC) Programme is an ESA optional programme subscribed to by its Member States and complemented by a significant

financial contribution of the EC. The EC contribution (670 million euros, 2008 e.c.) is governed through a delegation agreement under which the EC contribution is spent according to ESA rules and procedures with specific provisions to take account of the EU FP7 budget rules. These specific provisions include, in particular, that those items that are co-funded with EC budgets are open to all FP7 participants and that no geo-return targets are applicable in these cases. The agreement was signed in February 2008 and amended in January 2009 to regulate the contributions to Segments 1 and 2 of the GSC programme, respectively.

4.2.2.6. The EC GMES Bureau and ESA GMES Space Office

Good cooperation between the EC and ESA is key to the advancement of GMES. For this reason both organisations have established dedicated interfaces, the EC GMES Bureau within DG Enterprise and Industry and the ESA GMES Space Office within the Directorate of Earth Observation Programmes. The day-to-day interaction ensures this coordination; it also prepares decisions within both organisations at Director and Director General level and by the respective decision bodies of the two organisations.

4.3. Challenges of today's governance

Today's governance poses a number of challenges for the advancement of GMES. The most significant is how key decisions that are being made are split between the EC and ESA. Although there is good cooperation on a day-to-day basis, the coordination of decisions for both bodies poses a major issue due to different time schedules and composition of the ESA PBEO and EC FP7 PC.

Other challenges arise from the different nature of funds provided to GMES. Taking the case of the GMES Space Component, the contributions from ESA Member States and those of the EC are governed by different procurement rules. While ESA's rules are generally optimised for the procurement of space infrastructure, EC FP7 funds are primarily foreseen to support European research and development activities of a more general nature. It has been a significant achievement to negotiate and conclude the EC-ESA agreement on GMES which necessarily combines these two funding sources with their respective rules into a harmonised procurement of space infrastructure. The difference of the two rules imposes some restrictions on procurements which lead to the separation of ESA funded elements from those where EC funds are used.

Different procedures and time schedules for budgetary decisions pose another challenge. In particular, the major ESA decisions are taken at Ministerial Conferences which take place approximately every three years, and are mostly driven by the need for future funding of all ESA programmes, EU funding decisions are generally aligned with multi-annual financial perspectives. For example, the next period will cover 2014–2020, hence requiring the preparation of funding lines before the start of the period. For this reason, no firm commitments can be made until such decisions are made, which in this case will be towards the end of 2013. However, the needs of the GSC programme, in particular the finalisation of its build-up as well as the operation of the overall system, are dependent on EU funding. The late availability of operational funds can bring uncertainty into the procurement approach which ultimately reduces efficiency and increases cost. A case in point is the ability to procure several recurrent Sentinel spacecraft that are required for an operational system. This can not be performed as a one-off order but has to be aligned with available budget decisions.

It is fair to say that until now it has been possible to manage the above challenges because of the remarkably strong willingness by both organisations and the respective Member States to collaborate on GMES and ensure the most efficient implementation within the boundaries of the respective administrative environments.

4.4. GMES Space Component

Based on observation needs expressed by user communities in the environment and security domain, in 2005 ESA Member States decided to invest in the build-up of the necessary space infrastructure as a capacity complementary to those available in its Member States. The space infrastructure component for GMES, or GMES Space Component (GSC), will initially address service data needs for land/ice observations and systematic observations for the oceans and the atmosphere. Information that is generated will additionally feed activities in the areas of emergency response and citizens' security.

Current services mainly concentrate on the following observation techniques:

- SAR sensors, for all weather day/night observations of land, ocean and ice surfaces;
- Medium-low resolution optical sensors for wide area information on land cover, for example agriculture indicators, ocean monitoring, coastal dynamics and ecosystems;
- High Resolution (HR) or Medium Resolution optical sensors, panchromatic and multi-spectral, for regional and national land monitoring activities;

- Very High Resolution (VHR) optical sensors for targeting specific areas especially in urban areas for emergency and security applications;
- High accuracy Radar Altimeter systems for sea level measurements and climate applications;
- Radiometer to monitor the impact of climate change on land and ocean temperature;
- Spectrometer measurements for air quality and atmospheric composition monitoring.

As mentioned earlier, the GSC includes five families of dedicated satellite missions, called Sentinels, specifically developed by ESA to meet the needs of GMES, and a set of Contributing Missions, already in orbit or planned, that are owned and operated by national agencies or commercial entities of Member States, EUMETSAT or other third parties, including their ground segment, to fulfil the above observations.

All the dedicated Sentinels, Contributing Missions and Ground Segment infrastructure will be part of the overall GMES Space Component architecture coordinated by ESA. ESA is therefore responsible for establishing a mechanism to integrate, harmonise and coordinate access to all the relevant data from the multitude of missions contributing to GMES.

4.4.1. GMES satellites

4.4.1.1. Synthetic Aperture Radar missions

4.4.1.1.1. Sentinel-1

The Sentinel-1 European Radar Observatory is a polar-orbiting satellite system for the continuation of Synthetic Aperture Radar (SAR) operational applications. Sentinel-1 is a C-band imaging radar mission consisting of a pair of satellites aimed at providing an all-weather day-and-night supply of imagery for GMES user services. The first Sentinel-1 satellite is envisaged to be launched in 2012 and will be followed by the second satellite a few years later.

Sentinel-1 data will benefit numerous services. For example, services that relate to the monitoring of Arctic sea-ice extent, routine sea-ice mapping, surveillance of the marine environment including oil-spill monitoring and ship detection for maritime security, monitoring land-surface motion risks, mapping of land surface for forest, water and soil management and mapping in support of humanitarian aid and crisis situations.

Fig. 13: *Artist's impression of Sentinel-1, the first Earth observation satellite to be built for Europe's Global Monitoring for Environment and Security (GMES) programme (source: ESA/P. Carril).*

4.4.1.1.2. Sentinel-1 type contributing missions

Satellites carrying SAR (Synthetic Aperture Radar) instruments contributing to GMES include ESA's ERS-2 and Envisat missions, the Italian COSMO-SkyMed mission, the Canadian RADARSAT-2 mission and the German TerraSAR-X and TanDEM-X missions. COSMO-SkyMed operates at X-band and provides high-resolution imagery, as do the TerraSAR-X and TanDEM-X missions. RADARSAT operates at C-band and will augment Sentinel-1's revisit capabilities.

4.4.1.2. Optical missions

4.4.1.2.1. Sentinel-2

Sentinel-2 polar-orbiting satellites will provide systematic global acquisitions of high-resolution multi-spectral imagery with a high revisit frequency. The Sentinel-2 mission is envisaged to fly as a pair of satellites with the first planned to launch in 2013. Each Sentinel-2 satellite carries a Multi-Spectral Imager (MSI) with a swath of 290 km. It provides a versatile set of 13 spectral bands spanning from the visible and near infrared (VNIR) to the shortwave infrared (SWIR) imaging, featuring four spectral bands at 10 m, six bands at 20 m and three bands at 60 m spatial resolution. Data from Sentinel-2 will benefit services in areas such as land management by European and national public institutes, the agricultural industry and forestry as well as disaster control and humanitarian relief operations.

Images of extreme events such as floods, volcanic eruptions and landslides will also be acquired by Sentinel-2.

4.4.1.2.2. Sentinel-2 type contributing missions

High-resolution optical imagers and multi-spectral radiometers provide detailed images of the Earth's surface and atmosphere. GMES Contributing Missions carrying optical imagers will complement the multi-spectral imagery from Sentinel-2 and the Ocean Land Colour Instrument and Sea Land Surface Temperature Radiometer to be carried on Sentinel-3. They have many application areas such as agriculture, damage assessment associated with natural hazards, urban planning, land-cover mapping. They are, however, limited to fair weather, day-time only operation. Measurements may be used to infer a wide range of parameters, including sea and land surface temperature, snow and sea-ice cover, cloud cover. They supply an important source of data on processes in the biosphere, providing information on global vegetation and its variation through the seasons, important for identifying areas of drought and early warning of food shortages.

GMES Contributing Missions carrying high-resolution imagers and multi-spectral radiometers include: EnviSat, DMC, EnMap, EROS-A/-B, Pléiades, RapidEye, SeoSat/INGENIO, SPOT and TopSat.

4.4.1.3. Global land and ocean missions including altimetry

4.4.1.3.1. Sentinel-3

Sentinel-3 mission's main objective is to determine parameters such as sea-surface topography, sea- and land-surface temperature as well as ocean- and land-surface colour with high-end accuracy and reliability.

Benefiting from a heritage of ESA missions, Sentinel–3 carries several instruments:

- A surface temperature system called Sea Land Surface Temperature Radiometer (SLSTR), which is an evolution of Envisat's Advanced Along Track Scanning Radiometer (AATSR). The SLSTR uses a dual viewing technique and operates across eight wavelength bands providing better coverage than AATSR because of a wider swath width.
- An Ocean Land Colour Instrument (OLCI), which is based an evolution of Envisat's Medium Resolution Imaging Spectrometer MERIS instrument. The OLCI operates across 21 wavelength bands from ultraviolet to near-infrared and uses optimised pointing to reduce the effects of sun glint.

- A topography system, which includes a dual-band Ku- and C-band altimeter based on technologies used on ESA's Earth Explorer CryoSat mission, a microwave radiometer for atmospheric correction and a DORIS receiver for orbit positioning.

Sentinel-3 is primarily a mission to support services relating to the marine environment, with capability to serve numerous land, atmosphere and cryospheric-based application areas. The first Sentinel-3 satellite is expected to launch in 2013, followed by a second so that they work together to provide maximum coverage.

4.4.1.3.2. Sentinel-3 type contributing missions

Radar altimeters use the ranging capability of radar to measure the surface topography profile along the satellite track. They provide precise measurements of a satellite's height above the ocean by measuring the time interval between the transmission and reception of very short electromagnetic pulses.

A variety of parameters may be inferred using the information from radar altimeter measurements, such as time-varying sea-surface height (ocean topography), the lateral extent of sea ice and altitude of large icebergs above sea level, as well as the topography of land and ice sheets, and even that of the sea floor. Satellite altimetry also provides information for mapping sea-surface wind speeds and significant wave heights.

The following GMES Contributing Missions carry altimeters that will complement the altimetry instrument that will be carried on ESA's Sentinel-3 mission: ERS, EnviSat, Jason-1 and -2 (OSTM) and AltiKa.

4.4.1.4. Atmospheric Missions

4.4.1.4.1. Sentinel-4 and -5

The Sentinel-4 and -5 payloads will be carried on meteorological satellites operated by EUMETSAT. The Sentinel-4 payload will be deployed on the two Meteosat Third Generation-Sounder (MTG-S) satellites in geostationary orbit (planned to launch in 2017 and 2024). The Sentinel-5 payload will be carried on the post-EUMETSAT Polar System (EPS) spacecraft series (planned to launch starting in 2020). According to this concept, a Sentinel-5 precursor mission is planned to launch in 2015, to avoid data gaps between Envisat (SCIAMACHY data in particular) and Sentinel-5.

Both Sentinel-4 and -5 will be devoted to atmospheric composition monitoring for the GMES Atmosphere Service. This service will provide coherent information on atmospheric variables in support of European policies and for the benefit of European citizens. Services proposed will cover air quality, climate change and stratospheric ozone and solar radiation.

4.4.1.4.2. Sentinel-4 and -5 type contributing missions

Atmospheric chemistry instruments based on various techniques are used to measure the composition of the Earth's atmosphere. For example, each atmospheric gas is characterised by its absorption and emission spectra, which describe how the molecules respond to different frequencies of radiation. Remote sensing instruments exploit these signatures to provide information on atmospheric composition, using measurements over a range of wavelengths, between UV and microwave. The capability of providing a global picture of the atmosphere, and how it is changing on a daily, seasonal and geographical basis, ensures demand for these instruments in a wide range of applications, such as pollution monitoring and climatology, including studies of the carbon cycle and support to policy-making processes such as the Kyoto Protocol, volcanic eruption monitoring and operational meteorology.

GMES Contributing Missions carrying atmospheric instruments complement the Sentinel-4 and -5 missions and include Meteosat Second Generation (MSG), MetOp and GOSAT.

4.5. Sentinel Data Policy

The term "Sentinel Data" relates to all data and products that will be generated on the basis of observations acquired by Sentinel mission sensors and delivered by the GMES Ground Segment[568] during the exploitation phase of the Sentinel missions.

The Sentinel Data Policy, which governs the provision of Sentinel data, will be part of the overall GMES Data and Information Policy under the EC's responsibility and has the following objectives:

- Promoting the use and sharing of GMES information and data;
- Full and open access to information produced by GMES services and data collected through GMES infrastructure, subject to relevant security restrictions;
- Strengthening Earth observation markets in Europe, in particular the downstream sector, with a view to enabling growth and job creation;
- Contributing to the sustainability of the provision of GMES data and information;
- Supporting European research communities.

These objectives apply to the Sentinel Data Policy jointly established by the EC and ESA, including full and open access. Such an approach aims at maximising the beneficial use of Sentinel data for the widest range of applications and intends to stimulate the uptake of information based on Earth Observation data for end users. Thus it responds directly to the increasing demand for Earth Observation data in the context of climate change initiatives and in support of the implementation of environmental policies, also resulting in humanitarian benefits.

The principles of the Sentinel Data Policy are:

- Anybody can access acquired Sentinel data; in particular, no difference is made between public, commercial and scientific use or between European and non-European users;[569]
- The licenses for the Sentinel data itself are free of charge;[570]
- Sentinel data will be made available to the users via a "generic" online access mode, free of charge. Generic online access is subject to a user registration process and to the acceptation of generic terms and conditions;
- Additional access modes and the delivery of additional products will be tailored to specific user needs, and therefore subject to tailored conditions;
- In the event of security restrictions applying to specific Sentinel data that affect availability or timeliness, specific operational procedures will be activated.

The Sentinel Data Policy definition started in 2008. Several documents and their updates have been presented to ESA PBEO and GAC delegates. An industry information day took place in September 2009.

Agreement on data policy is led jointly by the EC and ESA and is reached through consultation between the respective decision bodies, which for ESA is the PBEO and for the EC the EU Council and European Parliament, as data policy is part of the Regulation on a GMES Initial Operations Programme that requires a decision process.

Final decision on the Sentinel data policy is expected by end 2010.

4.5.1. Other space data policies

There are certain common trends in the current data policies of various international and national organisations even though differences in position are present in some key areas.

One area of difference concerns the policies for data access of different organisations. The position of some of the most prominent international and national agencies such as ESA, EUMETSAT and NASA, is to provide data to all users on a non-discriminatory basis, from their funded and owned systems.

With respect to pricing policy there are clear differences among the different organisations. Data from NASA and its U.S. government partners, including NOAA, are made available at a reasonable price (i.e. cost of dissemination or COFUR, Cost Of Fulfilling User Requests) or, in cases where such pricing unduly inhibits use, below that cost.

ESA either provides data for free or at cost of reproduction (for Research and Development projects) or delegates the distribution of data (for commercial/operational use) to external entities that sell these data on the open market.

EUMETSAT delivers products and services on a free basis if they are considered "essential data". "Other" data and products are encrypted and their reception requires a licence agreement and a decryption unit and, depending on the type of user and use, a fee may be involved.

Canadian EO data policy varies from satellite to satellite. The launch of Radarsat-2 marked a complete shift from public sector ownership to a private sector enterprise, and the data started to be sold to recover costs of operations. The follow-on satellite mission, the 3-satellite Radarsat Constellation, reverts to a data policy whereby they are provided at low cost or cost of reproduction. However, details are still to be finalised in this regard.

Several other data suppliers, in particular in the very high resolution optical domain (up to about 5 m ground resolution) apply a commercial policy which uses the revenues to cover operations and/or development costs for the next satellites.

In the medium resolution domain (5–30 m) the general trend seems to be towards a free and open approach in which data are supplied on an unrestricted, open and non-discriminatory basis, and, in many cases, free of charge (or at dissemination cost).

As of 31 December 2008, a recent USGS (U.S. Geological Survey) policy change enabled Landsat data to be distributed at no charge via the Internet. There is evidence that more data have been distributed in the first six months since the change in the data policy than in the entire first 36 years of the Landsat missions combined.

As another example, at the end of the 1980s, Brazil began the development of a civilian remote sensing satellite programme (CBERS) with China. After the CBERS-2 launch, Brazil adopted the free of charge CBERS data distribution policy whenever data were requested in electronic format. Initially adopted for Brazilian users, it was extended at a later stage to other countries and thousands of images were distributed at no cost to users.

This is in line with a European directive called INSPIRE (Infrastructure for Spatial Information in Europe), which entered into force in May 2007, and whose goal is to establish an infrastructure for spatial information in Europe to support community environmental policies, whereby environmental data should be collected at the same standards and scales across Europe and be freely available to all.

4.6. Key issues on GMES and lessons learned

Good cooperation between the EC and ESA is mandatory for the success of GMES. This is a challenging task, as could be expected from a multi-organisation European space programme. In order to meet this challenge, a close relationship between the two major actors is required.

As leader of the overall initiative, the EC have a strong role in the policy domain.

The establishment of user requirements, under the responsibility of the EC, has been a relatively complex process. However, it is an essential step as it provides the basis for the definition of the space infrastructure. It is clear, therefore, that sufficient political attention and funding need to be given to the Services Component.

ESA will take on four main tasks within the GMES Space Component. As such, ESA will be:

- Coordinator of the overall GMES Space Component;
- Development agency for new GMES space infrastructure;
- Procurement agency for recurrent elements and
- Ad-interim operator of a sub-set of Sentinel missions (i.e. Sentinel-1, -2 and land part of -3).

While the development and procurement roles are well defined and practised since many decades, including the experience in the context of the ESA-EUMETSAT cooperation, the first and fourth roles in the above list have not been performed within GMES before. It is important that these roles are adequately defined and agreed with all stakeholders concerned.

Regarding the overall GMES context, the success of GMES requires that a number of challenges are met, some of which are listed below:

- The need to establish clear principles for the governance of GMES, which is aligned to the main funding streams;
- The need to prepare the operational programme of GMES including long-term sustainable funding lines;
- Procurement rules for joint EC-ESA programmes need to be defined, preferably at Space Council level, covering ideally the development and operational phases;

- The decision process for the Space Component is currently well defined for the build-up phase; however, for the operational programme, including the transition from the build-up to the operational phase, still requires further discussion and agreement;
- A long-term scheme for access to data from Member States and EUMETSAT missions needs to be established;
- The role of some key players, e.g. EUMETSAT, EEA and EUSC should be better defined with regard to their contribution to GMES;
- The agreement on a free and open Sentinel data policy is a key success criteria for GMES; preliminary decisions are made, however the final decisions are expected by end 2010;
- The In-situ Component needs to be further strengthened and integrated with the services and space component;
- Definition of the S (Security) of GMES has just started and will require further studies as well as an important political dialogue.

In summary, progress of the GMES programme is good despite the complexity of the initiative. In the services domain, first services are becoming pre-operational or operational. The Space Component was very successful in organise the necessary funding for the build-up phase. The challenge now will be to ensure sufficient funding for the operational programme including the smooth transition from the build-up to the operational programme.

Note by the Authors: Opinions expressed in this article are those of the authors and not necessarily those held by the European Space Agency.

[568] Direct access to the downlink of data produced by Sentinel sensors by any data reception system outside the GMES Ground Segment is subject to the conclusion of dedicated agreements.
[569] Restrictions may apply based on applicable security rules and regulations.
[570] As provided from the GSC Core Ground Segment; Priorities will be set in case of constraints in providing the data.

5. Integrated applications: a new way forward for Europe – Some legal thoughts

Frans G. von der Dunk

5.1. Europe and the integration of Earth observation, telecommunications and navigation

At the most recent ESA Ministerial Council, it was decided amongst others to establish an Integrated Applications Promotion (IAP) programme, essentially combining in one infrastructure Earth observation, telecommunications and navigation applications. ESA's website itself announced in somewhat greater detail: "The Integrated Applications Promotion (IAP) programme (ARTES element 20, Phase 1) will foster the use of integrated space systems and technologies (telecommunications, Earth observation, meteorology, etc.) alone or in combination with a variety of terrestrial systems, in a wide range of operational services for society and public policies (natural disaster monitoring and mitigation, search and rescue). The programme is based on two elements: *basic activities* (raising awareness of the potential users, identifying potential new services and preparing new projects for demonstration) and *demonstration activities* (projects that will lead to pre-operational services). Service providers, industry and user institutions will be involved from the outset with a view to their taking over the service when the activity is mature enough *to lead to sustainable operational services.*"[571]

This initiative may well bring rather novel perspectives to space applications, possibly shaping them across Europe or even globally in the near to mid-term future – but obviously will also raise legal issues in addition to the broader technological, economic and policy-related ones. Without claiming to be exhaustive, the present contribution seeks to provide a first inventory of such legal issues in order to help understanding where appropriate and timely solutions need to be developed, in particular in the law- and policy-making areas, to ensure that Europe would indeed reap the full benefit of such integrated applications.

Here, the key lies in the term "integrated": by integrating various existing applications, the legal frameworks existing for each of those applications effectively become relevant at the same time – without, however, being integrated themselves to any appreciable extent as of yet. The novelty, not only of the integration of the three application sectors, but to a considerable extent also of down-to-Earth applications of at least two of those sectors themselves (the navigation and Earth observation sectors) further exacerbates these problems.

Thus, the question would often arise whether it would be necessary or desirable to *either* develop a new legal regime almost as if from scratch, in the process carving out this special area from applicability of all such other regimes, *or* merely try to mend the inevitable gaps, overlaps and inconsistencies in such other regimes as appropriate. Finally, it may be noted that "Europe" as a concept in itself represents a study in "integration", too. Whether it concerns integration by means of ESA or by means of the Union, issues of individual Member State sovereign discretion versus the overarching greater common good of Europe arise all the time.

For the sake of analysis as well as for reasons of convenience, relevant issues will be categorised along three different lines: the ESA legal framework, the EU legal framework as relevant to space, and the "external" parameters, encompassing such issues as the ITU framework and the UN space treaties.

5.2. The ESA legal framework and integrated space applications

The aforementioned summary of the IAP initiative indicates that essentially basic activities and demonstration activities are envisaged, whilst the ultimate aim is to generate downstream value for European society and economies. Such activities clearly fit in rather well in general terms with the central purpose of ESA, defined as "cooperation [...] in space research and technology and their space applications, with a view to their being used for scientific purposes and for operational space applications systems".[572] At this – relatively early – stage of IAP, only two issues would seem to arise here.

The first issue essentially concerns the choice of options available within ESA when it comes to developing European space projects and space programmes. On the one hand, especially the "basic activities" as defined in the IAP would logically become "mandatory activities", meaning that once such a programme would be duly decided upon, all Member States would contribute their share of funds, as per a predetermined scale-based on average national income.[573]

With regard to the "demonstration activities" on the other hand, the situation is less clear. Following the rough outline offered on ESA's website, they might encompass the ensuring of "the elaboration and execution of a scientific programme including satellites and other space systems" as well as "the design, development, construction, launching, placing in orbit, and control of satellites and other space systems; [and/or] the design, development, construction, and operation of launch facilities and space transport systems", the former set under the ESA Convention invoking the label of "mandatory activity", the latter set one of "optional activity".[574]

Once a project would be undertaken as an "optional activity", it would still require a majority decision in the ESA Council as well as a unanimous decision on the level of resources,[575] but at the same time would allow individual States to determine their own level of contribution or even opt out altogether.[576] Thus, a determination that a project under IAP would have to be developed as an "optional activity" may result in considerable additional risk of debate and delays as a consequence of the need to get sufficient financing effectively committed by the Member States.

The second issue concerns the formal requirement for ESA to undertake and conduct its activities, projects and programmes "for exclusively peaceful purposes".[577] Rather than going into a detailed discussion on the exact meaning of that phrase here, it should suffice to note that ESA and its Member States have traditionally interpreted this clause as a requirement to refrain from any involvement in *military* activities, even for defensive or other "clearly" peaceful purposes. In particular in areas such as Earth observation and navigation – note that these form two of the three main pillars under the IAP! – this meant a serious handicap

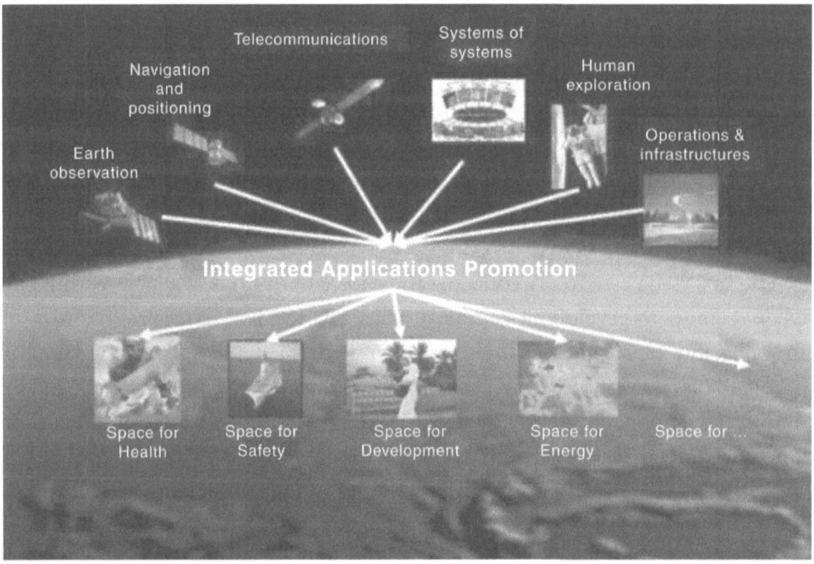

Fig. 14: *The Integrated Applications Promotion (IAP) initiative (source: ESA).*

for ESA; in the first case because of the obvious "spying" capabilities part of Earth observation, in the second case because of the equally obvious missile-guiding capabilities involved in satellite navigation applications.

With hindsight, it seems inevitable that the reality of space, where almost all hardware, software and know-how seem to have at least potential dual-use applicability, would someday catch up with this traditional effort to steer clear of political controversy, whether with individual Member States, with other European institutions such as the WEU[578], or with NATO[579] and the United States. In many respects, the current two major European space projects, Galileo and GMES, represent the first careful steps in the direction of involving ESA and the European Union (formally equally confined to dealing with non-defence issues[580]) into security and defence issues.

As for Galileo, though it is emphatically labelled a civil system and, unlike the U.S. and Russian GNSS, will be principally owned and operated by a civil governance structure, the possibility of adversary use of its signals should still be dealt with.[581] Even more pertinently, the envisaged Public-Regulated Service (PRS), whilst painstakingly avoiding any reference to "military", "defence" or "security", was modelled in many respects on the GPS Precise Positioning Signals. It consists of encrypted signals allowing only a limited group of users to access them, with technical features trying to prevent unwarranted and unwanted users from (ab)using or interfering with them.

Whilst the PRS is officially to be made accessible to all governmental services (as opposed to only the military ones, as was the case with the GPS PPS), and even to some non-public key infrastructure operations (telecommunications; energy), a debate has already arisen about whether such governmental uses should also include uses by the military of the respective Member States. To those familiar with the political history of the last half century, it will come as no surprise that France is most adamant in seeing no obstacle to such use, whereas at least until recently the United Kingdom was most adamant in emphasising that such military uses were never contemplated – neither should they be.

GMES in this sense represented the next step, since this project for the first time did prominently refer to the concept of "security" – interestingly in the process extending its scope, as GMES was originally meant to stand for Global Monitoring for *Environmental Security*, before the latter part was almost surreptitiously changed to "Environment *and* Security". GMES, being tasked to provide Europe with its own independent and comprehensive satellite Earth observation infrastructure for data and information generation on a comprehensive range of subjects, will indeed contribute to the inclusion of defence, security and military matters into the broader civil European governance structures.[582]

As GMES constitutes the most visible and central element of European Earth observation efforts for the years to come, and Galileo similarly is the most visible and central element of European satellite navigation efforts, this means two of the three main pillars of the IAP initiative are at stake in this fundamental discussion regarding the inroads of civil space efforts into the military, defence and security domain in Europe. Moreover, precisely the integration of highly valuable Earth observation data with highly accurate positioning and navigation information will raise the value from the security perspective: one and one adds up to considerably more than two here. Any comprehensive implementation of IAP can thus hardly avoid bringing this discussion to a more broad, visible and formal conclusion – inevitably taking shape at least partly through legislative documents.

The Rubicon has not yet been crossed here, as no one dared to redefine the reference to "exclusively peaceful purposes" so as to unequivocally include defence, military and security issues, but clearly Europe – and ESA with it – stands at the crossroads. From a broader angle, also in international legal circles, an increasing acceptance of the use of force (read "military" force) can be noted, as unfortunately often being necessary to enforce peace, or even to create the necessary prerequisites for peace, in particular as long as sanctioned by the United Nations. An interesting point in case was represented by INMARSAT, which under its statutory documents was obliged to comply with the mandatory requirement that all its activities were to be for "exclusively peaceful purposes".[583] The question then arose whether INMARSAT was entitled to offer its services in the context of the UN-sanctioned operations in Iraq in 1991 or whether such an offer would conflict with its peaceful purposes-obligation; a question answered in the first sense with that interpretation not being disputed by anyone since.

5.3. The European Union legal framework and integrated space applications

Formally speaking, at least for the time being, the IAP initiative does not involve the European Union or its quasi-executive arm, the European Commission. However, that does not mean that the EU legal framework can be completely ignored by the activities to follow on from the initiative. Almost all ESA Member States are members of the Union, and therefore the large majority of their industries and service providers fall under the sway of the Union's legal framework.

Moreover, as already indicated, Earth observation and navigation provide two of the three main pillars under the IAP. Since GMES within the former and Galileo within the latter constitute the key overarching European efforts, the fact that both are being driven at least on the political level by the European Commission already makes clear that at least in those two areas the Union's policy-making and legal framework, even if indirectly as of yet, may exert considerable impact on any further implementation of IAP.

Also in the context of the EU legal framework the security-paradigm first strikes the eye. The European Union, at least formally, was not supposed to deal with military, security and defence issues. The relevant clauses on tasks and objectives of the Union do not offer any, even indirect, reference to defence or military matters, whilst also indicating that it "shall act within the limits of the powers conferred upon it by this Treaty and of the objectives assigned to it therein".[584]

However, for the Union, that traditional perspective has steadily eroded over the past years more clearly than in the case of ESA. Already at the entry into force of the Treaty on European Union in 1993, the Common Foreign and Security Policy (CFSP) was established as the EU pillar to deal with common foreign and security policy, to which increasingly also defence and military issues were being relegated.

It should not be lost sight of, however, that the CFSP is a straightforward intergovernmental construction.[585] In this context, there is at best a marginal role for the European Commission as the guardian of the overarching European interest – for example, the Commission only "shall give its *opinion* particularly on whether the enhanced cooperation proposed [by EU Member States] is consistent with Union policies".[586] No role whatsoever arises for the elaborate legislative, adjudicative and enforcement jurisdiction (European Parliament and European Court of Justice) that was developed in the context of the EC Treaty. Thus, the cooperation under the CFSP can result in the "implementation of a joint action or a common position", but can never "relate to matters having military or defence implications"; those issues remain exclusively reserved for national governments to deal with as they see fit.[587]

Even with the recent entry into force of the Lisbon Treaty, formally deleting the separate pillar and integrating the CSFP into the revamped EC Treaty, this situation did not fundamentally change. It may, at best, open more opportunities to Member States so willing to "communitarise" some elements at the former CFSP, but ultimately it is the general political will which determines progress in that context.

At the same time, the aforementioned Western European Union (WEU), the vehicle for matters of international cooperation in the areas of defence and security in Europe, is now being integrated into the Union structurally, albeit still as part of the intergovernmental CFSP. The Treaty on European Union itself refers to the role of the WEU in somewhat ambiguous terms: security policies in the context of the CFSP pillar "shall not prevent the development of closer cooperation between two or more Member States on a bilateral level, in the framework of the Western

European Union (WEU) and NATO, provided such cooperation does not run counter to or impede that provided for" through the CFSP.[588]

The complications arising from this development for the Union's involvement in Galileo and GMES, and thus indirectly for any full-fledged implementation of the IAP initiative, are difficult to gauge. The best that can be said is that the Union, through the mechanisms available in the context of the CFSP, will likely try to assert some control over any IAP implementation in order to ensure that the interests of the CFSP, the Union and those of Europe in a wider sense will not be unduly interfered with. At the same time, in terms of the downstream operational services to spring from IAP, from a quite different angle especially the Commission might insist on applying as much of existing EC law as it can, to ensure in particular fair competition and an optimal operation of the Internal Market.

The most visible area where this may play out currently would likely be that of security sensitive exports of dual-use goods. Since 2000, the Commission has taken the lead within Europe in trying to transform the rather voluntary international export coordination and consultation arrangements existing under the Missile Technology Control Regime (MTCR)[589] and Wassenaar Arrangement[590] into a more binding, coherent and comprehensive system, at least for Europe. This was achieved in first instance by Regulation 1334/2000[591], which was updated regularly in order to take new developments in the MTCR and Wassenaar Arrangement context into consideration.[592]

Essentially, the Regulation established a system obliging the EU Member States to mutually inform and consult each other on potentially sensitive exports of any hardware, software or know-how to non-EU Member States (as the decision on allowing export remains a national one), whilst trying to harmonise and simplify the process of export authorisation – most notably by the introduction of a Community General Export Authorization (CGEA) for the lesser sensitive exports.[593]

In view of the large measure of potential dual-use applicability of much of the hardware, software and know-how involved in space activities and the consequently broad sweep of this regime, the effects thereof – much as it is still ultimately based on the sovereign discretion of individual Member States to allow or refuse relevant export – might have a considerable impact on IAP implementation, once it would involve some measure of transfer of relevant hardware, software or know-how to outside (or to a certain extent even within) the European Union.

The above analysis has especially focused on the Earth observation and navigation pillars under the IAP, in particular in their combination. When it comes to the third pillar, that of telecommunications, however, the situation is quite different.

On the one hand, this is the area within the space environment in which the EU legal framework has made truly large inroads also in a legal sense. Since the 1994 Satellite Directive[594] the Commission has set in motion a process of applying the Internal Market principles, still at the heart of the drive of the EU organs in the legal field, to the specific area of satellite communications. Even though progress is slow and often haphazard, the main principles – the break-down of intra-state trade barriers, and principled prohibitions with exceptions of cartel-like or monopolistic behaviour, as much as of State aid for commercial companies unless duly justified[595] – have now been applied, sometimes through legal proceedings, in the satellite communications sector. Obviously, any commercial activity in the telecommunications sector, including satellite communications, now has to comply with that regime, whether it occurs in the context of IAP or an implementation project thereof or not.

On the other hand, it does not seem that the public interest-driven IAP would have major problems with keeping within the boundaries of this regime, essentially abiding by the fair and open competition-rules that form the heart of it. The main *caveats* stemming from the applicability of the competition regime to IAP-related efforts would be that (1) any value derived from it by commercial ventures at the expense of the taxpayer should be made available without discrimination to *all* interested commercial ventures and (2) any solicited contribution from the private sector should be tendered in such a manner that *all* commercial ventures should be eligible as long as compliant with non-discriminatory and objective criteria.[596]

The above, to be sure, applies to all three pillars of the IAP initiative; yet, due to the level of commercialisation of the telecom sector as compared to that of the other two, it is by far the most relevant here. As the involvement of IAP applications in commercial sectors grows, EU policy-making and law-making may impact the IAP process more and more along the lines sketched above, even as the level of abstraction at which the IAP initiative currently operates may cause such issues to remain largely theoretical for the time being. Since the whole ultimate justification of IAP lies in resulting in downstream "sustainable operational activities" as per the website, further analysis on this issue – or at least some recurring monitoring for the purpose – would be warranted here.

5.4. External legal parameters: the ITU framework and the UN space treaties

Naturally, any substantive implementation of the IAP initiative would likely trigger external implications in addition to those following on from the two

European legal frameworks discussed. For the present purpose, only the parameters resulting from the ITU framework (in view of its relevance for all international telecommunications including satellite communications) and those resulting from the main UN space treaties (in view of the overriding space component of the IAP) will be dealt with, being the two most obviously important.

As for the ITU framework, the first problem for IAP that meets the eye may well lie with the process of allocation of frequency bands, as the first step in the system established in that framework's current version on the basis of the ITU Constitution[597] and Convention[598] to coordinate the international use of frequency spectrum for all wireless services. More precisely, specific allotment of frequencies

Fig. 15: *ITU Plenipotentiary Conference (Geneva, 1992). General view. From left to right: Mr. D. Maclean (ITU) and B. de Riedmatten, chairman of the conference (Switzerland) (source: ITU).*

available within an allocated frequency band to a particular satellite or satellite system of the IAP, the next step in the ITU coordination system,[599] can only occur if the satellite (system) is to be used for the particular category of service to which the frequency band was allocated in the first place.

Allocation is defined as reserving a particular range of frequencies, called a frequency band, "for the purpose of its use by one or more terrestrial or space radiocommunication *services* or the radio astronomy service under specified conditions".[600] The concept of "services" in turn is generally used to cover relatively broad and sometimes loosely defined categories – but not broad enough to encompass a satellite that might undertake *both* Earth observation *and* telecommunication-proper *and* navigation activities for IAP purposes.

In other words, an IAP satellite performing activities under current ITU regulation falling under more than one heading with a view to the allocation of frequency bands, may run into trouble already in this first stage, as its hybrid functionality might cause complications as regards the frequency band within which frequencies are to be requested and allotted. Further analysis of the ITU experience with the few hybrid satellites formally coordinated so far may provide some clues in this respect, but European authorities would do well to be prepared to deal with such a contingency (as well as others resulting from the ITU framework) by analysing the ITU categories and related process in detail, and trying to fit in as much as possible at the outset.

A further complication (though certainly not exclusive to the IAP project) concerns the fact that allotment of specific frequencies cannot be requested by, nor granted to intergovernmental organisations – not even the European Union, in spite of its political weight and legislative framework. Only States are entitled to put forward such requests, and then in turn would have to assign the frequencies, once allotted, to the actual operator in case that operator is not a State (body) itself.[601] Usually, in cases of intergovernmental operators, the host State is charged by internal arrangement to take it upon itself to ensure appropriate allocation and assignment. For IAP satellite frequencies, essentially similar solutions will have to be found.

As for the main UN space treaties, there are essentially three that merit being briefly surveyed here: the Outer Space Treaty,[602] the Liability Convention[603] and the Registration Convention.[604] In the last resort, however, those treaties seem to have very little impact specific to IAP as opposed to, for example, the individual areas of space applications it intends to integrate – satellite communications, satellite Earth observation and satellite navigations.

Either of those, at least in principle, complies with the general requirement of space activities under the Outer Space Treaty's terms to be "for the benefit and in the interests of all countries" and "in the interest of maintaining international peace and security and promoting international cooperation and understanding", and any beneficial combination of those as per the IAP would only enhance such principled benefits.[605] Also, IAP satellites would have to comport with clauses such as concerning international State responsibility for the lawfulness of their own as well as other "national activities in outer space" and the duty to inform and consult in case of threats of serious harm to space activities of others just as much as any other satellite.[606]

This is similar with respect to the Liability Convention. Those States qualifying under Article I(c) as launching states of respective IAP satellites are liable for damage caused by their satellites, absolutely so in case the damage is sustained somewhere on Earth and in principle without limit as to compensation.[607] From the perspective of the IAP, of particular note is that the definition of compensable damage under the Convention seems to exclude damage resulting from flaws in, or even total absence of, information generated by satellites.[608]

The Registration Convention finally, is clearly applicable to IAP satellites, calling for their registration in national registers as well as with the UN involving a minimum set of information parameters, without however posing issues specific to IAP so far.[609] Most peculiar perhaps is the requirement made under Article IV(1)(e) to indicate the "general function of the space object", which clause has been interpreted so loosely as to basically allow any general description.

The main, overarching issue under the UN space treaties thus, again, becomes the fairly limited range of possibilities in particular in legal terms for ESA (or even the Commission on behalf of the Union, though not formally involved in IAP (yet)) to protect and serve their interests independently from their Member States. As with the ITU, ESA and the Commission could at best enjoy an observer status, cannot undertake a formal legal role in altering any of the treaties' provisions – and can enjoy only limited status as "parties" to them.[610]

5.5. Concluding remarks

The one immediate conclusion that arises from the above, preliminary, analysis is that at least there are no "show-stoppers" for IAP resulting from the various legal frameworks surveyed above. In a sense, IAP is even furthering several key aspects of international (space) law, such as optimising the benefits space and space activities can bring for mankind.

Beyond that, however, several aspects of those frameworks may present cause for concern in that the feasibility, efficiency and occasionally even legality of specific options may be considerably impaired if those aspects and those frameworks are

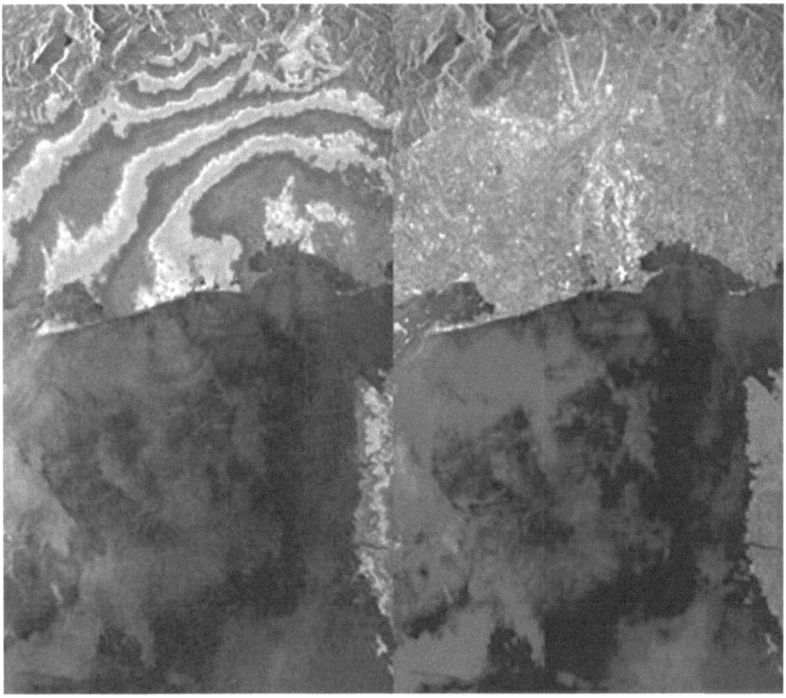

Fig. 16: *SAR Interferometry from the ERS-2 mission. ESA defined a new Data Policy to maximise the beneficial use of the Earth observation data for the ENVISAT satellite and to stimulate a balanced development of scientific and public and private sector applications, consistent with the mission objectives (source: ESA).*

not taken into due consideration in a timely manner. The present contribution in this regard touched upon the security- and frequency-related issues for the upstream part of IAP (the envisaged satellites and attendant infrastructure) and commercial issues for the downstream part of IAP (the "sustainable operational activities" it should generate) as per, in particular, EC and EU law.

Further issues might arise in the context of the governance structure to be ultimately developed for IAP, where it makes a considerable difference whether a single State, an intergovernmental organisation or a more or less private operator is going to be in charge – in terms of, for example, frequency coordination or liability. Once again, however, this does not amount so much to a "show-stopper" as it simply cautions to take such legal/institutional aspects into account as early in the process as possible.

IAP would promise large benefits at least for Europe, so the positive perspective should not be lost sight of. Addressing such issues, as well as other legal/institutional ones which doubtlessly will come up, in a timely manner would also considerably enhance the opportunities for IAP to bear fruit, as gaps, overlaps and other inconsistencies can be properly addressed. For an integrated programme like IAP, some measure of concomitant integration of the legal and institutional framework in the last resort is not only inevitable but also very helpful, and considering the time it costs to draft international treaties or "lesser" types of agreements, we should start tackling those issues soon as well.

---

[571] "Ensuring Competitive and Innovative Industries. Telecommunications and the ARTES Programme." 18 Nov. 2008. ESA. 12 July 2009. http://www.esa.int/SPECIALS/Ministerial_Council/SEMUIWSHKHF_0.html. Emphasis added.

[572] Art. II, Convention for the Establishment of a European Space Agency (hereafter ESA Convention), Paris, done 30 May 1975, entered into force 30 October 1980; 14 ILM 864 (1975); Space Law – Basic Legal Documents, C.I.1.

[573] Art. V(1.a), esp. sub (i) & (iii), ESA Convention; see further Art. XI, on decision-making by the Council, & Art. XIII, on financial contributions.

[574] Art. V(1.a), sub (ii), resp. Art. V(1.b), sub (i) & (ii), ESA Convention.

[575] See Art. XI(5.a), sub (i) & (ii), ESA Convention.

[576] See Art. XIII(2), ESA Convention; even though the clause is phrased the other way around, it essentially amounts to an *à la carte* system of participation by Member States to optional activities.

[577] Art. II, ESA Convention.

[578] The Western European Union (WEU), established by means of the Treaty of Economic, Social and Cultural Collaboration and Collective Self-Defence, Brussels, done 17 March 1948, entered into force 25 August 1948, was originally envisaged to be the 'defence version' of the European communities being developed in the late 1940s and 1950s. Currently, it is in the (long-drawn, and still-disputed) process of being integrated into European Union structures.

[579] The North Atlantic Treaty Organisation (NATO) was established by the North Atlantic Treaty, Brussels, done 4 April 1949, entered into force 24 August 1949; 34 UNTS 243; TIAS No. 1964; 63 Stat. 2241. NATO was designed to take care of all defence issues of Europe in the context of the Cold War in close alliance with, read under the leadership of, the United States.

[580] See however further *infra*, para. 3.

[581] This was essentially taken care of by involving a 'Galileo security centre' in the overall governance scheme for the Galileo system, as well as specific security-related regulations; see European Parliament and Council of the European Union. Regulation of the European Parliament and of the Council on the Further Implementation of the European Satellite Navigation Programmes (EGNOS and Galileo). Doc. (EC) No 683/2008 of 9 July 2008. Brussels: European Union. preambular para. 16, Artt. 7, 13, 14, 16; Council of the European Union. Council Joint Action on Aspects of the Operation of the European Satellite Radio-navigation System Affecting the Security of the European Union. Doc. 9569/04 of 19 May 2004. Brussels: European Union.

[582] See again further *infra*, para. 3, briefly discussing the European Union's building of its Common Foreign and Security Pillar in this respect.

[583] See Art. 3(3), Convention on the International Maritime Satellite Organization (INMARSAT), London, done 3 September 1976, entered into force 16 July 1979; 1143 UNTS 105; TIAS 9605; 31 UST 1; UKTS 1979 No. 94; Cmnd. 6822; ATS 1979 No. 10; 15 ILM 1052 (1976).

[584] Art. 5 Treaty establishing the European Economic Community (hereafter EC Treaty), Rome, done 25 March 1957, entered into force 1 January 1958; 298 UNTS 11; as amended by the Treaty on European Union, Maastricht, done 7 February 1992, entered into force 1 November 1993; 31 ILM 247 (1992); OJ C 191/1 (1992); see further Artt. 2–4.

[585] See Artt. 11–28, Treaty on European Union.
[586] Art. 27c, Treaty on European Union, as inserted by the Treaty of Nice amending the Treaty on European Union, the Treaties establishing the European Communities and certain related acts (hereafter Treaty of Nice), Nice, done 26 February 2001, entered into force 1 February 2003; OJ C 80/1 (2001). Emphasis added.
[587] Art. 27b, Treaty on European Union, as inserted by the Treaty of Nice.
[588] Art. 17, Treaty on European Union, as amended by the Treaty of Nice.
[589] The MTCR is a voluntary arrangement amongst currently 34 States (including 19 of 27 EU and all ESA Member States) to consult and coordinate exports to third States of missile and related technologies, based on the Agreement on Guidelines for the Transfer of Equipment and Technology Related to Missiles), done 16 April 1987; 26 ILM 599 (1987); amended in 1993.
[590] The Wassenaar Arrangement on Export Controls for Conventional Arms and Dual-Use Goods and Technologies, Wassenaar, done 19 December 1995, effective 12 July 1996, is a voluntary arrangement amongst currently 40 States (including 26 of 27 EU and all ESA Member States) essentially extending the scope of the MTCR to all dual-use goods with security-sensitive ramifications.
[591] Council of the European Union. Council Regulation Setting up a Community Regime for the Control of Exports of Dual-use Items and Technology. Doc. 1334/2000/EC of 22 June 2000. Brussels: European Union.
[592] For the latest update see: Council of the European Union. Council Regulation setting up a Community regime for the control of exports, transfer, brokering and transit of dual-use items. Doc. 428/2009/EC of 5 May 2009. Brussels: European Union.
[593] For the CGEA, see Council of the European Union. Council Regulation Setting up a Community Regime for the Control of Exports of Dual-use Items and Technology. Doc. 1334/2000/EC of 22 June 2000. Brussels: European Union. Art. 6(1) & Annex II.
[594] Commission of the European Union Communities. Commission Directive Amending Directive 88/301/EEC and Directive 90/388/EEC in Particular With Regard to Satellite Communications. Doc. 94/46/EC of 13 October 1994. Brussels: European Union.
[595] As to the latter, cf. esp. Artt. 81, 82 & 87, EC Treaty.
[596] As for the second, it may be noted that to this end EC law knows a set of regulations dealing with fair and transparent procurement of taxpayer-paid services from the commercial sector; see e.g.: European Parliament and Council of the European Union. European Parliament and Council Directive Amending Directives 92/50/EEC, 93/36/EEC and 93/37/EEC Concerning the Coordination of Procedures for the Award of Public Service Contracts, Public Supply Contracts and Public Works Contracts Respectively. Doc. 97/52/EC of 13 Oct. 1997. Brussels: European Union; European Parliament and Council of the European Union. Directive of the European Parliament and of the Council Coordinating the Procurement Procedures of Entities Operating in the Water, Energy, Transport and Postal Services Sectors. Doc. 2004/17/EC of 31 March 2004. Brussels: European Union; European Parliament and Council of the European Union. Directive of the European Parliament and of the Council on the Coordination of Procedures for the Award of Public Works Contracts, Public Supply Contracts and Public Service Contracts. Doc. 2004/18/EC of 31 March 2004. Brussels: European Union.
[597] Constitution of the International Telecommunication Union, Geneva, done 22 Dec. 1992, entered into force 1 July 1994; 1825 UNTS 1; UKTS 1996 No. 24; Cm. 2539; ATS 1994 No. 28; Final Acts of the Additional Plenipotentiary Conference, Geneva, 1992 (1993), at 1; substantially amended in 1994 and 1998.
[598] Convention of the International Telecommunication Union, Geneva, done 22 Dec. 1992, entered into force 1 July 1994; 1825 UNTS 1; UKTS 1996 No. 24; Cm. 2539; ATS 1994 No. 28; Final Acts of the Additional Plenipotentiary Conference, Geneva, 1992 (1993), at 71; substantially amended in 1994 and 1998.
[599] See Section 1.17, Radio Regulations, defining 'allotment' as the "entry of a designated frequency channel in an agreed plan, (. . .) for use by one or more Administrations for a terrestrial or space communication service in one or more (. . .) countries or (. . .) areas".

[600] Section 1.16, Radio Regulations; emphasis added.
[601] See Section 1.18, Radio Regulations, defining 'assignment' as the "authorization given by an Administration for a radio station to use a radio frequency or by an Administration for a radio station to use a radio frequency or radio frequency channel under specified conditions"; and 1st bullet, Annex to the ITU Constitution, defining 'Administration' as "Any governmental department or service responsible for discharging the obligations undertaken in the Constitution of the International Telecommunication Union, in the Convention of the International Telecommunication Union and in the Administrative Regulations".
[602] Treaty on Principles governing the Activities of States in the Exploration and Use of Outer Space, including the Moon and Other Celestial Bodies (hereafter Outer Space Treaty), London/Moscow/Washington, done 27 Jan. 1967, entered into force 10 Oct. 1967, 610 UNTS 205, 6 ILM 386 (1967).
[603] Convention on International Liability for Damage Caused by Space Objects (hereafter Liability Convention), London/Moscow/Washington, done 29 March 1972, entered into force 1 September 1972, 961 UNTS 187, 10 ILM 965 (1971).
[604] Convention on Registration of Objects Launched into Outer Space (hereafter Registration Convention), New York, done 14 Jan. 1975, entered into force 15 September 1976, 1023 UNTS 15, 14 ILM 43 (1975).
[605] Art. I, resp. Art. III, Outer Space Treaty.
[606] Art. VI, resp. Art. IX, Outer Space Treaty.
[607] Cf. Art. II, resp. Art. XII, Liability Convention.
[608] Cf. Art. I(a), Liability Convention.
[609] Cf. Artt. II, III & IV, Registration Convention.
[610] Cf. Art. XXII, Liability Convention, & Art. VII, Registration Convention; to an even more limited extent also Artt. VI & XIII, Outer Space Treaty.

6. The past and future of security-related satellite missions in Europe

Sascha Lange

Alongside the Galileo satellite navigation system, European States are currently developing and launching a whole new generation of Earth observation and communication satellites. This lively activity is driven by deep changes in European security priorities, technical progress and cost-reducing commercialisation.

The security priorities of European States have shifted considerably since the end of the Cold War and the events of 9/11. The clear, sharp confrontation between two superpower blocs has given way to a global web of different conflicts. The military aspects of European security policy have in the meantime been joined by phenomena such as global trade and climate change. This thematic and geographical expansion has greatly increased the information needs of political decision-makers and society, especially concerning questions of security.

Europe and its constituent States are responding to these strategic shifts by increasing their use of a domain that is global by its very nature: space and space-based assets. Given that satellites are predestined for global navigation, information gathering and communication applications (e.g. Earth observation or data transfer), many European States are actively increasing the number of security-relevant satellites they operate.

Technological advances are increasing both the spectrum of capabilities and the amount of information gathered and conveyed. This applies to civilian satellites (scientific and commercial) even more than to military ones.

As the number of satellites launched rises, it becomes easier to cut costs through series production. Gradual deregulation is intensifying competition among suppliers and causing further downward pressure on prices.

The result of technical progress, series production and gradual market opening is falling prices. As threshold prices fall, demand increases and the market grows. The dynamic unleashed in commercial space systems is so great that the speed of technical progress there currently outstrips the one in the military field. This in turn means that the military has increasing access to commercial providers. With only minor adaptations to their special needs, falling costs can be combined with increased performance.

6.1. Dual use

The original separation between civil and military applications is rapidly dissolving. The possibility of dual use of data and communication systems is growing in many fields. Whereas in the last century military satellites were generally considerably superior in performance to civil satellites, the latter have caught up in many fields over the past fifteen years. New commercial companies bundle the capacities of their satellites and offer these services to private-sector and State clients alike. Growing dual use increasingly brings together the security and defence aspects, because non-military clients now have access to data that would have been the privilege of the superpowers' armed forces even just a decade ago.

The United States especially is making increasing use of purchased commercial services alongside its already dominant military satellite constellations. Eighty percent of the data traffic of the U.S. armed forces for example, is transmitted by commercial providers. The telecommunications services of Iridium, Intelsat, Inmarsat and Telstar are joined by Earth observation products from DigitalGlobe and GeoEye. Satellite navigation and launch vehicles are further fields where a global market opening is under way. As global markets are established and competition increases between a growing number of providers, the costs of the technologies fall.

European States and the European Union (EU) are also increasingly joining this trend. Alongside standalone national programmes, there is an increasing trend to combine and formalise cooperation based on complementary capacities by expanding data exchange or financial participation.

The European Commission is increasingly becoming a driving force. Apart from the Galileo satellite navigation system, the Global Monitoring for Environment and Security (GMES) Earth observation project is one of the EU's best-known initiatives. Because GMES is based on the experience of numerous research satellite programmes of the European Space Agency (ESA), it is no surprise that cooperation between EU and ESA has emerged here.

A clear differentiation of the different satellite programmes has developed at ESA. The future programmes are rooted in the accomplishments of their forerunners. The three most important fields for the future are the Meteosat programme (in cooperation with EUMETSAT), the various research satellites of the Earth Explorer series and the Sentinel satellites of the GMES programme (in cooperation with the EU).

6.2. National programmes of European States

6.2.1. Reconnaissance

Imagery intelligence (IMINT) continues to make an important contribution to strategic reconnaissance. This type of Earth observation was one of the first objectives of low-orbit space exploration. In 1959, just two years after the launch of USSR's Sputnik, the United States launched its first Earth reconnaissance satellite. The first reconnaissance satellites still used chemically processed films. After exposure of the photographs, the satellites re-entered the atmosphere, revealing their valuable information only after the recovery of their payload. Image transmission by radio signal began in the 1970s, reducing the time-lag in acquiring reconnaissance results from several days to less than 24 hours.

Long-term Earth observation is fundamental for much military planning. It is used to create precise maps, and in recent years increasingly also for topographic databases. This represents an enormous aid to operational planning for air, sea and ground forces. On that basis – depending on the available resolution – strategic, operational and tactical targets can be observed. As resolution increases, targets can be increasingly better detected, identified, described and investigated.

These days observation is not restricted to the conventional optical window through the atmosphere. Because simple photography was unable to provide useful results at night or through cloud or fog, other parts of the spectrum have come to be used too. Infra-red sensors provide night-time imaging, while radar sensors even penetrate cloud. Significant processes and functionalities are Synthetic Aperture Radar (SAR) and Moving Target Indicator (MTI).

At the operative level of maritime surveillance and reconnaissance, significant capability leaps are expected in the coming years. Satellites are already effective at cueing Maritime Patrol Aircraft (MPA) searching for large ships. With resolutions under seven metres, convoys and ships can be roughly identified. By improving resolution down to under one metre, increasing revisit rates and expanding the quantity of data that can be sent to base stations, the Europeans will probably gain the certain operative abilities in the coming five years. Future constellations – after 2015 – could turn out to have even greater potential.

In land warfare, at least against strategic targets, imaging satellites are practically indispensable. As the performance of reconnaissance satellites increases, growing operational and even tactical applications will also be observed.

Alongside these imaging systems, there is also another category of reconnaissance satellite. Early warning satellites use frequency ranges primarily in the infra-red range to scan the Earth for missile launches. The conspicuous infra-red signature of missile booster plumes in the launch phase is captured by infra-red sensors and passed directly to the corresponding command centre.

Data gathered in this way is crucial for defence against long-range ballistic missiles, because it allows the fire-control radars of interceptor missile systems to be cued more quickly and accurately. Ultimately this expands the range within which interceptors can effectively hit their targets.

In the United States the Defense Support Programme (DSP) constellation is of central importance, while Russia operates satellites in the OKO and PROGNOS series. These systems originate from the Cold War era and are so complex and costly that no other power has comparable systems yet.

Reconnaissance is also strongly supported by another group of intelligence satellites, the signal intelligence systems (SIGINT). This field is still relatively weakly developed, but its potential is all the greater, because it allows important radio communication and radar signatures to be scanned without danger, even deep in an adversary's hinterland.

6.2.1.1. SAR-Lupe

The SAR-Lupe satellites, launched between 2006 and 2008, provide the German Bundeswehr with a system of five radar satellites offering resolution of well under 1 m. The 250 million euros TerraSAR-X radar satellite, launched in 2007 and operated by a German-based commercial consortium, supplies reconnaissance data – also with resolution down to well under 1 m – that is attractive for military users all over the world.

6.2.1.2. COSMO-Skymed

Since 2007, the Italian Space Agency (ASI) has been setting up the COSMO-Skymed system, which will ultimately comprise four radar satellites and cost 775 million euros. The resolution of this system should be under 1 m. As with TerraSAR-X, dual use is an explicit aim, in order to profit from both civil and military customers. Together with the PLEIADES satellites (described below) they form the Franco-Italian ORFEO system. Spain is also interested in acquiring its own radar imaging satellites (PAZ), but any launch date is unlikely to be before 2015.

6.2.1.3. SPOT

The launch of the French SPOT 1 Earth observation satellite in 1986 gave Europe its own photo-reconnaissance capability. SPOT 1 – like its two immediate successors – offered resolution down to 10 m. Through various improvements in optical instruments and optimised data processing, SPOT 5, which went into operation in 2002, achieves resolutions down to 3 m.

France improved the original capabilities of the SPOT system to develop the Helios 1 and 2 reconnaissance satellites. The experience gained in building and operating military satellites also flowed into the enhanced SPOT 4 and 5 spacecraft.

6.2.1.4. Helios

Since 1995, the French Armament Procurement Agency (DGA) has launched two Helios satellites, the first of which is still operational. The first of two Helios 2 successors was launched in 2004; the second is scheduled for a launch date in 2009 or 2010. Apart from the two satellites, the programme budget of 3 billion euros also includes the associated ground installations. Alongside its considerably improved optical resolution of about 30 cm, this 4200 kg spacecraft is characterised by its infra-red imaging capacity. This allows night-time images, as well as the detection of thermal activities associated with vehicles and buildings.

6.2.1.5. RapidEye

The German-led RapidEye system started with five optical satellites in 2008. Although this dual use system only allows resolutions down to somewhat over 6 m, the number of satellites makes it possible to revisit targets relatively often. This allows a rough initial optical assessment of an area of interest and speedy cueing of more precise reconnaissance by higher resolution satellites.

6.2.1.6. PLEIADES

In the coming years France plans to launch two PLEIADES satellites as successors to SPOT 5. With a resolution close to that of Helios 2, they also offer possibilities for military use.

6.2.1.7. INGENIO/PAZ/SEOSAT

In addition to the radar Satellite PAZ, Spain is is also interested in developing its own capacity for optical imaging satellites and is studying various options through the Ingenio programme. A possible launch date could be 2011. Together with the PAZ radar satellites, this project goes under the name of SEOSAT.

6.2.1.8. SPIRALE

In order to observe infra-red radiation events from space, the French Armament Procurement Agency (DGA) launched two SPIRALE demonstrator satellites in 2009. These two units, costing 124 million euros, represent France's first step towards the establishment of an early warning system for ballistic missile launches.

6.2.1.9. ESSAIM

France has been developing capacity for detecting, measuring and monitoring electromagnetic sources since 1995 with the technology demonstrators CERISE (COMINT) and CLEMENTINE (ELINT). The four-satellite ESSAIM system, costing more than 80 million euros, went into orbit at the end of 2008. Focused on detecting and mapping signals intelligence, ESSAIM succeeds CERISE. The analogous ELINT system that will go into orbit for at least three years from the end of 2009 will represent a considerable improvement over CLEMENTINE. The cost is currently estimated at 100 million euros.

6.2.2. Communication

Network-based operations (NBO) is increasingly speeding up the comprehensive exchange of data. Satellites are the pre-eminent communication platform for data in areas with underdeveloped infrastructure. Because armed forces increasingly operate in smaller and lighter units – spread out increasingly widely in more remote and inaccessible areas – this trend will continue.

Satellite-based communication has proven its value in major operations in Iraq and Afghanistan. Conventional radio communication is often impossible in Afghanistan, because many radio frequencies require a more or less direct line of sight. Radio relay aircraft can overcome radio shadow caused by mountains, but only in a relatively restricted range. In Iraq, where fighting units advanced relatively quickly and were spread over a large field of operations, conventional radio systems designed for a relatively high spatial concentration were often no longer able to ensure communication within a dispersed division.

6.2.2.1. Skynet

The U.K. launched its first Skynet satellites back in 1969, and the system has been transmitting data for the British Ministry of Defence and NATO since 1970. By 1993 a total of seven units of different versions had been put into orbit. The three satellites of the current Skynet 4 generation went into service in 1998. Their successor, Skynet 5, began its operations in 2006. Alongside NATO, which has a fifteen-year contract with Skynet operator Paradigm, the Portuguese defence ministry has also become a customer of this system

6.2.2.2. SYRACUSE

Since 1984 France has been launching communication satellites that are partly reserved for military communication. By 1996 seven SYRACUSE 1 and 2 satellites had been launched. In 2005 and 2006 two satellites of the more powerful SYRACUSE 3 series were added.

For the time after 2010 France is already developing dual use successor projects, named ALPHASAT and REMUS, featuring improvements in bandwidth and especially mobile tactical communications.

6.2.2.3. SICRAL

Italy put its SICRAL satellites into orbit in 2001. The Franco-Italian ATHENA-FIDUS system is to begin launching from 2009 to provide additional low-cost communication channels, potentially useable by the military. NATO will in future rely on a combination of Skynet, SYRACUSE and SICRAL for data transmission.

6.2.2.4. SatComBW

For a long time Germany used only leased data channels run by various private-sector providers such as Inmarsat. Starting in 2010 SatComBW Stage 2 will supply two dedicated satellites to guarantee the independence of military data traffic.

6.2.2.5. XTAR

Spain launched the XTAR-EUR and XTAR-LAND satellites in 2005 and 2006. The latter is also known as SPAINSAT 1. Both are Spanish-American joint ventures offering data channels to the U.S. and Spanish defence ministries, as well as to other States and commercial customers. The first outside customer was the U.S. State Department, which has secured data capacity for five years for a sum of up to 137 million U.S. dollars, followed closely by the Danish defence ministry as the first third-party nation.

6.3. Cooperation between ESA and EUMETSAT in Earth observation

At the operational and tactical level, weather observation (e.g. detecting sandstorms) makes an important contribution to supporting military operations of air, sea and land forces. This field of application is precisely a good example of the dual-use capability of data gathered in orbit.

The United States has dedicated military weather satellites such as the Defense Meteorological Satellite Programme, but data from primarily civilian systems such as the Geostationary Operational Environmental Satellites is also used. In Europe data from the Meteosat weather satellites is provided to both civil and military users.

6.3.1. Meteosat

ESA launched the first METEOSAT in 1977. In 1986, operational responsibility was taken over by a specially established body, the European Organisation for the Exploitation of Meteorological Satellites (EUMETSAT). Two of the first-generation satellites (Meteosat 1–7) are still in operation. Meteosat 8 and 9 from the more powerful Meteosat Second Generation (MSG) are already in orbit. The Meteosat Third Generation (MTG) is scheduled to begin launching in 2015.

6.4. ESA programmes

The overwhelming majority of ESA programmes serve pure research purposes. Especially through the Earth Explorer programme, ESA is creating a broad long-term Earth observation database that will benefit numerous applications. These programmes also collect data with dual use potential. Alongside weather, other data about environmental conditions is naturally of great importance for military planners. Soil humidity data for example can be of military interest. In the long term, findings that flow into climate modelling are important for the security of whole societies and regions, and this in turn influences the long-term structuring of armed forces.

6.4.1. ERS 1 and 2

The GMES Earth observation programme builds on scientific projects launched many years ago by France and other European States, including the European Remote Sensing Satellite (ERS). ERS, in turn, was a development of the pioneering French SPOT programme, (Système Probatoire d'Observation de la Terre) established by France in the 1980s. ERS 1 was launched in 1991; ERS 2 launched in 1995 is still operational in orbit. Military uses were not central to either of these missions – the low resolution of 30 m was inadequate for most military purposes.

6.4.2. ENVISAT

Europe also possesses scientific Earth observation capacities through the ENVISAT system based on ERS 1 and 2. The ASAR radar offers another rudimentary capability for observing the Earth's surface and shipping movements regardless of weather conditions.

6.4.3. GOCE

The Global Ocean Circulation Explorer satellite (GOCE) was launched in February 2009, complementing the existing Challenging Mini-Satellite Payload (CHAMP) and GRACE programmes. GOCE measures the Earth's gravitational field in high resolution, collecting data which will improve modelling of the Earth's core. Any security relevance is indirect, through the acquisition of fundamental knowledge.

6.4.4. SMOS

The Soil Moisture and Ocean Salinity satellite (SMOS) launched in 2009 measures the salt content of the oceans, moisture in soil and the water content of vegetation to investigate the natural water cycle. The security relevance is indirect, through gaining certain knowledge about soil conditions.

6.4.5. CryoSat2

CryoSat2 will be launched in 2009 to complete the mission of the CryoSat spacecraft lost during launch in 2005. The radar on board will record the planet's ice coverage and measure the thickness of the ice sheet. This information could be relevant for naval deployments. In particular, the condition of the Arctic ice sheet is of growing security importance. In the long term the exploitation of mineral resources in the Arctic is expected to increase enormously.

6.4.6. ADM/Aeolus

The Atmospheric Dynamics Mission (ADM-Aeolus) starting in 2010 will observe wind profiles at different levels of the atmosphere. Findings relating to future climate changes resulting in droughts or flooding could be relevant for predicting future resource conflicts.

6.4.7. SWARM

Building on the German CHAMP research satellites (Challenging Mini-Satellite Payload), the SWARM system starting in 2010 will provide an even more precise picture of the Earth's changing magnetic field. The security relevance is similar to that of GOCE.

6.4.8. EarthCare

Together with the Japanese Space Agency JAXA, ESA will launch the EarthCare (Earth Clouds, Aerosols and Radiation Explorer) Earth observation satellites in 2013. With a wide array of instruments, they are designed to measure humidity and radiation in the atmosphere in order to research cloud-formation processes.

6.5. Cooperation between ESA and EU/EC on Earth observation

6.5.1. Global Monitoring for Environment and Security (GMES)

The GMES project is designed to improve coordination of existing and future European capacities for ongoing Earth and environmental observation through space-based systems. Integration of data providers and users in the field of environmental and security-relevant capabilities will be improved in order to make more efficient use of existing resources.

This is factually an institutionalisation of the scientific Earth observation missions from ERS 1 through to the multi-satellite Earth Explorer programme. This programme for monitoring the state of the Earth is aptly named Earth Watch, and the individual satellite missions or measuring instruments are appropriately called Sentinels. They are called "satellite missions" rather than satellites because sometimes there will be several identical satellites, and sometimes instruments will piggy-back on EUMETSAT satellites to cut costs.

6.5.1.1. Sentinel-1

The Sentinel-1 mission is scheduled to launch in 2011. The satellite will be equipped with a SAR radar in the C-band frequency range that will allow considerably higher resolutions than its two purely scientific predecessors ERS and ENVISAT. With a maximum resolution of 5 m, however, the military uses will be limited to tasks such as locating large ships.

6.5.1.2. Sentinel-2

The second satellite in the Sentinel mission series is dedicated to Earth observation and budgeted to cost 195 million euros. Although its resolution is not very high (between 10 and 20 m) Sentinel-2 is designed primarily to ensure continuity of data collecting with its predecessors SPOT and LANDSAT. The launch is scheduled for 2012. A second satellite could be launched later to increase the revisit rate.

6.5.1.3. Sentinel-3

The estimated cost of Sentinel 3 is 305 million euros. The task of these satellites is to observe land and sea masses. The instruments it carries allow it to continue the data collection mission of its functional predecessors ENVISAT and Cryosat. The launch is scheduled for 2012.

6.5.1.4. Sentinel-4 and -5

Because the payloads of the Sentinel-4 and -5 missions are relatively small, they will fly as piggy bags on EUMETSAT missions rather than having satellites of their own. This will lower costs, and the simultaneous measurements and images will be easier to compare and combine.

6.6. Navigation

The applications of navigation systems now extend far beyond their original function, which was to position submarines armed with intercontinental ballistic missiles with absolute precision. In all fields of land, sea and air transport, positioning signals sent from space have become a vital and highly valued instrument for precise spatial orientation.

The first navigation-supporting satellites were launched into orbit by the United States in 1960. The GPS system which has become so widely known was declared fully operational in 1993. In 1996 the Russian Global Navigation Satellite System (GLONASS) was also declared operational. Both systems struggle to a greater or lesser extent with loss of performance through the failure of older satellites. The United States succeeds significantly better at maintaining overall operability.

6.6.1. Galileo

The European Union and ESA plan to have the civilian-run Galileo navigation fully functioning in the course of the coming decade. The navigation signals are to be compatible with the GPS system, and will produce greater reliability and precision by combining Galileo and GPS.

Civilian administration is a delicate security issue, because the European armed forces will probably not be able to use the very precise Galileo system (or the even better Galileo/GPS combination) to their own maximum advantage without additional cost.

The Galileo navigation system already has a turbulent history behind it. The first test satellite, Galileo In-Orbit Validation Element A (GIOVE A), was launched in 2005. But full operational launch of the system is already years late, following numerous industrial policy-related problems. Now that the final contract has been signed, four pre-series In-Orbit Validation (IOV) satellites are to be produced, later to be joined by 28 series satellites. Whether the currently estimated cost of more than 4.5 billion euros will be sufficient to put the system into service remains to be seen. Uncertainty also applies to financing the running costs.

Further coordination could be needed with the Russian GLONASS and Chinese Beidou systems to avoid signal interference impairing function.

6.7. Financial threats?

The world is in the midst of the worst economic crisis since the Great Crash of 1929, whose full impact will not hit public budgets until 2010. The consequences for State finances will be felt over several years in the form of immense budget deficits and pressure on departments to make cuts. In Europe, the Maastricht Treaty limits annual budget deficits to 3% of GDP.

Growing uncertainty about the future funding of multinational space programmes threatens to spark a trend towards national programmes. In order to avoid risking the loss of substantial resources when a partner decides to quit a major multinational programme, there will probably be a shift to the more predictable framework of national sovereignty.

The expensive multinational projects will be the first to come under scrutiny, because cost overruns and delivery delays increase with the number of partners. Pressure is looming for large and highly-integrated projects like Galileo and GMES. A manned European space programme – which would demand investment of more than 5 billion euros – is unrealistic under these circumstances.

6.8. Outlook – complementary strengths

Space has already a central role in the economy and society. Phenomena like global warming and the ensuing consequences for security make it even more important. These great challenges demand the expansion of satellite constellations that provide information (navigation) gain information on the global situation (Earth observation) and can pass it on quickly (communication).

Large States like the United States and China will be in a position to run their own national programmes, which will then often be used for international cooperation in the soft power context.

Big multinational programmes like GMES and Galileo are plagued by enormous delays and cost overruns, caused by their complexity and economic and political frictions. These programme structures have yet to prove their worth as successful models.

Another way forward could be offered by combining complementary programmes of individual States. Much can be learned from the Franco-German cooperation in SAR-Lupe and Helios 2. The European Multinational Space-based Imaging System (MUSIS), in particular, can be set up initially through complementary combinations of national programmes. Satellites – especially in the low Earth orbits (LEO) that are important for Earth observation – have only a limited lifetime, so if cooperation between particular partners works well, system integration can be expanded in following satellite generations. Where different States have different technological profiles (e. g. Germany has excellent expertise with radar-satellites) it makes sense to combine complementary strengths.

7. Latest trends in the national authorisation and regulation of space activities in Europe

Irmgard Marboe and Florian Hafner

7.1. Introduction

In the past few years, more and more States have become aware of the need to regulate and control outer space activities. One of the reasons for this increasing interest is the growing importance of private space activities. While twenty years ago only States were able to conduct space operations, nowadays private companies all over the world are actively involved in this growing branch of business. Today, privately operated satellites are no longer a rarity, and it is even possible to buy "vacation trips" into outer space via travel agencies.[611] The future development of technical progress and the growing experience will allow private space activities to expand further. As a result of this, the number and complexity of rules and regulations in the field of space law will multiply.

On the other hand, international law imposes on the States themselves a number of obligations and responsibilities. The most important ones are contained in the five space treaties and in the principles adopted by the United Nations General Assembly.[612] In order to comply with their obligations, States see the need to supervise and control private space activities.

Several States in Europe have already decided to enact national space legislation. One of the most important issues in this respect is the authorisation of space activities by the State. This paper aims at analysing the latest trends in the national authorisation and regulation of space activities in Europe by examining the most recent pieces of national space legislation which entered into force in Belgium (2006), in the Netherlands (2008) and in France (2008). The situation under EU law will also be briefly addressed. Moreover, the first-year findings of the working group of the Legal Subcommittee of the UN Committee for the Peaceful Uses of Outer Space on national legislation relevant to the peaceful exploration and use of outer space (2009) shall be presented.

7.2. Is there a need for national space legislation?

At the outset, the pertinent international legal framework will be briefly analysed and it will be examined to what extent it contains an obligation to enact national space legislation. In 1967, when the basis of international space law – the Outer Space Treaty[613] – was established, probably none of the States Parties to the Treaty had borne in mind that private entities would carry out space activities. Nevertheless, Article VI of the Outer Space Treaty states that

> *"States Parties to the Treaty shall bear international responsibility for national activities in outer space, including the Moon and other celestial bodies, whether such activities are carried on by governmental agencies or by non-governmental entities, and for assuring that national activities are carried out in conformity with the provisions set forth in the present Treaty."*

Furthermore, the second sentence of Article VI of the Outer Space Treaty stipulates that States should authorise and supervise space activities of non-governmental entities:

> *"The activities of non-governmental entities in outer space [. . .] shall require authorization and continuing supervision by the appropriate State Party to the Treaty."*

It has to be noted that this provision does not impose an obligation to enact a national space law.[614] Nevertheless, there is general agreement that a national system of authorisation, supervision and registration, whether being regulated by one particular space act or not, is necessary to fulfil the obligations contained in the UN space treaties.[615]

7.3. Latest trends in the national authorisation and regulation of space activities in Europe

It is interesting that after a longer period of stagnation[616] three European countries enacted national space laws within a short period of time, two of them in the past year. This shows that national space legislation has gained increasing practical relevance. In particular, the new trends in national authorisation regimes in Europe are worth being looked at more closely. The cases of the Belgium, the Dutch and the French space legislation, which are the most recent examples, shall in the following briefly be introduced.

7.3.1. Belgium

The Belgian Law on the Activities of Launching, Flight Operations or Guidance of Space Objects[617] was promulgated on 17 September 2005.[618] It entered into force on 1 January 2006. It has been supplemented by a Royal Decree implementing certain provisions of the Law.[619] This decree was adopted on 19 March 2008 and entered into force on 11 April 2008.[620] It is notable that the Belgian space legislation leaves regulatory powers to the Belgian King, particularly with respect to the limitation of the operator's liability. The Belgian Space Act has three main targets.[621] First, the authorisation and supervision of space activities performed under Belgian jurisdiction. Secondly, the establishment of a national registry for space objects and thirdly, the avoidance of liability which may arise according to Article VI of the Outer Space Treaty. Belgium stimulated the development of a Belgian space law in order to provide a legal framework to existing or emerging activities in Belgium, as well as to possible activities performed by Belgian citizens. Thirdly, the new Belgian space law should create compliance with the obligations of the Belgian State deriving from international law. Besides, as an ESA Member State, Belgium considered itself as a (co-) launching State of space objects launched by ESA.

7.3.2. The Netherlands

In the Netherlands, the Law Incorporating Rules Concerning Space Activities and the Establishment of a Registry for Space Objects[622] was officially enacted by the Dutch Parliament on 25 January 2007 and entered into force on 1 January 2008.[623] Similar to the case of Belgium, the Netherlands enacted national space legislation due to the fact that Dutch companies had started to be actively involved in space activities.[624] Concerning the territorial scope of application of the Dutch Space Activities Act, it has to be noted that the Kingdom of the Netherlands consists of its European part and of six islands in the Caribbean, Aruba and the five Netherlands Antilles. The Kingdom had ratified the five United Nations treaties on outer space also on behalf of its overseas territories. Therefore, it was necessary to decide if the Dutch space legislation should become effective in the Kingdom as a whole. Eventually, the overseas territories declared that they preferred to retain their autonomy with respect to the regulation of space activities. Thus, the Dutch Space Activities Act only entered into force with effect in the European part of the Netherlands.[625] The Dutch Act provides for the installation of an authorisation regime and gives the possibility of redress in cases where the Netherlands are held liable under the international legal regime.[626]

7.3.3. France

The most recent piece of space legislation in Europe was enacted in France, namely the Law No. 2008-518 of 3 June 2008 regarding space operations.[627] It was adopted by the French Senate on 22 May 2008 and published in the French Republic Official Journal on 4 June 2008. The purpose of the French Space Operations Act is to set up a national regime to authorise and control space operations in accordance with the French government's international commitments.[628] Before the enactment of the law of 3 June 2008, French space activities had been administered on the basis of administrative practices with no unified legal basis. The new French space legislation deals with three main topics: the installation of an authorisation regime, the distribution of liability between the State and non-governmental entities and the introduction of a sanction regime for non-compliance with the authorisation requirements.[629]

France has major space launching facilities on its territory, a large and growing space industry, and is the host of the headquarters of the European Space Agency, a major actor in space activities. It follows that national legislation in this area is of specific interest to the industry and has important practical relevance also across the French borders. The French Space Operation Act shall be accompanied by a general implementing decree which has not been enacted yet at the time of writing the present article.

Fig. 17: *Ariane 5 enclosing Herschel and Planck on its launch pad on 13 May 2009 (source: ESA/S. Corvaja).*

7.4. The contents of the authorisation regimes

In order to implement the obligation to authorise space activities as required by international law, several issues have to be clarified. The most important questions are: who authorises, what kinds of activities require an authorisation, what conditions are required for an authorisation, and how is compliance controlled. In the following, the way these questions are addressed in the Belgium, the Dutch, and the French space acts will be analysed.

7.4.1. Who authorises?

According to Article VI of the Outer Space Treaty, the authorisation is incumbent on the "appropriate" State. The meaning of this term is subject to different interpretations.[630] In accordance with the rules of treaty interpretation contained in Article 31 of the Vienna Convention of the Law of Treaties,[631] the "appropriate" State should in principle be equal to the "responsible" State. Otherwise the State would be responsible for an activity which it is deprived to authorise and supervise.[632] How do the three recent space acts interpret this term? In which cases do the States consider themselves to be the "appropriate" State?

First, territorial or "quasi-territorial" jurisdiction is established in the three acts under consideration. The Dutch Space Activities Act applies to "*space activities that are performed in or from within the Netherlands or else on or from a Dutch ship or Dutch aircraft*".[633] France has made clear that the French Space Operations Act applies to activities carried out within the boundaries of French territory or French jurisdiction.[634] Belgium explicitly mentions "zones under the jurisdiction and control of the State".[635]

Secondly, the citizenship of the person carrying out the space activities plays a certain role. This personal jurisdiction is, however, much more differentiated in the three acts under consideration. The law in France also refers to the nationality of the operator, be it a natural or a legal person.[636] Belgium does this only subsidiarily, depending on international agreements. The Dutch Act holds that its regulations can also be declared applicable to activities performed by Dutch natural or juridical persons in countries not party to the Outer Space Treaty.[637]

The Outer Space Treaty only refers to the "appropriate State" but it is not more specific on the question who – within the State – should carry out the obligation to authorise. It follows that the States are free to select the appropriate State organ or entity. The competent authority may vary from country to country, but in the three countries under review a Ministry has the competence to grant or deny authorisation. In Belgium, the Minister competent for the space sector is responsible for granting licences.[638] In the Netherlands the Ministry of Economic Affairs[639] is the competent organ. During the authorisation and/or supervision process the Belgian Minister may be assisted by an ad hoc committee of technical experts if it seems to be useful.[640] In France authorisations are under the administrative authority of the Ministry of Research in charge of outer space affairs. According to Article 28 of the French Space Operations Act, CNES has to assure, by delegation of the Minister in charge of space, that the systems and procedures implemented by a space operator are compliant with the technical regulations.

So far, the ministries mentioned in the space acts have not delegated their powers of authorisation. This competence is still kept within the area of responsibility of the public administration. It has to be seen if this remains as it is or if the competence to authorise will be delegated to another organisation or agency, as it is the case in other countries. In particular, the implementing decree of the French law will be interesting in this regard.

7.4.2. What kind of activities require authorisation?

Article VI of the Outer Space Treaty only requires authorisation for "activities in outer space". The international obligation stemming from this provision is thus limited to the authorisation of activities that actually take place in outer space.[641] However, the national laws under consideration do not limit themselves to regulate only those activities.

The launch of a space object, although taking place on Earth, is subject to authorisation in all of the above mentioned national space laws. The Belgian space law has it already in its title.[642] In the Netherlands, the launch of a space object falls under the scope of "space activity".[643] The French law requires authorisation for space operators. It prescribes that launching activities or even the attempt to launch are covered by the term space operation and therefore shall obtain authorisation from the administrative authority.[644] Surprisingly, the "launch procurement" of an object into outer space, although important for registration and liability matters under the space treaties,[645] is not mentioned in any of the countries under review.

Furthermore, in Belgium and France, a license for "space operations" is needed. The Belgium space legislation states that flight operations are covered by its area of application[646] and defines a flight operation as an "operation relating to the flight conditions, navigation or evolution in outer space of the space object, such as the control and correction of its orbit or its trajectory".[647] In France, the term "space operations" is used as an umbrella term and includes "any activity to launch or attempt to launch an object into outer space or to assure the control of a space object during its stay in outer space [. . .] as well as, where appropriate, on its return to

Earth".[648] In the Netherlands "flight operations" of space objects in outer space require authorisation.[649] It follows that, generally, "space operations" or "flight operations" fall under the authorisation regime.

In addition, the Belgium and the Dutch space laws explicitly mention the "guidance of space objects".[650] This usually includes operations related to the flying conditions, navigation or evolution in outer space of space objects, such as the control and correction of orbit or trajectory.

The comparison of the coverage of the three space act shows that the scope of application as regards "space activities" is rather similar. It covers the launch of space objects, space operations and the guidance of space objects. This is, on one hand, more than Art. VI of the Outer Space Treaty demands. On the other hand, it is less than other countries which have chosen a much broader scope for their authorisation regimes like, for example, Ukraine.[651]

7.4.3. What are the conditions?

According to Art. VI of the Outer Space Treaty, the States Parties bear international responsibility for national activities in outer space and for assuring that national activities are carried out in conformity with the treaty provisions. This means that the States must ensure – by way of authorisation – that the activities by non-governmental entities do not violate the norms contained in the Outer Space Treaty.[652]

However, it has to be kept in mind that the obligations in the Outer Space Treaty and other treaties are international obligations and thus directed to States and not to private entities. The norms are defined and formulated as obligations of States. They include the obligation to pursue space activities for peaceful purposes on the basis of equality and non-discrimination, for the benefit of all mankind and in accordance with the Charter of the United Nations. They contain the principles of non-appropriation, avoidance of harmful contamination, non-interference and international cooperation.[653] Some of these obligations can be easily transferred to private actors, others not.

There is, in particular, disagreement on whether the obligation to use outer space for peaceful purposes can be binding upon private actors. Some commentators say that "war" is per definition a situation of violence between countries.[654] The prohibition of the use of force under Article 2/4 of the UN Charter only prohibits the threat or the use of force between States. Others say that it is obvious that States must ensure that private actors also use outer space exclusively for peaceful purposes and do not place nuclear weapons and weapons of mass destruction in outer space and/or test any type of weapon.[655] It is

Tab. 1: *Obligations in national space laws transferred to private entities.*

	Belgium	France	Netherlands
Economic interest of the State	X		
Ensuring optimal use of outer space	X		
Financial capability of the applicant	X	X	X
Fulfilment of international obligations	X	X	X
Protection of national/international security		X	X
Protection of public health		X	
Protection of the environment	X	X	X
Protection of the public order			X
Safety of people and property	X	X	X
Space debris mitigation		X	
Strategic interest of the State	X	X	X
Technical capability of the applicant	X	X	X

interesting to note that none of the national space laws under review mentions such restrictions.

The national space acts may contain general clauses or a list of specific conditions explicitly mentioned. An overview of the obligations incumbent on private entities incorporated in the three national authorisation regimes is shown in Table 1.

It is notable that six out of twelve conditions mentioned above have to be fulfilled in all of the three countries under review. In Belgium, France and the Netherlands, the applicant has to prove his financial and technical capability. Moreover, he must not endanger the safety of people and property. All the three States have the possibility to deny a licence if the fulfilment of international obligations, the protection of the environment or the strategic interest of the State is endangered. The protection of the environment is also mentioned as a necessary condition in all of the three acts. The economic interest of the State and the optimal use of outer space play an explicit role when granting a license only in Belgium. The protection of public health and the mitigation of space debris are conditions which are specifically mentioned in France. The protection of the public order is a condition appearing only in the Dutch law. The Belgian law, in contrast to the other two, does not mention explicitly national security among the prerequisites for a license.

It follows from the above that the conditions imposed on applicants for a licence for a space activity are largely concurrent under the three laws. However, they all provide rather general guidelines and leave considerable discretion to the authorising body. This discretion may be narrowed down by an implementing decree and by reference to existing standards, for example in technical matters. Specific emphasis on certain aspects which are different in the three laws may be regarded as a reflection of the areas of concerns of the three States. These may result from the different types of space activities envisaged and from different political priorities.

7.4.3.1. How is compliance controlled?

National space legislation varies considerably as regards the control and enforcement of the authorisation regime. The most usual sanction in case of non-compliance or of an infringement is the non-issuance of the license or the suspension or withdrawal of the license.[656] In addition to fines on the basis of ordinary administrative law, there are extra fines under the Belgian, the Dutch and the French provisions. According to the Belgian Space Act, a fine between 25 and 25,000 euros may be imposed.[657] In France,[658] the overall fine is 200,000 euros (not up to 200,000 euros) and in the Netherlands[659] there is a fine up to 450,000 euros or 10% of the relevant annual sales of the company, whichever is more. Furthermore, the Belgian space legislation stipulates that any person carrying out space activities without authorisation, shall be liable to a period of imprisonment of eight days to one year.[660]

7.5. European perspective

The European Union (EU), as the major political and economic actor in Europe, doesn't have explicit competences on space matters so far. Nevertheless, on the European level, a number of initiatives have already taken place to address the issue of national space legislation.[661] They were aimed at exploring the perspective of a coordinated approach on the European level. They have prepared the ground for the development of common principles and approaches, but so far remained merely of academic interest. An explicit competence on space matters in the EU is foreseen in the Treaty of Lisbon of 13 December 2007,[662] whenever it will enter into force.

In the Treaty on European Union as amended by the Treaty of Lisbon, space appears as a "shared competence" between the Union and the Member States.

Article 4 of the Treaty on the Functioning of the European Union explains that the Union shall share competence with the Member States where the Treaties confers on it a competence which does not relate to areas of exclusive competence or to areas where the Union only has the competence to carry out actions to support, coordinate or supplement the actions of the Member States. Article 4(3) mentions space, together with research and technological development, as a specific category of such a shared competence:

> *"In the areas of research, technological development and space, the Union shall have competence to carry out activities, in particular to define and implement programmes; however, the exercise of that competence shall not result in Member states being prevented from exercising theirs."*[663]

The specific areas of competence of the Union in space matters are contained in Article 189 under Title XIX on "Research and Technological Development and Space". The first two paragraphs of Article 189 read:

> *"1. To promote scientific and technical progress, industrial competitiveness and the implementation of its policies, the Union shall draw up a European space policy. To this end, it may promote joint initiatives, support research and technological development and coordinate the efforts needed for the exploration and exploitation of space.*
>
> *2. To contribute to attaining the objectives referred to in paragraph 1, the European Parliament and the Council, acting in accordance with the ordinary legislative procedure, shall establish the necessary measures, which may take the form of a European space programme, excluding any harmonisation of the laws and regulations of the Member States."*

The changes of this article in relation to the prior version on the Draft Constitution[664] are striking. The first two paragraphs in Article III-254 under Section 9 under the title "Research and Technological Development and Space" said:

> *"1. To promote scientific and technical progress, industrial competitiveness and the implementation of its policies, the Union shall draw up a European space policy. To this end, it may promote joint initiatives, support research and technological development and coordinate the efforts needed for the exploration and exploitation of space.*
>
> *2. To contribute to attaining the objectives referred to in paragraph 1, European laws or framework laws shall establish the necessary measures, which may take the form of a European space programme."*

In the new version, the possibilities of harmonisation of national space legislation are removed in two ways: first, the possible adoption of "European laws or framework laws" is taken away and replaced by "the European Parliament and the Council, acting in accordance with the ordinary legislative procedure". The Treaty on the functioning of the European Union does not contain the concepts of "European Law" or "European framework laws" anymore, but returns to the traditional concepts of legislative acts, namely regulations, directives and decisions. Secondly, it contains an explicit exclusion of "any harmonisation of the laws and regulations of the Member States" in Article 189(2). Due to this explicit exclusion of harmonisation in the new version of the Treaty on the functioning of the European Union, an *e contrario* argument in favour of harmonisation can not be made.[665]

It becomes evident that the recent version of the space related provisions emphasises the subsidiary or complementary role of the European Union. Space appears, therefore, as an area where the Union has the competence only to carry out actions to support, coordinate or supplement the actions of the Member States whereas, according to Article 2(5), legally binding acts shall not entail harmonisation of Member States' laws or regulations.

7.6. Latest trends from the Legal Subcommittee: The working group on national legislation relevant to the peaceful exploration and use of outer space

Due to the growing interest of its Member States, the Legal Subcommittee of the Committee on the Peaceful Uses of Outer Space established a working group under agenda item 11, entitled "General exchange of information on national legislation relevant to the peaceful exploration and use of outer space" during its 783rd meeting, on 23 March 2009.[666] The discussion and the results of this working group can also be regarded as an indicator for new trends in the field of national space legislation and authorisation of space activities.

At the outset, the working group noted that every national regulatory framework represented a different legal system and that States had adapted their national legal frameworks according to their specific needs and to practical considerations.[667] Newer pieces of space legislation are usually unified acts; older provisions are often a combination of different national legal instruments. National legal requirements depend to a high degree on the range of space activities conducted and on the level of involvement of the private sector.[668]

Fig. 18: *A COPUOS session meeting at the U.N. being led by current Chairman of the Committee, Ambassador Ciro Arévalo of Colombia (source: R. Williamson).*

In considering the reasons for States to enact national space legislation, the working group noted that common grounds for national space legislation were the need to fulfil obligations under treaties to which a State had become a party, the need to achieve consistency and predictability in the conduct of space activities under the jurisdiction of the State and the need to provide a practical regulatory system for private sector involvement.[669] With regard to the competence of national authorities to authorise and supervise private national space activities, the working group found out that there were different national authorities involved in those procedures, ranging from space agencies and other similar authorities up to ministerial-level authority.[670]

Concerning the issue of the scope of space activities targeted by national regulatory frameworks, the working group noted a broad variety of activities. The launching of objects into outer space, the operations of a launch or re-entry site, the operation and guidance of space objects, in some cases the design and manufacturing of spacecraft, the application of space science and technology such as that used for Earth observation and telecommunications have for example been qualified as space activities.[671]

As for the conditions to be fulfilled in order to obtain an authorisation, the working group noted that ensuring the safety of space activities was an important policy underpinning most national space laws, in particular laws governing the launch of objects into outer space.[672] Most launch-licensing regimes included measures to ensure that the launch did not create a significant risk of personal injury, environmental damage or damage to property. Conditions concerning safety and technological standards were also closely linked to the States' concern about meeting space debris mitigation requirements. Other conditions related to the professional and financial qualifications of the applicant. In addition, national security and foreign policy interests were usually involved in authorisation and licensing procedures.[673]

Regarding the control and enforcement of the authorisation regime, the working group found out that most States applied procedures for the supervision and monitoring of licensed space activities. It appeared that most national regulatory regimes operated with a set of administrative measures for minor violations and a sanctions regime, including penal sanctions in some cases, for more serious offences.[674]

7.7. Conclusion

The analysis of the three most recent pieces of national space legislation, the Belgian, the Dutch and the French space act, has shown that, despite some differences, they are constructed in a rather similar way. The distinct motivations and scope of space activities envisaged for regulation are not recognisable at first sight. The laws are constructed in a general way leaving rather large discretion to the authorising bodies. This discretion may be reduced or streamlined by way of implementing decrees or through precedents and practice over time.

Nevertheless, such general laws also fulfil the aim of providing a legal basis for the authorisation procedure under national law as required by Article VI Outer Space Treaty. In addition, the legal procedures aim at upholding European standards on legal certainty and transparency. Some of the divergences, such as the different ministries responsible for the authorisation, derive from the internal structures and the division of competences and are of no significant importance. The same can be said for the divergence in the systems of compliance monitoring.

On the other hand, the existing divergences concerning the interpretation of the "appropriate" State in authorisation matters are a little more troublesome. They reflect a different approach as to the applicability of the authorisation regime. Of the three laws under consideration, the French law has the broadest scope of application, the Belgian the smallest. This is due to the different application of the personal jurisdiction on space operators. While the French system is applied to natural and legal persons having the French nationality and is open also to others, the other two laws apply the nationality of the operator only subsidiarily and have their focus on territorial application. In this respect, a more coordinated approach on the European level seems to be desirable.

Other differences concern the conditions imposed for acquiring a licence. It depends on their implementation, to what extent they contribute to the convergence or the divergence of the authorisation practice in Europe. In the worst case, they might lead to a distortion of competition in the space sector.

However, as it has been shown under the new regime of the Treaty of Lisbon, harmonisation of laws regulating space matters is excluded. Yet, this does not mean that European States, when enacting and implementing space legislation, are prevented from adopting a coordinated approach. Even on the international level, as the recent discussions in the working group on national space legislation of the UN Legal Subcommittee show, a coordinated approach is considered to be desirable.

---

611 The operator of the online travel agency German Unister GmbH for example, offers a suborbital flight at the price of 102,000 dollars. This suborbital flight is operated by Space Adventures Ltd. and shall be conducted from 2010 onwards. See Ab-in-den-urlaub.de website. 28 Aug. 2009. http://www.ab-in-den-urlaub.de/weltraumreisen.htm.

612 This concerns (1) the Treaty on Principles governing the Activities of States in the Exploration and Use of Outer Space, including the Moon and Other Celestial Bodies (hereafter Outer Space Treaty), London/Moscow/Washington, done 27 January 1967, entered into force 10 October 1967, 610 UNTS 205, 6 ILM 386 (1967); (2) the Agreement on the Rescue of Astronauts, the Return of Astronauts and the Return of Objects Launched into Outer Space (hereafter Rescue Agreement), London/Moscow/Washington, done 22 April 1968, entered into force 3 December 1968, 672 UNTS 119, 7 ILM 151 (1968); (3) the Convention on International Liability for Damage Caused by Space Objects (hereafter Liability Convention), London/Moscow/Washington, done 29 March 1972, entered into force 1 September 1972, 961 UNTS 187, 10 ILM 965 (1971); (4) the Convention on Registration of Objects Launched into Outer Space (hereafter Registration Convention), New York, done 14 January 1975, entered into force 15 September 1976, 1023 UNTS 15, 14 ILM 43 (1975); and (5) the Agreement Governing the Activities of States on the Moon and Other Celestial Bodies (hereafter Mon Agreement), New York, done 18 December 1979, entered into force 11 July 1984, 1363 UNTS 3, 18 ILM 1434 (1979).

613 See Outer Space Treaty.

614 Kayser, Valérie. "Commercial Exploitation of Space: Developing Domestic Regulation." Annals of Air and Space Law 17.1 (1992): 190. Some others are, however, of another opinion. See Bourély, Michel. "Quelques Réfléxions au Sujet des Législations Spatiales Nationales." Annals of Air and Space Law 16.1 (1991): 247.

615 See Von der Dunk, Frans G. "Implementing the United Nations Outer Space Treaties – The Case of the Netherlands." National Space Law – Development in Europe – Challenges for Small Countries. Eds. Christian Brünner and Edith Walter. Graz: Böhlau, 2008: 83–84; Hermida, Julian. Legal Basis for a National Space Legislation. Dordrecht: Kluwer Academic Publishers, 2004: 29–30.

616 Norway enacted its space act in 1969, Sweden in 1982, and the U.K. in 1986.

617 Hereafter Belgian Space Act. For the text of the law see Böckstiegel, Karl-Heinz, Benkö, Marietta, and Hobe, Stephan. Space Law – Basic Legal Documents. Volume 5. Utrecht: Eleven International Publishing, 2008. E.X.; National Space Law – Development in Europe – Challenges for Small

Countries. Eds. Christian Brünner and Edith Walter. Graz: Böhlau, 2008: 183–194; The Belgian Federal Science Policy Office. 28 Aug. 2009. http://www.belspo.be/belspo/res/rech/spatres/loispat_en.stm.

[618] Mayence, Jean-François. "Introduction to the Belgian Law on the Activities of Launching, Flight Operations or Guidance of Space Objects." Space Law – Basic Legal Documents. Volume 5. Eds. Karl-Heinz Böckstiegel, Marietta Benkö and Stephan Hobe. Utrecht: Eleven International Publishing, 2008. E.X.

[619] Royal Decree implementing certain provisions of the law of 17 September 2005 on the activities of launching, flight operations and guidance of space objects (hereafter Royal Decree). For the text of the decree see National Space Law – Development in Europe – Challenges for Small Countries. Eds. Christian Brünner and Edith Walter. Graz: Böhlau, 2008: 195–200; Belgian Federal Science Policy Office. 28 Aug. 2009. http://www.belspo.be/belspo/res/rech/spatres/loispat_en.stm.

[620] Mayence, Jean-François. "Introduction to the Belgian Law on the Activities of Launching, Flight Operations or Guidance of Space Objects." Space Law – Basic Legal Documents. Volume 5. Eds. Karl-Heinz Böckstiegel, Marietta Benkö and Stephan Hobe. Utrecht: Eleven International Publishing, 2008; Belgian Federal Science Policy Office. 28 Aug. 2009. http://www.belspo.be/belspo/res/rech/spatres/loispat_en.stm.

[621] Mayence, Jean-François. "Introduction to the Belgian Law on the Activities of Launching, Flight Operations or Guidance of Space Objects." Space Law – Basic Legal Documents. Volume 5. Eds. Karl-Heinz Böckstiegel, Marietta Benkö and Stephan Hobe. Utrecht: Eleven International Publishing, 2008.

[622] Hereafter Dutch Space Activities Act. For the text of the law see Böckstiegel, Karl-Heinz, Benkö, Marietta, and Hobe, Stephan. Space Law – Basic Legal Documents. Volume 5. Utrecht: Eleven International Publishing, 2008. E.XIV; National Space Law – Development in Europe – Challenges for Small Countries. Eds. Christian Brünner and Edith Walter. Graz: Böhlau, 2008: 201–210.

[623] De Brabander-Ypes, Heleen. "Introduction to the Law Incorporating Rules Concerning Space Activities and the Establishment of a Registry for Space Objects." Space Law – Basic Legal Documents. Volume 5. Eds. Karl-Heinz Böckstiegel, Marietta Benkö and Stephan Hobe. Utrecht: Eleven International Publishing, 2008. E.XIV.

[624] De Brabander-Ypes, Heleen. "The Netherlands Space Law – An introduction to contents and dilemma's." Presentation. 47th session of the Legal Subcommittee of the Committee on the Peaceful Uses of Outer Space. Vienna, Austria. 28 Aug. 2009. http://www.oosa.unvienna.org/pdf/pres/lsc2008/pres-02.pdf.

[625] Aruba declared that it would ban space activities completely and the Dutch Antilles will draft their own space legislation on the basis of the Dutch Space Activities Act. De Brabander-Ypes, Heleen. "Introduction to the Law Incorporating Rules Concerning Space Activities and the Establishment of a Registry for Space Objects." Space Law – Basic Legal Documents. Volume 5. Eds. Karl-Heinz Böckstiegel, Marietta Benkö and Stephan Hobe. Utrecht: Eleven International Publishing, 2008. E.XIV.

[626] Ibid.

[627] Hereafter French Space Operations Act. For the text of the law see United Nations General Assembly. Conference Room Paper Containing Information on the National Legislation of France Relevant to the Peaceful Exploration and Use of Outer Space. UN Doc. A/AC.105/C.2/2009/CRP.18. Vienna: United Nations.; National Space Law – Development in Europe – Challenges for Small Countries. Eds. Christian Brünner and Edith Walter. Graz: Böhlau, 2008: 211–224.

[628] Clerc, Philippe. "The French Space Operation Act." Presentation. 48th session of the Legal Subcommittee of the Committee on the Peaceful Uses of Outer Space. Vienna, Austria. 28 Aug. 2009. http://www.oosa.unvienna.org/pdf/pres/lsc2009/pres-04.pdf.

[629] Couston, Mireille. "Introduction to the French Space Operation Act." Space Law – Basic Legal Documents. Volume 5. Ed. Karl-Heinz Böckstiegel, Marietta Benkö and Stephan Hobe. Utrecht: Eleven International Publishing, 2008. E.XVI.

[630] Van Traa-Engelman and Louise Hanneke. Commercial utilization of outer space. Dordrecht: Martinus Nijhoff Publishers, 1993: 62–63; Reijnen, Bess C.M. The United Nations Space Treaties

Analysed. Gif-sur-yvette: Editions Frontières, 1992: 114; Cheng, Bin. "Article VI of the 1767 Space Treaty Revisited: "International Responsibility", "National Activities", and the "Appropriate State"." Journal of Space Law 26.1 (1998): 14.

631 According to Article 31 of the Vienna Convention of the Law of Treaties, a treaty should be interpreted in good faith in accordance with the ordinary meaning to be given to it in its context and in the light of its object and purpose.

632 See Reijnen, Bess C.M. The United Nations Space Treaties Analysed. Gif-sur-yvette: Editions Frontières, 1992: 114–115; Wassenbergh, Henri A. Principles of Outer Space Law in Hindsight. Dordrecht: Kluwer Academic Publishers, 1991: 25.

633 Dutch Space Activities Act. Chapter 1, Section 1.

634 French Space Operations Act. Title 2, Chapter 1, Article 1.

635 Belgian Space Act. Chapter 1, Article 2.

636 French Space Operations Act. Title 2, Chapter 1, Article 2.

637 Dutch Space Activities Act. Chapter 1, Section 1.

638 At the moment this is the Minister for Federal Scientific Policy. See Mayence, Jean-François. "The Belgian Space Law." Presentation. 48th session of the Legal Subcommittee of the Committee on the Peaceful Uses of Outer Space. Vienna, Austria. 28 Aug. 2009. http://www.oosa.unvienna.org/pdf/pres/lsc2009/pres-08.pdf.

639 Dutch Space Activities Act. Chapter 1, Section 1.

640 Royal Decree. Chapter 2, Article 2.

641 Hermida, Julian. Legal Basis for a National Space Legislation. Dordrecht: Kluwer Academic Publishers, 2004: 34.

642 Belgian Space Act. Chapter 1, Article 2.

643 Dutch Space Activities Act. Chapter 2, Section 3.

644 French Space Operations Act. Title 1, Article 1 and Title 2, Article 2.

645 See Article 1 Registration Convention and Article 1 Liability Convention.

646 Belgian Space Act. Chapter 1, Article 2.

647 Ibid. Chapter 1, Article 3.

648 French Space Operations Act. Title 1, Article 1.

649 Dutch Space Activities Act. Chapter 1, Section 1; De Brabander-Ypes, Heleen. "The Netherlands Space Law – An introduction to contents and dilemma's." Presentation. 47th session of the Legal Subcommittee of the Committee on the Peaceful Uses of Outer Space. Vienna, Austria. 28 Aug. 2009. http://www.oosa.unvienna.org/pdf/pres/lsc2008/pres-02.pdf.

650 See Belgian Space Act. Chapter 1, Article 2; Dutch Space Activities Act. Chapter 1, Section 1.

651 Law on Space Activity of Ukraine of 15 November 1996 as amended 2000 and 2006, Section 1. For the text of the law see Böckstiegel, Karl-Heinz, Benkö, Marietta, and Hobe, Stephan. Space Law – Basic Legal Documents. Volume 5. Utrecht: Eleven International Publishing, 2008. E.VIII.

652 Cheng, Bin. "Article VI of the 1767 Space Treaty Revisited: "International Responsibility", "National Activities", and the "Appropriate State"." Journal of Space Law 26.1 (1998): 13–14.

653 Kopal, Vladimír. "International Legal Regime on Outer Space: Outer Space Treaty, Rescue Agreement and the Moon Agreement." Proceedings of the United Nations/Nigeria Workshop on Space Law – Meeting international responsibilities and addressing domestic needs. Ed. United Nations Office for Outer Space Affairs. Vienna: United Nations, 2005: 8–10.

654 Gerhard, Michael. Nationale Weltraumgesetzgebung. Köln: Carl Heymanns Verlag, 2002: 81.

655 Bourbonnière, Michel and Ricky J., Lee. "Legality of the Deployment of Conventional Weapons in Earth Orbit: Balancing Space Law and the Law of Armed Conflict." European Journal of International Law 18.5 (2007): 883–884; Gorove, Stephen. "Implications of International Space Law for Private Enterprise." Annals Air & Space Law 7.1 (1982): 321.

656 This is the case in Belgium where Chapter 3, Article 11 of the Belgian Space Act provides for the suspension or withdrawal of the license. See Belgian Space Act. Also the Netherlands allow the withdrawal of authorisation by their competent authorities. See Dutch Space Activities

Act. Chapter 2, Section 7. For France see the French Space Operations Act. Chapter IV, Article 9.

[657] Belgian Space Act. Chapter 7, Article 19.

[658] French Space Operations Act. Title 1, Chapter 4, Article 11.

[659] Dutch Space Activities Act. Chapter 5, Section 15. It has to be noted that the Dutch Space Activities Act deals with the enforcement-topic in a very comprehensive way. 11 out of the total 28 Sections deal with the question of enforcement.

[660] Belgian Space Act. Chapter 7, Article 19.

[661] See, for example, Gerhard, Michael and Kai-Uwe Schrogl. "Report of the "Project 2001" Working Group on National Space Legislation." "Project 2001" – Legal Framework for the Commercial Use of Outer Space. Eds. Karl-Heinz Böckstiegel. Köln: Carl Heymanns Verlag, 2002: 529; Hobe, Stephan, Bernhard Schmidt-Tedd and Kai-Uwe Schrogl, eds. "Project 2001 Plus" – Global and European Challenges for Air and Space Law at the Edge of the 21st Century. Vol. 4. Köln: Carl Heymanns Verlag, 2004.

[662] Treaty of Lisbon amending the Treaty on European Union and the Treaty establishing the European Community, signed at Lisbon, 13 December 2007. OJ No. C 306 of 17 December 2007.

[663] Treaty on the Functioning of the European Union, Article 4(3).

[664] Treaty establishing a Constitution for Europe, signed in Rome on 29 October 2004. OJ No. C 310.

[665] This e contrario argument was put forward by Sergio Marchisio with regard to the Draft Constitution. See Marchisio, Sergio. "Potential European Space Policy and its Impact on National Space Legislation." "Project 2001 Plus" – Global and European Challenges for Air and Space Law at the Edge of the 21st Century. Vol. 4. Eds. Hobe, Stephan, Bernhard Schmidt-Tedd and Kai-Uwe Schrogl. Köln: Carl Heymanns Verlag, 2004: 145, 150.

[666] The working group on National Legislation Relevant to the Peaceful Exploration and Use of Outer Space was chaired by Irmgard Marboe (Austria) and held six meetings, from 31 March to 3 April 2009. For the detailed report of the working group see United Nations General Assembly. Report of the Legal Subcommittee on its Forty-Eight Session, Held in Vienna from 23 March to 3 April 2009. UN Doc. A/AC 105/935 of 20 April 2009. Vienna: United Nations.

[667] Ibid: 35.

[668] Ibid.

[669] Ibid.

[670] Ibid.

[671] Ibid.

[672] Ibid.

[673] Ibid: 36.

[674] Ibid.

8. Iridium-Cosmos collision and its implications for space operations

Ram S. Jakhu

8.1. Introduction

The physical condition of the environment in which space activities take place must be conducive to the safe and sustainable development and implementation of all space operations. Rapidly increasing debris in space is posing serious risks to space activities of all nations. Such risks are real, as evidenced by an unprecedented collision between a defunct Russian satellite weighing about 900 kg (Cosmos 2251) and an active about 500 kg U.S. commercial satellite (Iridium 33). The theory of "big sky" or "vastness of space" is being questioned as space operators face new challenges primarily due to the enhanced risks posed by an ever increasing number of space debris, particularly in the region of space that is being used extensively for meeting important earthly needs, like communications, Earth observation, weather forecasting, reconnaissance, navigation and global positioning services, early warning, etc. The availability and risk-free utilisation of space is becoming difficult. The accident seems to have sounded a wakeup call for all countries, especially the space-faring nations and the States having an interest in the exploration and use of space. From that perspective, this paper addresses some relevant questions associated with the legal and policy aspects of the Iridium-Cosmos collision. Specifically discussed are the issues concerning possible liability of the States directly involved in the collision under currently applicable international law; the general space environment; added risks and costs due to the debris created by the collision; and, finally, the concerns of space operators and some governments and the efforts they are making to achieve and maintain safe and sustainable development and use of space. Undoubtedly, these issues are highly complex. However they are discussed here in a general fashion and no attempt is made to carry out a thorough critical analysis in this paper.

8.2. Determination of liability for damage

Since the Iridium-Cosmos collision involved satellites launched by at least two different States, it is international in nature and scope. As such, the law applicable to the accident essentially stems from two important international space treaties: i.e. the 1967 Outer Space Treaty[675] and the 1972 Liability Convention,[676] to which both the U.S. and the Russian Federation were parties at the time of the collision. These treaties impose liability on a "launching State", whose space object has caused damage.[677] Thus, it is imperative first to determine the status of each country as a "launching State". Under both treaties, the term "launching State" is defined similarly.[678] Applying the common definition, it is evident that the Russian Federation was the launching State for Cosmos 2251 satellite, since it was launched on 16 June 1993 by Russia with its Cosmos rocket from its Plesetsk cosmodrome.[679] The U.S. was a "launching State" for Iridium 33 commercial communications spacecraft since the U.S. can be deemed to have "procured" the launch of the satellite which was owned by an American private company, Iridium Satellite LLC (though the launch agreement seems to have been privately negotiated by the Iridium Corporation).[680] Interestingly, the Russian Federation and Kazakhstan were also launching States for Iridium 33, which was launched on 14 September 1997 with a Russian Proton K rocket from the Russian leased Tyuratam (Baikonur Cosmodrome) facility in the territory of Kazakhstan.[681] The issue concerning the possible liability of Kazakhstan is not addressed here.

Iridium 33, which "was part of a planned commercial communications network comprised of a constellation of 66 LEO spacecraft using L-Band to provide global communications services",[682] was licensed by the U.S. under its 1934 Communications Act (as amended),[683] partly a space legislation. Thus, the U.S. exercised jurisdiction and control over Iridium 33,[684] as it is entitled and indeed obliged under Article VIII of the Outer Space Treaty,[685] though it did not register the satellite with the UN as required by the Registration Convention[686] to which the U.S. is also a party.

Since the collision occurred in space, under Article III of the Liability Convention, a launching State is "liable only if the damage is due to its fault or the fault of persons for whom it is responsible"; i.e. it creates a rule of fault-based liability if the damage is caused in outer space. The burden of proving fault is cast upon the State that makes a claim; i.e., the State that is claiming compensation is under obligation to establish or to prove "fault" on the part of other State(s). What legal criteria are to be used for the determination of the precise meaning and scope of "fault"? Since the Liability Convention does not define the term "fault", one must look for its ordinary meaning in law, specifically under international law. Normally, "fault" in law is equated with negligence; and negligence is determined on the basis of duty of care. After tracing the origin of the international law principle of "fault" back to *The Jamaica Case of 1798*, Bin Cheng thoroughly discussed this principle and reached the conclusion that "[. . .] contrary to what is assumed by a majority of writers, fault in modern jurisprudence is no longer identified exclusively with negligence or malice. [. . .] it means any act or inaction which violates an obligation [duty]."[687] Thus, the questions that need to be addressed are: (a) whether any of the States concerned was under a duty, and if so what was the nature of that duty; (b) whether leaving one's dead satellite in orbit is a negligent act and (c) whether inaction or failure to take action to avoid collision amounts to negligence (fault)?

It is necessary to recall what actually happened in the Iridium-Cosmos collision in order to determine whether any of the parties could be considered to have been at fault. The collision occurred on 10 February 2009 at 11.56 AM EST, at an altitude of about 800 km over Northern Siberia.[688] Both satellites were orbiting at a speed of about 17,000 mph (7.5 km/s); i.e. about 9–10 times the speed of a bullet fired from a rifle. At the time of collision, Cosmos 2251 was a "free flying" dead space object; i.e. without any station-keeping or manoeuvring capability. On the other hand, Iridium 33 was a normal operational satellite providing telecommunication services. It was fully equipped and capable of making manoeuvres, if desired; i.e. its behaviour could be controlled and its orbit could be adjusted to avoid any collision with other space object(s).

It is known that Russia had abandoned its Cosmos 2251 satellite when it became non-operational about 10 years ago. Russia did not take any action to remove this space object (actually debris)[689] from its orbit in LEO. Such inaction is considered contrary to Russia's obligation under the Outer Space Treaty,[690] particularly its Article IX, which, in part, specifies that States must conduct all their activities in outer space with due regard to the corresponding interests of all other States parties to the Treaty. Thus, such inaction and disregard of the interests of other States on the part of Russia amounts to Russia's fault and Russia should be considered liable to the U.S. for the destruction of the Iridium 33 spacecraft belonging to an American company. However, since the dawn of the space age, the practice among launching States, though unfortunate, has been the routine abandonment of dead satellites. Thus, one may assert that due to this practice of over 50 years, a new rule of customary international law has evolved that modifies a State's general obligation under Article IX of the Outer Space Treaty,[691] and States are therefore entitled to leave their dead satellites in orbit without any responsibility for keeping outer space free of hazards. However, this assertion is not tenable since the two essential elements or requirements that must exist in order for a practice to become a rule of customary international law (i.e. steady space practice and *opinio juris*)[692] are missing. The practice has not been followed regularly by States. Neither have space-faring States expressed a legal right in the past to abandon their dead satellites in space. For several years, most space-faring States have adopted regulations and policies which require their respective agencies and commercial operators to de-orbit or relocate their near-death (near end-of-life) satellites in all orbits.[693] There are many examples of the routine implementation of such regulations and policies, the result being that numerous satellites have been de-orbited.

As the State having jurisdiction over and responsibility for actions of Iridium Corporation under the above-mentioned Article IX of the Outer Space Treaty and the Liability Convention,[694] the U.S. is obliged not to cause damage to, or interfere with the operation of, satellites belonging to other States. Nevertheless, it appears that the U.S. suffered loss due to its own inaction/inability to predict and/or avoid collision. Rationally speaking, one must expect, or lay the responsibility on, the owner/operator of a functional satellite (i.e. Iridium 33) to carry out appropriate manoeuvre(s) to avoid a collision with those satellites that have the right to operate under Article 1 (2) of the Outer Space Treaty.[695] Though capable of implementing its collision-avoidance procedures, the Iridium Corporation did not take any action in this regard since it had no prior warning about the possibility of the Iridium-Cosmos collision. According to Jeffrey White, Regional Director EMEA and Russia for Iridium Satellite LLC, "Iridium does not utilize any standard processes for collision risk monitoring/collision avoidance".[696] The Iridium Corporation, like any other satellite operator, is aware of the position-situation of its own satellites but not of other satellites in the neighbourhood. For monitoring and conjunction (possibility) predictions about collisions between objects in space, several operators rely upon civil and commercial tools and resources like SOCRATES (developed and operated by Analytical Graphics Inc. (AGI)). However, they are limited in their capabilities and are unable to monitor all satellites all the time. Moreover, their conjunction predictions are not precise and highly reliable. Jeffrey White reiterated the statement of AGI that "the Iridium 33/Cosmos 2251 conjunction was not even in the top 150 most probable predicted conjunctions for that particular day... [In fact]... 16 other Iridium satellites had higher probability conjunctions that day than did Iridium 33".[697] The other entity involved in monitoring and conjunction predictions has been the Space Surveillance Network (SSN), an official establishment of the U.S. Government, which operates the world's most advanced and extensive multi-billion U.S. dollar system for carrying out space situational awareness (SSA) activities and conjunction predictions. It disseminates the coordinates and velocity of the satellites through the internet free of charge. However, due to lack of "computing power, trained personnel and sensor capabilities"[698] the SSN's priority has been the monitoring and conjunction predictions of U.S. military satellites and manned spacecraft. The U.S. military does not analyse the collision probabilities of all satellites. It is clear that the SSN (i.e. the U.S. Government) did not monitor Iridium 33 before it collided with Cosmos 2251. Consequently, it failed: (a) to predict its possible collision with Cosmos 2251 and (b) to alert or instruct Iridium Corporation management to undertake appropriate manoeuvre(s) for changing its orbit in order to avoid the collision.

In April 2009, the U.S. House of Representatives Committee on Science and Technology's Subcommittee on Space and Aeronautics conducted hearings regarding the challenges posed (particularly by space debris) to space operators. In her opening statement, the Subcommittee Chairperson, Gabrielle Giffords, expressed the view that it was "difficult to believe that nothing could have been done to prevent the (Iridium-Cosmos) collision, given that one of the satellites was active and by all accounts would have had the capability to manoeuvre out of harm's way".[699]

Inaction or even inability on the part of the U.S. cannot relieve it of its international responsibility not to cause harm on the basis of the decision of the International Court of Justice in *The Corfu Channel Case*. In that case, the Court held Albania responsible for laying mines in its territorial waters over which it had exclusive jurisdiction, as a result of which British ships suffered damage. Albania pleaded that it did not know about the existence of the mines and thus could not prevent the accident. The Court dismissed Albania's plea stating "nothing was attempted by Albania to prevent the disaster, and these grave omissions involve her international responsibility."[700] It is well-known that the U.S. has been (and continues to) monitoring thousands of satellites (including pieces of debris), and has successfully avoided impending accidents by undertaking appropriate collision-avoidance manoeuvres numerous times. Even if there are technical limitations on the capabilities of the SSN, it is believed that the U.S. could or should have known about the possibility of the Iridium-Cosmos collision had it monitored these satellites in the same manner as it was (and is) surveying other satellites, like the International Space Station and the Space Shuttle space transportation system, with the regular monitoring of which the U.S. did avoid several potential accidents with space debris. Since the U.S. omitted, ignored, failed, or did not attempt to prevent the Iridium-Cosmos collision, it remains internationally responsible for the consequences of the accident.

As mentioned above, the U.S. registered Iridium 33 on its national registry but failed to register it with the UN. Thus, the U.S. is in violation of its international obligation under the Registration Convention. However, non-registration with the UN does not relieve the U.S. of its responsibility and possible liability for being the State that exercised jurisdiction and control over Iridium 33 and also for being its launching State. It may be noted here that, interestingly, neither did the Russian

Federation deregister its decayed Cosmos 2251 with the UN as required under the Registration Convention.[701] Similarly, although the Russian Federation too violated its obligation under the Registration Convention, the legal consequences of such inaction are not directly related to its liability under the Liability Convention.

If the U.S. is at fault in this case, is Russia entitled to claim compensation for the destruction of Cosmos 2251? For this, Russia needs to prove not only fault on the part of the U.S. but also the quantum and nature of the damage it suffered. In this regard, it may be difficult to accept that Russia suffered any damage (as understood under the Liability Convention)[702] as a result of the destruction of its dead and abandoned satellite.

In sum, it is submitted that the Russian Federation had no right under customary international law to abandon its decayed Cosmos 2251 in space. Neither was the U.S. (for and on behalf of Iridium Corporation) free not to take any action to prevent the collision which occurred on 10 February 2008. The case of Iridium-Cosmos collision seems to have been resolved by mutual understanding since there is no information available in the public domain about any legal action (option) that is being considered or pursued by either party. It is contemplated that, perhaps the absence of an ensuing claim for any compensation is due to the fact that the parties have realised the weaknesses of their respective legal positions; i.e. Russia did not suffer any real damage and the U.S. was at fault (or at least was contributorily negligent) primarily due its own lack of knowledge of, or inaction or inability to prevent, the accident.

8.3. Consequences and implications of the collision

8.3.1. Regulatory uncertainty

Since the issue of the Iridium-Cosmos collision has not been litigated in any international or national court of law, it did not authoritatively clarify the applicable law (i.e. the concept of fault) to cases of accidents in space.

The incident demonstrated, however, that States, particularly major space-faring nations, are not regularly and faithfully observing the binding obligations they have voluntarily assumed under the Registration Convention. This unfortunate observation raises the following two concerns: (a) small and emerging space-faring nations might follow the unfortunate practice of the U.S. and the Russian Federation and (b) States might be expected not to seriously comply with non-binding principles[703] related to the prevention of space debris.

In addition, the collision unmistakably indicates the lack of prompt availability of sufficient, precise and timely information, both from commercial and governmental (security) sources that could be used by civil and commercial space operators to formulate advance warnings and take any collision-avoidance actions or procedures in a timely manner. Moreover, the capabilities for monitoring space situation and conjunction predictions were essentially within the U.S. alone. In such a case, it will be extremely difficult, if not impossible, to collect full, accurate, and unbiased evidence comprising of all the appropriate facts and circumstances of an accident in space should one occur and the case is brought before a judicial tribunal. The collection of such evidence would, a priori require: (a) the presence of fairly detailed and agreed upon "rules of the road" or "space navigational rights, responsibilities and procedures" (b) a significant amount of financial resources needed for advanced capabilities for precise monitoring, cataloguing and analysing extensive amounts of data about complete space situation and traffic in space and finally, (c) appropriately trained human resources possessing the necessary expertise and working within or through an independent international body. Thus, the Iridium-Cosmos collision signifies the urgent need of rules, procedures, resources and a neutral international body (fully equipped with the appropriate mandate, resources and regulatory infrastructure) for the collection and determination of relevant evidence.

8.3.2. Increasing risks and cost of space operations

Initially, the Iridium Corporation and Russia downplayed the problem caused by the Iridium-Cosmos collision. However, several experts, space entrepreneurs and government spokespersons expressed serious concerns related to national security, cost of space operations and even the very viability of space programmes in the future. For example, a day after the accident, the U.S. Marine Gen. James E. Cartwright, Vice Chairman of the Joint Chiefs of Staff, speaking on the national security ramifications of the collision, is reported to have said that "Many of the commercial and national security satellites, particularly communications satellites, rely on certain spacing between other objects in order to be effective. Losing a spot because of debris could have a financial or operational impact on anyone wanting to use the space. If that's going to be long term, that's a problem for us."[704] Similarly, Lieutenant General Larry James, head of the U.S. Strategic Command's Joint Functional Component Command for Space, said the Iridium-Cosmos collision, "which was not predicted by the U.S. military or private tracking groups, had a big impact on future U.S. military planning by tangibly demonstrating the vulnerability of our space assets."[705]

Today the global space industry generates over 250 billion U.S. dollars worth of economic activity that employs hundreds of thousands of high-tech, high-wage middle class workers.[706] A dire warning about the economic and operational implications of the Iridium-Cosmos collision has been well expressed by Launchspace Staff: "Should such an event occur, several bad things will happen to many satellite operators. If another Iridium satellite is involved the company would be forced to replace the lost satellite. The frequency of close encounters in orbits near that of Iridium's constellation would suddenly increase to levels that would cause several operators to reassess the viability of existing space applications. Satellite insurance providers might be forced to raise premiums on in-orbit performance to record high levels. Future launch plans for almost all low orbit satellites may be curtailed. Space-based services to the world would diminish over time. The economic impact is not even calculable. This is scary!"[707]

In order to avoid collision, an active satellite would have to change its orbital path with the use of onboard fuel. Such a manoeuvre depletes the necessary fuel and consequently shortens the operational life of the satellite. GeoEye, an American private satellite imaging enterprise, had to manoeuvre its Ikonos satellite several times in order to avoid collisions with pieces of space debris. Consequently, these manoeuvres resulted in the imposition of extra costs on the company in terms of wastage of fuel (thus shortening the satellite's lifespan) and personnel time to execute these manoeuvres.[708] In 2007, an American 1.3 billion U.S. dollars Terra satellite had to be manoeuvred to avoid collision with a piece of debris from the Chinese Fengyun-1 C satellite destroyed by China's ASAT test.[709]

Space debris poses risks not only to unmanned satellites but also to spacecraft with humans on board; e.g. the International Space Station (ISS) and the U.S. Space Shuttle. It has been reported that the crew of the ISS had to make several manoeuvres and sometimes take refuge in Soyuz spacecraft in order to avoid possible collisions with space debris.[710] The windshield windows of the Space Shuttle had to be replaced several times due to the damage caused by space debris present in the orbit used by the Shuttle. According to the U.S. House of Representatives Committee on Science and Technology, "the recent Iridium-Cosmos collision has added to the debris field in LEO and represents a 71% increase in the amount of threatening debris to STS-125 [Space Shuttle]."[711]

Due to the gravity of the Earth, pieces of space junk are expected to reenter the Earth's atmosphere and possibly burn up at the time of reentry. However, the life span of orbiting debris depends on the altitude of its orbit. It takes about one year for pieces of debris at an altitude of 400 km to come down. Nevertheless, if the orbit of debris is at an altitude of 1000 km, it could take over a thousand years for it to reenter the Earth's atmosphere.[712] Thus, the junk in the geostationary orbit (i.e. about 36,000 km above the Earth's equator) will remain there for millions of years;

i.e. forever. Also, all pieces of space debris, especially the bigger ones, do not burn up during their reentry into the Earth's atmosphere, and sometimes, they land on unpredictable places. Thus, space debris poses risks to humans or can cause damage or gives rise to environmental concerns on the surface of the Earth. Here are some examples: in 1978 the Soviet satellite Cosmos 954 disintegrated and scattered radioactive debris over a large area in Northern Canada.[713] Similarly, in 1979 the U.S. space laboratory Skylab broke down and showered hundreds of pieces of debris over Western Australia and the Indian Ocean. There were serious concerns expressed by the public over the possibility of Russia's Mir Space Station coming down. Fortunately, on 23 March 2001, it successfully de-orbited, broke apart at reentry into the atmosphere and the unburned pieces crashed over the South Pacific Ocean without any damage to anyone. In December 2008, NORAD's operation centre could not predict the precise time and place of the coming down of an earlier launched Russian Proton K rocket, which without prior warning, crossed through Canadian airspace and landed uneventfully in Labrador, Canada. This event was considered to be a "reminiscent of a 2005 incident, when a U.S. military rocket splashed down in the vicinity of the Hibernia oil platform, on Newfoundland's Grand Banks, shortly after its launch from Florida".[714]

Since the beginning of the space age in 1957, several thousand satellites and rockets have been launched by or on behalf of about 50 countries. Most of these satellites have become spent since, yet they remain in orbit. Consequently, space around the Earth is increasingly becoming polluted and contaminated with dead satellites (including those with nuclear reactors on board), break-ups of spacecraft and rocket bodies, leftover fuel or other reactive chemicals, debris caused by accidental collisions between orbiting objects or by deliberate destruction of satellites (e.g. ASAT tests), etc. The U.S. Department of Defense (DoD) has been tracking about 19,000 space objects 10 cm (4 inches) or larger in diameter, out of which only about 900 are operational and manoeuvrable satellites.[715] In addition, it is estimated that there are more than 300,000 untracked objects measuring between 1 and 10 cm in diameter orbiting the Earth and millions smaller than 1 cm. Their number is expected to increase fast in the near future.[716] As several new countries are entering the space field and the current crop of space-faring nations are launching more satellites, "a conservative estimate projects that the number of active satellites will jump from 1300 to 1500 over the next decade".[717] This will inevitably be accompanied by an equivalent increase in the rate of generation of space debris.

Pieces of space debris sometime travel at speed of about 35,000 km per hour. "At such high velocity, even small junk can rip holes in a spacecraft or disable a satellite."[718] No satellite can be reliably protected against this kind of destructive force.

Tab. 2: *The Known or Suspected Past Collisions between Objects in Space.*

Year	Event
1991	Inactive [Russian] Cosmos 1934 satellite hit by catalogued debris from [Russian] Cosmos 296 satellite
1996	Active [French] Cerise satellite hit by catalogued debris from [European] Ariane rocket stage
1997	Inactive [U.S.] NOAA 7 satellite hit by uncatalogued debris large enough to change its orbit and create additional debris
2002	Inactive [Russian] Cosmos 539 satellite hit by uncatalogued debris large enough to change its orbit and create additional debris
2005	Inactive U.S. rocket body hit by catalogued debris from Chinese rocket stage
2007	Active [European] Meteosat 8 satellite hit by uncatalogued debris large enough to change its orbit
2007	Inactive [U.S.] NASA OARS satellite believed hit by uncatalogued debris large enough to create additional debris
2009	Active [U.S.] Iridium 33 satellite hit by inactive [Russian] Cosmos 2251

The probability of inter-debris collisions increases as the debris population grows, thereby creating the potential to create even more debris. It is estimated that 40% of tracked debris results mostly from break-ups of spacecraft and rocket bodies.[719] There is a trend of a steadily increasing number of such break-ups.[720] Since 1991, there have been several major debris generating collisions in space that are believed to have significantly increased the space debris population. Table 2 above, which has been compiled by David Wright of the Union of Concerned Scientists, shows the known collisions in space, and the three italicised entries are those that resulted in the destruction of operational satellites.[721]

The extent of the debris clouds generated by the Iridium-Cosmos collision ranges from 200 to 1700 km. The SSN has identified about 1500 pieces larger than 10 cm that were created by the Iridium-Cosmos collision.[722] Half of these pieces might fall back to Earth within 5 years. Some will remain in orbit by the end of the century and even for "significantly longer periods"[723] afterwards. Sadly, after the Iridium-Cosmos collision, scientists predict the possibility of more accidents in the future.[724]

Due to rapidly increasing rate of production of space debris, several experts have expressed alarm about the shortage of "space in space", and scientists predict that '"close encounters" between satellites and debris in orbit will rise by 50% in the next 10 years, and by 250% by 2059, to more than 50,000 a week'.[725] If the number of spacecraft with nuclear reactors on board increase,[726] the chances of radioactive contamination of space and possibly the safety of the public and property on Earth will grow, if such satellites somehow collide in space with debris or an operational satellite.

Deliberate destruction of satellites, primarily by ASAT tests, has been another disturbing source of space debris. During the 1980s both the United States and the Soviet Union carried out several of such tests. The latest ASAT test was carried out by the Peoples Republic of China on 11 January 2007. With the use of its ground-based medium-range ballistic missile, China successfully destroyed a dead Chinese weather satellite orbiting at an altitude of about 800 km.[727] It is estimated that as a result of the test, the satellite "shattered into thousands of pieces that were thrown into a wide range of orbits ranging in altitude from 3800 km on the high end down to about 200 km at the lowest."[728] According to Nicholas Johnson, NASA's Chief Scientist for Orbital Debris, this test, "is by far the worst satellite fragmentation in the history of the space age, in the past 50 years"[729] as it created about 2500 new pieces of junk and thus added 25% more debris. In Table 3 below, Johnson shows significant increases in the space debris population caused by the 2009 Iridium-Cosmos collision as well as the 2007 Chinese ASAT test.[730]

Tab. 3: *Catalogued objects in Earth orbit.*

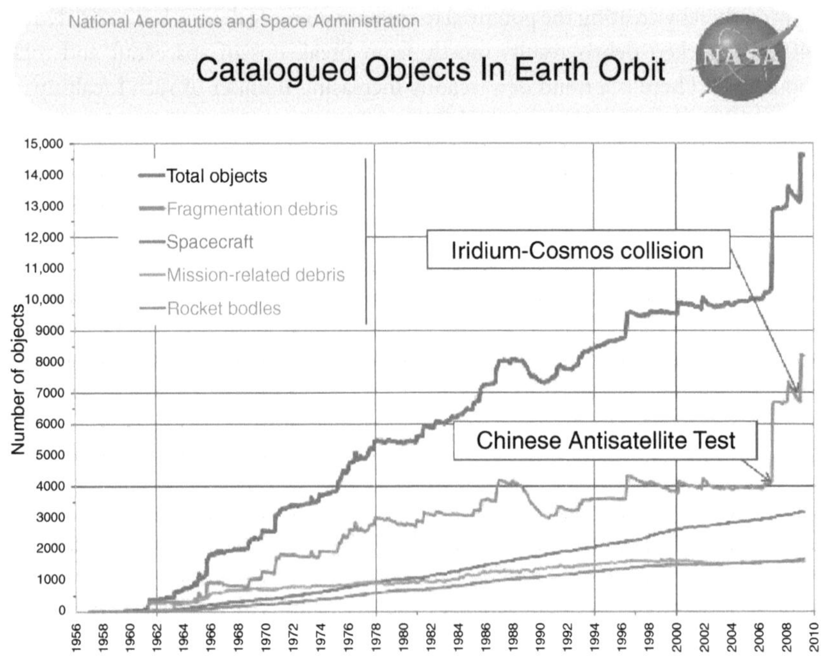

The region in space where the Chinese ASAT test and the Iridium-Cosmos collision took place is one that is heavily used by satellites belonging to several countries. Thus, the Iridium-Cosmos debris has worsened a situation that was already serious and enhanced the risks to satellites of several nations for a long time. It has correctly been pointed out by experts that "Orbital debris is the gravest threat to new and existing space systems."[731] Consequently, the Iridium-Cosmos collision has raised serious concerns and urgent calls signaling the need for appropriate policies and means to avoid the repetition of similar mishaps and to maintain space for safe and sustainable use by all nations.

8.4. Efforts to keep space for safe and sustainable development and use

8.4.1. Prevention: debris mitigation regulatory measures

The international space community, and even the general public, is becoming aware of and concerned about the danger posed by rapidly increasing space debris. A UN general circular, recently distributed around the world, highlighted the *Space Debris Problem as One of the Ten Stories the World Should Hear More About.*[732] There are often news and discussions about space debris in all sorts of media, at least in the Western world. Almost all major space-related academic and governmental conferences or seminars invariably deal with the space debris problem, either directly or indirectly. This growing awareness of the negative impact of space debris upon public safety on the Earth as well as the safety and sustainability of space operations seems to have encouraged some space-faring nations to take steps to mitigate the production of new debris through the development of national space debris mitigation measures.[733] However, such initiatives are useful only in the short term, since they apply exclusively to respective national space operations and consequently their effectiveness is limited as a single major accident could create hazards for the space activities of all States. It seems imperative that all space-faring and space-aspiring nations ought to be involved in the adoption and effective application of all sorts of measures in a coordinated manner to prevent the production of new pieces of debris and elimination of the existing space debris.

At the international level, there are no effective and binding rules, standards and procedures in existence that apply to space debris. After debating the issue since 1994, the UN Committee on Peaceful Uses of Outer Space (COPUOS) adopted Space Debris Mitigation Guidelines,[734] which were endorsed by the UN General Assembly in its resolution 62/217 of 21 December 2007. The Guidelines aim at curbing pollution in space, and at promoting international efforts for safe spacecraft operations. The Guidelines are based on and consistent with the Inter-Agency Space Debris Coordination Committee (IADC) guidelines. The COPUOS believed that the Guidelines would increase mutual understanding on acceptable activities in space and thus enhance stability in space-related matters and also decrease the likelihood of friction and conflict. Such preventive measures, if fully and effectively implemented by all space-faring nations, are very important but are not sufficient to keep the space environment safe enough to facilitate sustainable development of the space operations of all nations as the Guidelines do address the extensive amount of existing debris currently in orbit. They are aimed at mitigating the production of new debris only.

Unfortunately, the Guidelines are one of the lowest forms of any statements on highly serious space-related international concern by a UN body. They are voluntary, legally non-binding, and present only general recommendations in the form of seven principles (or goals) to be implemented (or achieved) through national legislation and regulatory mechanisms. Moreover, it is believed by some experts that the goals set in the Guidelines might prove more difficult to achieve than their negotiation, during a period of about fifteen years, primarily because of the different views, approaches and concerns within various national (regulatory and technical) institutions that are involved in the design, adoption and implementation of detailed domestic rules, procedures and standards for mitigating space debris.

Nevertheless, irrespective of their inherent weaknesses, it seems imperative that the Guidelines must be implemented effectively nationally as soon as possible rather than waiting for the adoption of perfect implementation mechanisms and standards. At the same time, the international community must continue seriously searching for means to strengthen them and to coordinate national space debris mitigation efforts. For this purpose, the COPUOS (actually its Scientific and Technical Sub-Committee) should make concerted efforts, in collaboration with neutral national or international bodies, like the IAASS (International Association for the Advancement of Space Safety), or any other neutral group of technical experts, to develop and adopt technical standards and regulatory procedures so that they could be followed uniformly by all nations and space operators.

8.4.2. Prevention: space situational awareness and space traffic management

Most of the satellites, especially civil and commercial satellites, "fly blind"; i.e. they are not aware of other spacecraft that are in their vicinity in space. The first and the most important practical step to prevent Iridium-Cosmos collision like accidents "is to determine what is in Earth orbit and where it is going: space situational awareness (SSA)."[735] According to the testimony of the Secure World Foundation, the "owner or operator of a particular satellite usually has excellent knowledge about the position of that satellite in space, but little to no information about the locations of other objects around them. This situation was the *root cause* behind the collision of two satellites in February – the owner of the Iridium satellite, which could have potentially manoeuvred it out of the way, did not know about the impending close approach."[736] As noted above, the U.S. Government Space Surveillance Network neither monitored Iridium 33 nor could it alert its operator about the possible collision. However, since the Iridium-Cosmos collision the U.S. "SSTRATCOM has been providing daily conjunction reports for the Iridium constellation."[737] In fact, the U.S. has started significantly enhancing its SSA capabilities (monitoring, conjunction prediction and information distribution) by adding new staff and facilities that track on a daily basis 800 new space objects and distribute conjunction reports to civil and commercial satellite operators through its DoD Commercial and Foreign Entities (CFE) pilot programme.[738] Several friendly foreign governments and their entities also benefit from the programme. In order to further improve its SSN, the U.S. government is planning for a new system called "space fence", which "will constantly report the motion of all objects 5 cm wide and larger in medium and low-Earth orbits."[739] However, the CFE programme is not open to all space operators from all countries since the DoD prefers not to divulge sensitive information, particularly about its military satellites. This pretext of the DoD is considered and not fully convincing since "many sensitive satellites are large and easily found, even by amateur astronomers."[740]

Because of the limitations of the CFE, three major commercial satellite operators (i.e. Intelsat, SES and Inmarsat) have recently decided to set up their own Satellite Data Association (SDA) and "have issued a request for proposals for a company to design and operate a database on satellite positions, planned manoeuvres and signal transmissions with a view to reducing the chance of orbital collisions."[741] It is too early to judge the viability and possible success of the SDA, but it certainly seems to be an appropriate and timely step in the right direction for the commercial space operators to protect their investments and interests. Preferably, a neutral international system (a space data center) mandated to carry out SSA activities and to provide regular and precise conjunction reports to all space operators is needed. Instead of clinching to outdated theories of absolute national sovereignty and freedom of action in a highly globalised and interdependent world, it is in the interest of all nations to collectively resolve the problem of space debris, which adversely affects all satellites in orbit without discrimination. China, France, Germany, India, Japan and Russia have their own, though limited, space surveillance facilities in one form or another. However, no country shares SSA information regularly with anyone. As major space-faring nations, they all should pool their resources together and set up a cooperative system, preferably under an international agreement, that will help them, their space operators and other interested public or private entities in the pursuit of safe and sustainable utilisation of space. Such capabilities will be indispensable for undertaking manoeuvres not only for collision avoidance but also for the purpose of de-orbiting satellites, when required. The operations of such a system need not be expensive since it should employ a small group of appropriately trained and experienced experts and acquire the necessary equipment, especially high powered computers. The most expense tool for carrying out the work of the system will be the raw satellite data, but that may be procured from (or provided by) participating national systems, like the U.S. Space Surveillance Network and other suppliers. Moreover, such a neutral international system could provide unbiased and accurate statements of facts and expert opinions which are necessary for the proper resolution of disputes and problems like the one created by the Iridium-Cosmos collision. The second important step to prevent space collisions is to have precise "rules of the road" for all space operators; i.e. space traffic management rules.[742] Simply speaking, as all users of roads on the Earth must know the rules for using the highway in order to avoid accidents, similarly spacecraft orbiting around the Earth and space transportation operators must follow predetermined, uniform, universal, precise, legally binding and effectively enforceable rules. There exist no such rules today, except those that are set through the International Telecommunication Union (ITU) for the satellites operating in the geostationary orbit. New modes of transportation, namely orbital and suborbital-orbital spaceflights, are becoming viable and will be used routinely in the near future for transporting people and cargo from point-to-point on the surface of the Earth through space, and also as a viable means of space transportation and travel, including space tourism. The safety of space operations and space travel will depend, among other things, upon the safe and secure construction and operation of all modes of space transportation and the rules they will follow in their flights through airspace and outer space. It is, therefore, logical that international space safety and traffic management regulations should be negotiated through the International Civil Aviation Organization (ICAO), which has extensive experience in such rules with respect to aviation. The rationale for this suggestion is the same as that followed by the U.S. in gradually extending the mandate of the Federal Aviation Administration (FAA) to set rules and procedures for space transportation systems and implementing them, as and when necessary.

8.4.3. Cure: removal of space debris

As noted above, without effective and timely efforts and means to physically remove space junk from space, particularly from the densely populated orbits, the space environment would not only remain hazardous for space operations but could also become more dangerous as additional debris would be created by collisions between pieces of existing debris (i.e. "domino effect" also known as "Kessler Syndrome"). For the last two decades, there have been in place some national regulations and policies requiring space operators to de-orbit their satellites when they approach the end of their lives and place them in the so-called graveyard orbit above the geostationary orbit. These policies have not been consistently followed and thus are correctly considered to "abjectly fail to remove risks imposed by derelict spacecraft, due in great part to the false sense of security created by the term [de-orbit]".[743] Nevertheless, since the Iridium-Cosmos collision, various concepts for space debris removal are being actively conceived, proposed, debated, and developed. Some of them include: (a) the development of specially designed "Debris Collection Spacecraft" that would be capable of rendezvousing with space object, manoeuvring and removing them[744] and (b) an idea to develop a spaceship to collect all the debris and push it closer to the Earth so that it burns up at the time of reentry into the Earth's atmosphere.

The following two proposals seem to be more serious: (a) on 11 October 2009, *The Observer* reported that German robots that are being designed "can rescue failing satellites and push "dead" ones into deep space. . . .[They might] be ready in four years. . . . and will dock with failing satellites to carry out repairs or push them into "graveyard orbits", freeing vital space in geostationary orbit".[745] This proposal appears to be interesting and somewhat valuable but tends to postpone the real and desired solution of removing pieces debris from space (not only from their respective orbits). (b) The second one is the initiative, taken in September 2009, by the Defense Advanced Research Projects Agency (DARPA), which is a part of the U.S. DoD. DARPA issued a Request for Information regarding the design and development of capabilities for removal of space debris. The programme "will involve all space-faring nations, tens of billions of dollars and decades of development, testing and operations".[746] The second proposal seems to be a good and possibly viable solution because of the involvement of all space-faring nations. However, the costs involved, the national security interests and long time to develop might prove challenging to overcome. In addition, both these proposals might also have to overcome the perception that the capabilities and tools developed for space debris removal might also be used as space weapons. Secondly, under current space law, there is no right of salvage; i.e. no right to collect and destroy satellites (including debris) belonging to others. Defunct satellites or their component parts remain under the jurisdiction and control of the owner, unless specifically disclaimed. Such right needs to be created before technical solutions, like the foregoing can be used to sufficiently eliminate risky space debris. Nevertheless, small pieces of debris (millions in number) orbiting at extremely high speed will remain a problem, especially for space travellers and for insufficiently protected satellites. The search for an appropriate cure will and must continue until safe and sustainable use of space is assured for all space operators.

8.5. Conclusion – message

A safe space environment is imperative for the sustainable development of the space operations of all nations for civil, commercial and security purposes. It is inevitable that as space becomes more crowded, Iridium-Cosmos collision like accidents would happen. The collision highlighted an urgent need and initiated intensive efforts to search for technical, policy, organisational and regulatory solutions, including expanded space situational awareness, safety standards, space traffic rules, removal of existing debris, etc. One can only hope that something positive might come out of the tragic accident of 10 February 2009. It is submitted that the COPOUS Space Mitigation Guidelines must be enhanced and strengthened at the international level in order to expressly include a clear obligation to remove defunct satellites and a right to salvage. I reiterate here my earlier statement, i.e. Space debris is primarily a global issue. Global problems need global solutions, which must be effectively implemented internationally as well as nationally. In addition, COPUOS must intensify efforts to ensure that the Registration Convention is adhered to by all parties to the treaty. It is believed that the proposed neutral international space data center could help all space operators in achieving safe and sustainable utilisation of space. It is unfortunate and ironic but true that events like the Iridium-Cosmos collision are sadly necessary for appropriate solutions to the problems which would not otherwise be resolved in a timely manner. *Nécessité fait loi.*

---

[675] Treaty on Principles governing the Activities of States in the Exploration and Use of Outer Space, including the Moon and Other Celestial Bodies (hereafter Outer Space Treaty), London/Moscow/Washington, done 27 January 1967, entered into force 10 October 1967, 610 UNTS 205, 6 ILM 386 (1967).

[676] Convention on International Liability for Damage Caused by Space Objects (hereafter Liability Convention), London/Moscow/Washington, done 29 March 1972, entered into force 1 September 1972, 961 UNTS 187, 10 ILM 965 (1971).

[677] Article VII of the Outer Space Treaty stipulates that a launching State is internationally liable for damage caused by its space object. Similarly, Articles II and III of the Liability Convention provide that a launching State is (a) absolutely liable to pay compensation for damage caused by its space object on the surface of the Earth or to aircraft in flight and (b) liable if the damage is caused in outer space and is due to its fault or the fault of persons for whom it is responsible.
[678] Article VII of the Outer Space Treaty and Article I (c) of the Liability Convention similarly specifies that a 'launching State' is a State that launches or procures the launching of an object and a State from whose territory or facility an object is launched.
[679] Information Furnished in Conformity with the Convention on Registration of Objects Launched in to Outer Space. "Note verbale dated 14 June 1994 from the Permanent Mission of the Russian Federation to the United Nations addressed to the Secretary General". UN Doc. ST/SG/SER.E/275 of 13 June 1994. Vienna: United Nations: 2. It should be noted that there is a typographical error in writing the number of the satellite; i.e. it is listed as 2551, but the number should have been 2251.
[680] According to Article VI of the Outer Space Treaty, States bear international responsibility for national activities in outer space, whether such activities are carried on by their governmental agencies or by non-governmental (private) entities, and for assuring that their national activities are carried out in conformity with the provisions set forth in the Treaty. The activities of non-governmental entities require authorisation and continuing supervision by the appropriate State. NASA National Space Science Data Centre, an official entity of the U.S. Government, enlists Iridium LLC of the United States as the funding agency of Iridium 33 satellite: see NASA National Space Science Data Centre. 2 Nov. 2009. http://nssdc.gsfc.nasa.gov/nmc/spacecraftDisplay.do?id=1997-051C.
[681] NASA National Space Science Data Centre. 2 Nov. 2009. http://nssdc.gsfc.nasa.gov/nmc/spacecraftOrbit.do?id=1997-051C. On 4 March 1998, Russia informed the UN that "On 14 September 1997, seven Iridium satellites were placed in Earth orbit by a single Proton carrier rocket from the Baikonur launch site. [...] The satellites are owned and operated by the Motorola company (United States of America)". See: Information Furnished In Conformity With The Convention On Registration Of Objects Launched Into Outer Space. "Note verbale dated 4 March 1998 from the Permanent Mission of the Russian Federation to the United Nations addressed to the Secretary-General". UN Doc. ST/SG/SER.E/332 of 19 March 1998. Vienna: United Nations.
[682] Ibid.
[683] "Communications Act of 1934". Federal Communications Commission. 3 Nov. 2009. http://www.fcc.gov/Reports/1934new.pdf. Also applicable are the U.S. Code of Federal Regulations: Title 47-Telecommunication; Chapter I-Federal Communications Commission; Part 25-Satellite Communications.
[684] The official U.S. Registry of Space Objects Launched into Outer Space, which is maintained by the U.S. Department of State's Bureau of Oceans and International Environmental and Scientific Affairs, enlists Iridium 33 (with International Code 1997-051 C and NORAD 24946), for which the USA is the flag state and affirms that the satellite was not registered with the UN by the U.S. See: U.S. Space Objects Registry. 2 Nov. 2009. http://usspaceobjectsregistry.state.gov/registry/dsp_DetailView.cfm?id=1517. Similarly, the SPACEWARN Bulletin Number 527 substantiates that Iridium 33 was an American communications spacecraft that was launched with a Russian Proton-K rocket. See: "SPACEWARN Bulletin Number 527". 1 October 1997. NASA National Space Science Data Centre. 2 Nov. 2009. http://nssdc.gsfc.nasa.gov/spacewarn/spx527.html.
[685] According to Article VIII of the Outer Space Treaty, a State "on whose registry an object launched into outer space is carried shall retain jurisdiction and control over such object".
[686] Convention on Registration of Objects Launched into Outer Space (hereafter Registration Convention), New York, done 14 January 1975, entered into force 15 September 1976, 1023 UNTS 15, 14 ILM 43 (1975). Article IV (1) of the Convention requires each State of registry to furnish to the Secretary-General of the United Nations information concerning each space object carried on its registry.
[687] Cheng, Bin. General Principles of Law as Applied by International Courts and Tribunals. Volume 2 of Grotius classic reprint series. Cambridge: Cambridge University Press, 2006: 225.

[688] NASA National Space Science Data Centre. 2 Nov. 2009. http://nssdc.gsfc.nasa.gov/nmc/spacecraftDisplay.do?id=1997-051C.
[689] The Cosmos satellite involved in this collision was a 'space object' under the Liability Convention, irrespective of the fact that it was non-functional or a piece of space debris. It should be kept in mind that Cosmos 954 too was a non-functional object when it crashed into Canadian territory and the Soviet Union did pay to Canada "the sum of three million Canadian dollars (C$ 3,000,000.00) in full and final settlement of all those matters [. . .] [connected with the disintegration of the Soviet satellite Cosmos 954 in January 1978], including the claim advanced by Canada in this respect". See: Disintegration Of Cosmos 954 Over Canadian Territory In 1978. 2 April 1981. Doc. Ref. Canadian Department of External Affairs. Communiqué No. 27.
[690] Article 1 (2) of the Outer Space Treaty specifies that, "Outer space [. . .] shall be free for exploration and use by all States without discrimination of any kind, on a basis of equality and in accordance with international law". By implication, this Article obliges States to refrain from causing damage to, and interfering with the similar use of space by, other States.
[691] According to the Deputy Legal Advisor to the United States Department of State, the emergence of a new customary rule may result in a modification in the operation of a prior treaty rule. To substantiate his opinion, he relies on Article 68(c) of the 1964 International Law Commission draft articles on the law of treaties, which stated that "the operation of a treaty may be modified by subsequent emergence of a new rule of customary law relating to matters dealt with in the treaty and binding upon all the parties". 2 ILC Yearbook 198 (1964); adopted unanimously, 1 ILC Yearbook 318 (1964). 2 Nov. 2009. http://www.state.gov/s/l/65626.htm. Also, the International Court of Justice recognised in the 1974 *Fisheries Case* that the right of a State to set up twelve-mile fishing zones had crystallised as customary international law notwithstanding the provisions in the 1958 High Seas Convention creating freedom of fishing on the high seas. See: *Fisheries Jurisdiction Case* (United Kingdom v. Iceland, Merits), Judgment of 25 July 1974 (1974) I.C.J. Rep. 3.
[692] A practice of States becomes a rule of customary international law only when it is followed with a sense of legal obligation (*opinio juris sive necessitatis*). *North Sea Continental Shelf Cases* (Denmark/The Netherlands v. Federal Republic of Germany) Judgment of 20 February 1969 (1969), I.C.J. Rep.:3. at page 44: "Not only must the acts [of States] concerned amount to a settled practice, but they must also be such, or be carried out in such a way, as to be evidence of a belief that this practice is rendered obligatory by the existence of a rule of law requiring it."
[693] For example, see U.S. Government Orbital Debris Mitigation Standard Practices. 2 Nov. 2009. http://www.orbitaldebris.jsc.nasa.gov/library/USG_OD_Standard_Practices.pdf.
[694] It may be also noted that normally, a State is not considered responsible for an illegal act of its non-State actors unless there is a genuine link between the act and the State; i.e. an illegal act must be somehow imputable to the concerned State. However, that principle of international law has been modified by Articles VI and VII of the Outer Space Treaty, which make a State responsible (and possibly liable) for the space activities of all its entities. Even under general International law, according to Richard S.J. Tol and Roda Verheyen, "as soon as an activity is permitted or licensed by a state (under the control of. . .), the resulting behaviour is attributable to the state because [the principle that] states must exercise due diligence in [the] control of private persons is an acknowledged principle". Tol, Richard S. J. and Roda Verheyen. "State Responsibility and Compensation for Climate Change Damages – a Legal and Economic Assessment". Energy Policy 32.9 (2004): 1111. As noted earlier, the operation of Iridium 33 was licensed by, and under the control of, the U.S. Government, thus the U.S. was (and will remain) responsible (and possibly liable) for any damage, harm or injury caused by or with this satellite (or its debris).
[695] Article 1 (2) of the Outer Space Treaty specifies that, "Outer space [. . .] shall be free for exploration and use by all States without discrimination of any kind, on a basis of equality and in accordance with international law".
[696] White, Jeffrey. "Iridium Satellite: Aviation – We are everywhere you want to fly". 5 Nov. 2009. http://www.eurocontrol.int/nexsat/gallery/content/public/Steering%20Group/Meeting10/IRIDIUM_JeffWhitePresentation32109.pdf.

[697] Ibid.
[698] "Vacuum in Space". 16 Feb. 2009. International Herald Tribune. http://www.iht.com/articles/2009/02/16/opinion/edspace.php.
[699] "Committee Examines Ways to Make the Space Environment Safer for Civil and Commercial Users". 28 Apr. 2009. Press Release from the Committee on Science and Technology of the U.S. House of Representatives. 23 Sept. 2009. http://science.house.gov/press/PRArticle.aspx?NewsID=2447.
[700] *The Corfu Channel Case* (Merits), Judgment of 9 April 1949 (1949) I.C.J. Rep.: 4. at p. 23.
[701] The Registration Convention, Article IV (3).
[702] According to Article I (a) of the Liability Convention, the term 'damage' "means loss of life, personal injury or other impairment of health; or loss of or damage to property".
[703] For example, the principles incorporated in the 2007 COPUOS Space Debris Mitigation Guidelines, discussed below.
[704] Stump, Adam M. "Satellite Collision Debris May Affect Space Operations, Cartwright Says". 12 Feb. 2009. American Forces Press Service. 19 Feb. 2009. http://www.defenselink.mil/news/newsarticle.aspx?id=53077.
[705] Shalal-Esa, Andrea. "Pentagon May Reach Satellite Analysis Goal Early". 28 Apr. 2009. Reuters. 2 Nov. 2009. http://www.reuters.com/article/idUSN2836780620090428.
[706] "Global Space Economy Revenues Reach 257 bln in 2008: Report". 31 Mar. 2009. Chinaview. 2 Aug. 2009. http://news.xinhuanet.com/english/2009-03/31/content_11106035.htm.
[707] "Space Debris – Problem Solved". 31 Aug. 2009. Space Daily. 30 Aug. 2009. http://www.spacedaily.com/reports/Space_Debris_Problem_Solved_999.html.
[708] Moskowitz, Clara. "U.S. 'Decades Behind' on Space Debris Threat, Official Says". 6 Nov. 2009. Space.com. 8 Nov. 2009. http://www.space.com/missionlaunches/091106-space-junk-risk-increase.html.
[709] "Space Debris: A Growing Challenge". Oct. 2009. Aerospace America. 2 Nov. 2009. http://www.gsb.stanford.edu/news/packages/2009/PDF/AIAA-SpaceDebris_OCT2009.pdf: 31.
[710] Ibid.
[711] Hearings on "Keeping the Space Environment Safe For Civil and Commercial Users". 28 Apr. 2009. U.S. House of Representatives Committee on Science and Technology Subcommittee on Space and Aeronautics.
[712] "Astronomy Question Of The Week: Is Space Debris Dangerous". 25 Aug. 2009. Spacemart. 25 Aug. 2009. http://www.spacemart.com/reports/Astronomy_Question Of The Week_Is_Space_Debris_Dangerous_999.html.
[713] See *supra* footnote 690.
[714] "Space Debris Poses Risk". 3 Aug. 2009. The Times & Transcript. 3 Aug. 2009. http://timestranscript.canadaeast.com/newstoday/article/748848.
[715] Glassman, Albert. "The Growing Threat of Space Debris". July 2009. Todaysengineers. 2 Nov. 2009. http://www.todaysengineer.org/2009/jul/space_debris.asp.
[716] "Committee Examines Ways to Make the Space Environment Safer for Civil and Commercial Users". 28 Apr. 2009. Press Release from the Committee on Science and Technology of the U.S. House of Representatives. 23 Sept. 2009. http://science.house.gov/press/PRArticle.aspx?NewsID=2447. Thomas, Brad. "STS 121". Apr. 2006. Space Center Roundup, Lyndon B. Johnson Space Center. 9 Nov. 2009. http://www.jsc.nasa.gov/roundup/online/2006/0406_p8_11.pdf: 8.
[717] "Space Debris: A Growing Challenge". op.cit. *supra* footnote 710.
[718] Lovgren, Stefan. "Space Junk Cleanup Needed, NASA Experts Warn". 19 Jan. 2006. National Geographic News. 19 Feb. 2009. http://news.nationalgeographic.com/news/2006/01/0119_060119_space_junk.html.
[719] Ibid.
[720] NASA, Orbital Debris Quarterly News 11.1 (2007): 2.
[721] Wright, David. "Colliding Satellites: Consequences and Implications". 26 Feb. 2009. Union of Concerned Scientists. 11 Oct. 2009. http://www.ucsusa.org/assets/documents/nwgs/SatelliteCollision-2-12-09.pdf.

[722] "United Nations' COPUOS Receives Update on Iridium-Cosmos Collision". NASA Orbital Debris Quarterly News 13.3 (2009): 2.
[723] Ibid.
[724] Swaine, Jon. and Stephen Adams, "Experts Warn of More Space Crashes After U.S. and Russian Satellites Collide: Space Scientists Issued a Stark Warning Yesterday Over the "Inevitable" Prospect of More Satellites Crashing into Each Other". 12 Feb. 2009. The Telegraph. 15 Mar. 2009. http://www.telegraph.co.uk/science/space/4603851/Experts-warn-of-more-space-crashes-after-US-and-Russian-satellites-collide.html.
[725] Moskowitz, Clara. "U.S. 'Decades Behind' on Space Debris Threat, Official Says". 6 Nov. 2009. Space.com. 8 Nov. 2009. http://www.space.com/missionlaunches/091106-space-junk-risk-increase.html.
[726] It has recently been reported that several nuclear-powered manned spacecrafts might be launched soon both by NASA and Russia. See: "Russia Goes All Out To Develop Nuclear-Powered Spacecraft". 16 Nov. 2009. Space Travel. 16 Nov. 2009. http://www.space-travel.com/reports/Russia_Goes_All_Out_To_Develop_Nuclear_Powered_Spacecraft_999.html; Kislyakov, Andrei. "Old Russian Nuclear Satellite Returns". 30 Jan. 2009. Spacedaily. 2 Nov. 2009. http://www.spacedaily.com/reports/Old_Russian_Nuclear_Satellite_Returns_999.html; Isachenkov, Vladimir. "Russia Hopes Nuclear Ship Will Fly Humans to Mars". 29 Oct. 2009. Associated Press. 2 Nov. 2009. http://www.google.com/hostednews/ap/article/ALeqM5jfhjVbCZfHYXlG0zJNKwYr8BtNqgD9BKPDOO0.
[727] "China Confirms Satellite Downed". 23 Jan. 2007. BBCnews. 2 Nov. 2009. http://news.bbc.co.uk/2/hi/asia-pacific/6289519.stm.
[728] Morring, Jr., Frank. "China Asat Test Called Worst Single Debris Event Ever". 11 Feb. 2007. Aviation Week & Space Technology. 10 Mar. 2007. http://www.aviationweek.com/aw/generic/story_generic.jsp?channel=awst&id=news/aw021207p2.xml.
[729] Ibid.
[730] Johnson, Nicholas L. "Preserving the Near-Earth Space Environment with Green Engineering and Operations". 30 Sept.–1 Oct. 2009. Presentation. NASA Green Engineering Masters Forum. 3 Nov. 2009. http://ntrs.nasa.gov/archive/nasa/casi.ntrs.nasa.gov/20090032041_2009032577.pdf.
[731] "Space Debris: A Growing Challenge". op.cit. *supra* footnote 710.
[732] "Space Debris: Orbiting Debris Threatens Sustainable Use of Outer Space". 2008. United Nations. 10 Oct. 2009. http://www.un.org/en/events/tenstories/08/spacedebris.shtml.
[733] See, UN Doc. A/AC.105/820/Add.1 of 15 December 2003; UN Doc. A/AC.105/820/Add.1 of 4 February 2004; Roscosmos. "Activity of Russian Federation on Space Debris Problem". Presentation. 46th Session of the Scientific and Technical Subcommittee of UNCOPUOS. Vienna, Austria. 17 February 2009. 2 Nov. 2009. http://www.unoosa.org/pdf/pres/stsc2009/tech-46.pdf; Wirt, Uwe. "UN-Space Debris Mitigation Guidelines – National Implementation Mechanism." Presentation. 48th Session of the Legal Subcommittee of UNCOPUOS. Vienna, Austria. 31 March 2009. 2 Nov. 2009. http://www.unoosa.org/pdf/pres/lsc2009/pres-06.pdf; Klinkrad, Heiner. "Space Debris Mitigation Activities at ESA". Presentation. 46th Session of the Scientific and Technical Subcommittee of UNCOPUOS. Vienna, Austria. 17 February 2009. 2 Nov. 2009. http://www.unoosa.org/pdf/pres/stsc2009/tech-40.pdf; NASA "USA Space Debris Environment and Policy Updates". Presentation. 45th Session of the Scientific and Technical Subcommittee of UNCOPUOS. Vienna, Austria. 11–22 February 2008. 2 Nov. 2009. http://www.unoosa.org/pdf/pres/stsc2008/tech-26E.pdf.
[734] United Nations General Assembly. Official Records of the General Assembly, Sixty-second Session. UN Doc. Supplement No. 20 (A/62/20) of 2007. New York: United Nations. The UN General Assembly in its Resolution endorsed the Space Debris Mitigation Guidelines of the Committee on the Peaceful Uses of Outer Space in 2007. See: United Nations General Assembly. Resolution adopted by the General Assembly on International Cooperation in the Peaceful Uses of Outer Space. Sixty-second session, Agenda item 31. UN Doc. A/RES/62/217 of 1 February 2008. New York: United Nations.
[735] "Keeping the Space Environment Safe for Civil and Commercial Users". Testimony of Secure World Foundation. U.S. House Committee on Science and Technology, Subcommittee on Space and Aeronautics. Hearing on 28 April 2009.

[736] Ibid. Emphasis added.
[737] Jeffrey White. op.cit. *supra* footnote 697.
[738] Taverna, Michael A. "Traffic Cop: U.S. Satellite Protection System is Entering Service, but Data Factors Remain an Issue". Aviation Week & Space Technology 20 Apr. 2009: 55.
[739] Marks, Paul. "Race is on for Best Space-junk Alarm System". 22 July 2009. NewScientist. 23 July 2009. http://www.newscientist.com/article/mg20327185.800-race-is-on-for-best-spacejunk-alarm-system.html.
[740] Moltz, James Clay. "Space Jam". 18 Feb. 2009. The New York Times. 19 Feb. 2009. http://www.nytimes.com/2009/02/19/opinion/19moltz.html?_r=1eta1.
[741] de Selding, Peter B. "Satellite Firms Moving Ahead on Orbital Database". 11 Nov. 2009. Space News. 12 Nov. 2009. http://www.spacenews.com/satellite_telecom/091118-satellite-firms-moving-ahead-orbital-database.html.
[742] For details discussion of various aspects of Space Traffic Management, see Schrogl, Kai-Uwe. "Space Traffic Management. The New Comprehensive Approach for Regulating the Use of Outer Space: an International Perspective". October 2007. ESPI Perspectives 3. 2 Sept. 2009. http://www.espi.or.at/images/stories/dokumente/flash_reports/stmflashrep3f2.pdf.
[743] "Space Collisions and Opportunity for Entrepreneurship". 3 Aug. 2009. Spacetalk now 10 Aug. 2009. http://spacetalknow.org/wordpress/?p=740.
[744] "Space Debris – Problem Solved". 31 Aug. 2009. Space Daily. 30 Aug. 2009. http://www.spacedaily.com/reports/Space_Debris_Problem_Solved_999.html.
[745] Day, Michael. "Rogue Satellites to Be Cleared From Earth's Orbit by German Robots". 11 Oct. 2009. The Observer. 14 Oct. 2009. http://www.guardian.co.uk/science/2009/oct/11/space-robots-clear-rogue-satellites.
[746] "Space Debris Gets Some Respect". 28 Sept. 2009. Spacemart. 2 Nov. 2009. http://www.spacemart.com/reports/Space_Debris_Gets_Some_Respect_999.html.

9. The Space Policy of the New U.S. Administration

Scott Pace

9.1. Introduction

The new U.S. Administration under President Obama is conducting a review of space policy and it is too soon to tell what major changes will result. However, it is possible to identify the major policy issues being considered and likely development based on past statements and actions taken by the Administration so far in 2009. Specific study groups that have been publicly identified include space protection, international cooperation, acquisition reform and national space strategy. Presidential policy remains in force until the President or his successor changes it and thus there is a high degree of interest in the on-going White House policy review being led by staff to the National Security Council.

During the 2008 Presidential election campaign, the Obama campaign released a detailed position paper on space policy issues which represented a great deal of continuity with the space policies of both Bush Administrations and the Clinton Administration. In particular, areas of continuity including completing the International Space Station and retiring the Space Shuttle, seeking to close (or at least reduce) the U.S. gap in human access to space after the retirement of the Space Shuttle, and continuing to conduct robotic missions. The position paper placed an increased emphasis on Earth science research and aeronautics, more international collaboration on climate change research and exploration, and the negotiation of international agreements on "rules of the road" to improve space security, protect space assets, and prevent the "weaponisation" of space.

New political appointees with responsibilities for space activity have been nominated by President Obama and confirmed by the Senate where needed. Most notably, retired Marine Corps General Charles Bolden and Lori Garver are the new Administrator and Deputy Administrator, respectively, for NASA; Dr. Jane Lubchenco is the new Administrator for the National Oceanic and Atmospheric Administration, and retired Air Force General Bruce Carlson is the director of the National Reconnaissance Office. Several senior military officers with space responsibilities have continued in their positions, these include General James Cartwright, the Vice Chairman of the Joint Staff, General Kevin Chilton,

Commander of U.S. Strategic Command, and General Robert Kehler of Air Force Space Command.

9.2. Civil Space Policy

In the course of his confirmation hearing, Charles Bolden stated his support for programmatic and policy priorities in-line with the Obama campaign positions on civil space policy. These included "building on investments made" in the International Space Station (and presumably not ending U.S. involvement in it by 2016 as current budget projections show), "accelerating the development of next-generation launch systems" (which may or may not be the Ares 1 and 5 vehicles currently in work), "enhancing the study of the Earth", lead space science to "new achievements", continue "cutting-edge technology" development, support "commercial, entrepreneurial innovation". and "inspire students" to study math and science.

Climate change research is an early and notable priority for the new Administration. The American Recovery and Reinvestment Act of 2009 was the largest single appropriations bill ever passed by the U.S. Congress. Part of the bill included an agreement to provide 400 million U.S. dollars (266 million euros) to NASA to "accelerate the development of the tier 1 set of Earth science climate research missions recommended by the National Academies Decadal Survey and to increase the agency's supercomputing capabilities". For NOAA, 170 million U.S. dollars (113.3 million euros) was made available to "address critical gaps in climate modelling and establish climate data records for continuing research into the cause, effects and ways to mitigate climate change". There will likely be renewed discussion over how to best coordinate the differing roles and responsibilities of NASA, NOAA, and other government agencies in Earth observations. In general, NASA tends to be best at R&D, demonstration of new scientific sensors, and spacecraft acquisition while NOAA is best at spacecraft operations, routine observations, and long-term data management.

Cost growth and delays with the National Polar-orbiting Operational Environmental Satellite System (NPOESS) have resulted in the total life cycle cost estimate approaching 14 billion U.S. dollars (9.3 billion euros) – about double the original estimate with inclusion of long-term launch, operations, and maintenance costs. The first launch has also slipped to 2014. A satellite known as the NPOESS Preparatory Project was supposed to be a test platform for NPOESS sensors but it is now planned to have an operational role when it is launched in 2011. An independent review panel reported to NPOESS management in March 2009 that the programme needed at least an additional one billion U.S. dollars (666 million euros) and creation of single decision-maker in place of the current tri-agency committee structure of NOAA, the Department of Defense, and NASA. White House officials from the National Security Council, Office of Management and Budget, and the Office of Science and Technology Policy are monitoring progress and will have input into upcoming budget decisions.

At the same time, NASA space science themes have been under increased stress due to cost overruns in flagship-class missions, notably Mars Science Laboratory. In the case of future robotic Mars exploration, NASA and the European Space Agency are engaged in potential cooperative discussions of ExoMars for a 2016 or 2018 launch opportunity. As might be expected, the definition of a 2018 mission is much more uncertain than what would be necessary for a 2016 launch. In June 2009, ESA agreed to create a Mars Exploration Joint Initiative (MEJI) that will provide a framework for the two agencies to define and implement their scientific, programmatic and technological goals at Mars.

The area of space policy that has been most subject to public debate has been the future of human space exploration beyond low Earth orbit, in particular human missions to the Moon and Mars. Current U.S. Space Exploration Policy embodies the direction in the Bush Administration's 2004 Vision for Space Exploration, a policy that has been supported by two successive NASA authorisation bills in 2005 and 2008. The future direction of U.S. space policy will be shaped by the interaction of Presidential and congressional priorities, international interests, the capabilities of NASA and its industrial partners, and the input of various external groups.

The most significant policy event for U.S. human space exploration in 2009 will likely prove to be the *Review of U.S. Human Space Flight Plans Committee*, also know as the Augustine Committee, after its chairman, retired Lockheed Martin Chief Executive Office Norm Augustine. Created at the request of the White House Office of Science and Technology Policy, the Committee is charged with conducting an independent review of ongoing U.S. human space flight plans and programmes, as well as alternatives, to "ensure the nation is pursuing the best trajectory for the future of human space flight – one that is safe, innovative, affordable, and sustainable". In effect, this is a review of NASA's Constellation programme for future U.S. missions to Low Earth Orbit, the Moon, and eventually Mars.

The U.S. Congress provided an additional one billion U.S. dollars (666 million euros) for NASA in fiscal year 2009 as part of one-time economic recovery funds, including 400 million U.S. dollars (266 million euros) for exploration, to which the Obama Administration allocated 150 million U.S. dollars (100 million euros) for commercial orbital transportation services

(COTS) from the exploration funds. The Obama Administration requested 18.686 billion U.S. dollars (12.4 billion euros) for NASA in fiscal year 2010, an increase of slightly over 5% over the prior years approved funding. This is helpful in the critical transition years now underway with the upcoming retirement of the Shuttle, but the additional funding is too late to accelerate development of Orion/Ares 1. Beyond fiscal year 2010, NASA's budget is effectively flat from FY2011-2014, and does not keep up with inflation. The nominal growth rate to compensate for inflation was projected at 1.36% compared to fiscal year 2009 and prior year projections of 2.4% annual growth rate. This further reduces NASA's expected long-term purchasing power.

The Administration has noted that approximately 3 billion U.S. dollars (2 billion euros) in the NASA Exploration budget for fiscal years 2011–2013 may be subject to change pending the outcome of the Augustine-led review. If these funds were to be fully removed, it would represent the continuation of a budget pattern since 2005. Through FY 2008, there has been a cumulative total of 11.7 billion U.S. dollars (7.8 billion euros) in reductions and costs absorbed within NASA's budget since the Vision for Space Exploration was announced and this has been the major cause of the growth of the "gap" between the Shuttle's retirement and the planned initial operating capability of Ares 1/Orion. Today, technical factors such as the time needed to develop the J2-X upper stage rocket engine for the Ares vehicles, prevent the complete elimination of the gap time even if more funds were spent. The Augustine Committee intends to provide a range of options to the Obama Administration, presumably consisting of possible spending increases, schedule delays, and changes in programme technical content, to better align national policy objectives with politically acceptable levels of spending.

The Augustine Committee requested a congressional perspective on human spaceflight policies, particularly those contained in the NASA Authorization Acts of 2005 and 2008 (Public Law 109–155 and Public Law 110–422, respectively). In a 17 July letter to Norm Augustine, Representative Bart Gordon, Chairman of the House Committee on Science and Technology argued that a durable congressional consensus had been established on the goals, objectives, and overall approach for the human and robotic exploration of the solar system.[747] As a result, Representative Gordon cautioned against major departures from that consensus as exemplified in the most recent NASA authorisation from 2008:

Sec. 402: "Congress hereby affirms its support for –

(1) the broad goals of the space exploration policy of the United States, including the eventual return to and exploration of the Moon and other destinations in the solar system and the important national imperative of independent access to space,

(2) *the development of technologies and operational approaches that will enable a sustainable long-term program of human and robotic exploration of the solar system,*
(3) *activity related to Mars exploration, particularly for the development and testing of technologies and mission concepts needed for eventual consideration of optimal mission architectures, pursuant to future authority to proceed with the consideration and implementation of such architectures, and*
(4) *international participation and cooperation, as well as commercial involvement in space exploration activities.*

The 2008 NASA Authorization Act responded to the observation of the Columbia Accident Investigation Board, created after the loss of the Space Shuttle *Columbia* in February 2003, that remaining in low Earth orbit was an inadequate justification for the risks of human spaceflight. The legislation placed an emphasis on developing systems to "support human missions beyond low Earth orbit" as show in Finding #8 of Public Law 110–422:

> *"Developing United States human space flight capabilities to allow independent American access to the International Space Station, and to explore beyond low Earth orbit, is a strategically important national imperative, and all prudent steps should thus be taken to bring the Orion Crew Exploration Vehicle and Ares 1 Crew Launch Vehicle to full operational capability as soon as possible and to ensure the effective development of a United States heavy lift launch capability for missions beyond low Earth orbit."*

International cooperation was recognised as making vital contributions to U.S. space efforts, perhaps most visibly in the durable partnership created by the International Space Station (ISS). The congress encouraged steps by the partnership to increase the "resiliency" of the ISS after the retirement of the Space Shuttle. Section 603 of P.L. 110–422 addresses the interplay of international and commercial contributions to ISS resupply:

> *"The (NASA) Administrator shall develop a plan and arrangement, including use of International Space Station international partner cargo resupply capabilities, to ensure the continued viability and productivity of the International Space Station in the event that United States commercial cargo resupply services are not available during any extended period after the date that the Space Shuttle is retired".*

In one of the last public meetings of the Augustine Committee on 12 August 2009 in Washington, D.C., Committee member Dr. Sally Ride made a presentation on the budgetary affordability of alternative exploration scenarios. While the analysis is ongoing, she concluded that there are no viable options for human exploration beyond low Earth orbit under the fiscal year 2010 budget currently proposed by the Obama Administration. This presents the U.S. Administration with a clear "disconnect" between stated policies, as noted above, and its budget proposals. Resolving the conflict between policies and budgets will require either additional funds for NASA or acceptance of less ambitious goals for human exploration, e.g., continuation of work in low Earth orbit or cancellation.

9.3. National security space policy

Moving to the national security space sector, there are several reviews and initiatives underway. In 1999, the 2000 Defense Authorization bill created a commission to look at national security space issues and threats. Formally called the "Commission to Assess United States National Security Space Management and Organization" it was also known at the Rumsfeld commission after its chairman, Donald Rumsfeld, who stepped down when he became Secretary of Defense in 2001. Its key message was that "If the U.S. is to avoid a 'space Pearl Harbour,' it needs to take seriously the possibility of an attack on U.S. space systems", said the commission, referring to the 1941 attack on U.S. naval forces in Hawaii's Pearl Harbour. "The U.S. is more dependent on space than any other nation. Yet the threat to the U.S. and its allies in and from space does not command the attention it merits", said the commission.

During the Bush Administration, the Department of Defense saw several reorganisation and personnel changes in response to the recommendations of the Rumsfeld Commission, concerns about the vulnerability of U.S. space assets have only grown and it can be argued that little in the way of fundamental progress toward improving space security has been made in recent years. The Obama Administration has said relatively little on national security space issues to date, but has acknowledged the crucial role of space systems. The White House website has a statement on "Guiding Principles to Rebalance Defense Capabilities for the 21st Century" in which space is mentioned specifically:

> **Space:** The full spectrum of U.S. military capabilities depends on our space systems. To maintain our technological edge and protect assets in this domain, we will continue to invest in next-generation capabilities such as operationally responsive space and global positioning systems.

> We will cooperate with our allies and the private sector to identify and protect against intentional and unintentional threats to U.S. and allied space capabilities.[748]

Space systems such as satellite communications, navigation and positioning constellations, and weather satellites are part of the critical infrastructure of all developed (and most developing) countries. One example of international engagement by the State Department on this topic this year was a statement on critical infrastructure protection made at the UN Institute for Disarmament Research conference in Geneva on 16 June 2009.[749] A particular point was made about the importance of engagement between the United States and its closest allies:

> Trans-Atlantic discussions on the protection of terrestrial infrastructure also can build from ongoing dialogues with the European Union and the European Space Agency on space situational awareness to address the increasingly congested and complex space environment. This trans-Atlantic collaboration can also serve as a foundation for expanded cooperation with other spacefaring nations around the globe who share our commitment to the safe and responsible use of space.

In 2009, the first notable national security space decision was an Administration decision to proceed with a next generation electro-optical intelligence satellite architecture. Quoting from a press release by the Director of National Intelligence: "Government-owned satellites would be developed, built and operated by the National Reconnaissance Office [. . .] [and] The Department of Defense and the Intelligence Community would increase the use of imagery available through U.S. commercial providers. This additional capability would provide the government with more flexibility to respond to unforeseen challenges [. . .] making them especially useful as a near-term supplement and backup to the government's existing imagery architecture".[750]

The Defense Authorization Act for Fiscal Year 2009 directed the Secretary of Defense and the Director of National Intelligence to jointly conduct a study of the policy, requirements, and objectives for space situational awareness, space control, space superiority, including defensive and offensive counterspace and protection, force enhancement and force application, space-based intelligence and surveillance and reconnaissance from space, and the integration of space and ground control and user equipment. Known as the "Space Posture Review", this effort is also directed to assess the relationships between military space policy, national security space policy, national security space objectives, arms control policy, export control policy and industrial base policy (presumably the space industrial base in particular as opposed to the entire U.S. industrial base). The report is due to the Congress by 1 December 2009.

The Department of Defense is also engaged in its Quadrennial Defense Review (QDR) for 2010. The QDR is a congressionally-mandated "wide-ranging review of strategy, programs, and resources" that is expected to link national defense strategy to the most recent National Security Strategy by defining force structures, modernisation plans, and a budget plan. Secretary of Defense Gates has already made some major decisions for Fiscal Year 2010 budget, such as cancelling a large military communications satellite programme known at T-SAT, reduction in the purchase of a Space-based Infrared System satellite for missile warning and stopping the testing and development of space-based kinetic energy weapons (KEW) for ballistic missile defense. Other decisions have yet to be made and the 2010 QDR is expected to have a major influence on national security space programmes and funding. A particular theme in the terms of reference for the 2010 QDR has been potential threats to what is termed the "global commons" e.g., areas like air, sea, space and cyberspace by both state and non-state actors. These threats seem to include both intentional acts as well as unintentional ones such as those posed by gaps in space situational awareness of orbital debris. Funding increases for improved space situational awareness are expected, including efforts to ease sharing of information on space objects with commercial and international space operators.

While the QDR, Space Posture Review, and Human Spaceflight Reviews are underway, the U.S. State Department has continued to work with the United National Committee on Peaceful Uses of Outer Space (COPUOS) on potential "rules of the road" guidelines to improve the "sustainability" of the space environment in the face of proliferating orbital debris and promote transparency and confidence-building measures in outer space to deter and discourage actions that would threaten peaceful space activities. In this regard, the United States has continued to support informal working groups that bring together experts from the public and private space sectors to explore additional measures to ensure the long-term sustainability of space activities, such as the initiative led by Gerard Brachet, the former chairman of COPUOS and past Director-General of the French space agency, CNES.

9.4. Looking Ahead

While it is still early in the tenure of the new Administration, a few predictions can be made on the direction of U.S. space policy. In scientific research, Earth science and climate monitoring will be emphasised in both budgetary terms and term of international cooperative initiatives. Continued participation in international efforts such as the Group on Earth Observations (GEO) will be supported in a continuing effort to build a Global Earth Observation System of Systems (GEOSS).[751] Given fiscal pressures, it is unlikely that general spending on space science will rise more than by inflation and cost overruns on major programmes, combined with Earth-oriented initiatives, will increase pressures for either more international cooperation or cancellations as is being seen in the robotic Mars exploration programmes.

In human space exploration, U.S. policy commitment to completing and utilising the International Space Station looks to have firm support from both the President and Congress despite the lack of U.S. funding projected by the White House Office of Management and Budget for beyond 2016. Similarly, support for retiring the Space Shuttle after it completed its current manifest for ISS assembly looks to be firm. Budget reductions in exploration budget lines, both past and current, have lengthened the gap between the end of Shuttle missions and the debut of a new generation of NASA vehicles to carry humans to low Earth orbit. It is possible that commercial services, such as those proposed by SpaceX, could arrive sooner but funding would need to be redirected from Ares development efforts that NASA has in the past said were of higher priority. The Augustine Review will likely have a major influence on whether the Obama Administration chooses to proceed with the current Constellation effort (Ares and Orion in particular) as is, to add funding or delay schedules, or to order development of a new architecture that would fit within funding levels the Administration is willing to support. It can be expected that the results of the Augustine report will be the beginning of a new round of debate in the Unites States on human space exploration.

In the national security space sector, budget constraints will limit the development of any major programmes. Combined with policy statements and reductions to ballistic missile defense, the Obama Administration will most likely not proceed with any efforts that would appear to be "space weapons" however that term is defined or intended. On the other hand, efforts to improve space situational awareness through greater multilateral cooperation with foreign governments and industry can be expected. Diplomatic efforts will support voluntary guidelines for satellite operations and information sharing to protect the space environment as part of a "global commons".[752]

In general, the rhetoric and inclination of the new U.S. Administration is and will continue to be more multilateral and less prone to statements of unilateral intent. This does not change, however, the enduring national interests pursued by the United States through space activities. The United States will likely increase efforts to ensure a benign and stable international system across all parts of the global commons, including space, cyberspace, aviation, and maritime sectors. Continuing globalisation and the proliferation of advanced technologies means the wider ranges of States (and non-state actors) are capable of exploiting space systems and are increasingly capable of creating their own space capabilities. On one had, this means there are more areas of potential common interest as the number of spacefaring States increase. On the other hand, it means the arrival of new participants (e.g., Iran, North Korea) that are markedly different from past space-capable States. In the particular case of Europe, there will likely be continuing debates on the relative roles and responsibilities for space-related security discussions involving both NATO and the European Commission as well as individual European countries with whom the United States has long hand security ties.

The next major space policy events to look for in 2010 will be the outcome of the National Space Policy review, any budgetary changes to the President's Fiscal Year 2010 submission, and the President's Fiscal Year 2011 Budget submission to the Congress in February of 2010.

---

[747] Gordon, Bart. "Statement to the Review of U.S. Human Space Flight Plans Committee." 17 July 2009. 6 Nov. 2009. http://democrats.science.house.gov/Media/file/AdminLetters/BG_to_Augustine%2007%2017%2009.pdf.

[748] White House website. 6 Nov. 2009. http://www.whitehouse.gov/issues/defense/.

[749] "Remarks of Richard H. Buenneke Deputy Director, Space Policy Office of Missile Defense and Space Policy. Bureau of Iternational Security and Nonproliferation. U.S. Department of State." 16 June 2009. United States Mission to the United Nations and Other International Organizarions in Geneva. 6 Nov. 2009. http://geneva.usmission.gov/CD/updates/2009/0616Buenneke.html.

[750] "DNI Blair Announces Plan for the Next Generation of Electro-optical Satellites." 7 Apr. 2009. Office of the Director of National Intelligence Public Affairs Office. 9 Nov. 2009. www.dni.gov/press_releases/20090407_release.pdf.

[751] GEO was launched in response to calls for action by the 2002 World Summit on Sustainable Development and by the G8 (Group of Eight) leading industrialised countries.

[752] See the article co-authored by Undersecretary of Defense for Policy Michele Flournoy: "The Contested Commons." July 2009. U.S. Naval Institute Proceedings. 9 Nov. 2009. http://www.usni.org/magazines/proceedings/story.asp?STORY_ID=1950.

10. China's space programme and Asia

Srikanth Kondapalli

10.1. Introduction

China's space programme in 2008/2009 exhibited several new features, while continuing to follow the guidelines enunciated at the turn of the century. While the recent emphasis on satellite, manned spacecraft and lunar mission launches continued to be reflected in the Chinese space programme, further modernisation of space technologies, research on new launchers, and "batch" production of these space systems was also pushed forward, testifying maturity in the domestic space industrial potential and capabilities. Among the significant achievements of China in this field in this year are stepped up satellite launches, the extra vehicular activity of Shenzhou 7 manned spacecraft in September 2008, the March 2009 realisation of the lunar mission launched in October 2007 and bidding for the international commercial launches.[753] Nevertheless, despite these successes, the failure to put into specified orbit a French-built Indonesian satellite on 31 August 2009 dents into these images of robust space programme of China.

Although quite advanced, China is not alone in the space faring efforts in Asia and beyond. Others, such as Japan, India, South and North Koreas, Iran, Israel, Taiwan, Malaysia, Singapore and Vietnam have expressed a keen interest in these efforts and some like Japan and India are in much advanced stages compared to China.[754] As a result, China is not yet a leader in Asia in several aspects of the space programme and needs the help of several countries including Russia, European States, the U.S. and some within Asia itself to occupy this position. As Japan, India and South Korea are also making efforts in this field, China is following several strategies to be on the upper hand in Asia. While the U.S. and the Soviet Union had conducted lunar missions in the 1960s, the three rising countries in Asia, Japan, China and India, are following a similar path for commercial, technological or even strategic/military reasons. Thus, Japan took the lead in this sector, followed by China and India. All three also have medium-term plans for unmanned/ manned missions to the Moon.

Commercial considerations are said to be the major triggers for these countries' space launches, given several surveys assessments that a 1 U.S. dollar investment could bring in profits of about 8–14 U.S. dollars to the society.[755] More than half of

China Aerospace Science and Technology Corporation's revenues came from the civilian sectors. In October 2007 and in April 2008, China established centres at Shanghai and Xian respectively for civilianising space technologies. By 2012, these centres are expected to produce nearly 3 billion U.S. dollars worth of goods. [756] However, the commercial satellite launch business is considered more seriously in China and India than in Japan.

Another consideration concern the military benefits, i.e. national security considerations in which China and India (and lately other countries as well) had shown much interest.[757] While the Chinese programme essentially started with military in the forefront (and still is), the Indian space programme (as well as most others in Asia), started with civilian and is now exhibiting strategic flavours.[758]

The other main feature – specific to the period under study – is the impact of the global financial crisis. 2008/2009 witnessed global financial meltdown triggered by the mortgage crisis, high energy prices and food stuffs and stock exchange decline. Reduction in domestic demand in the U.S., Europe and Japan affected the developing economies such as China and India as well. Yet, availability of credit as such had not affected China and India which witnessed higher allocations in the infrastructure projects as a part of their stimulus packages. Indirectly, this included the space programme as well where credit crunch was not the main factor for furthering space faring activities. While the per capita spending in China and India in the space programme is still far below the western allocations, we did witness the unveiling of new space programmes in these countries. In the case of China, sitting as it were on more than 2 trillion U.S. dollars in foreign exchange reserves, the global financial crisis is less of a negative influence. Indeed in 2008/2009, while China invested about 20 billion yuan to finance the first phase of the space programme, its "pulling effect" on the country's economic and technological sectors (specifically on electronics, raw material, component production, new materials, and new techniques) is estimated at ten times the investment. According to Zhang Jianqi, assistant director and chief engineer of the space programme and delegate of the National People's Congress:

> Our country's space program to take humans to space will not only not be affected by the international financial crisis, it will have a stimulus effect on promoting growth to better face the crisis . . . The space industry is also an industry that can raise the national cohesion of a country. It has an important effect on augmenting national pride and confidence . . . The spirit of taking humans to space can energize the strong faith of unity among hundreds of thousands of people to face a difficult time together.[759]

Thus, it could be argued that the financial crisis is not affecting the Chinese space programme, although the effort appears to be to convince the domestic audience on national unity and "skilful stage management". Selective release of information, hype created by Taikonauts tours of Hong Kong and Macau, etc still form part of the Chinese reality.[760]

10.2. Satellites

2008/2009 is a significant year for the Chinese space programme, specifically in the areas of satellite launches, manned spacecraft and lunar missions. China had stepped up satellite launches in the recent period according to the China Academy of Space Technology. While the number of launches in the 9th Five Year Plan was about 1.5 a year, China launched on an average about eight rockets in 2006–2008. In 2009 alone, China planned to launch as many as 15–16 satellites.[761] These figures are also reflected in the global percentages. In a survey of the expected global satellite launches in 2009–2013, while the U.S. is still the leader in this field, China is poised to acquire a share of about 8.3% of the global satellite launches, albeit mostly for domestic customers.[762] Significantly, for the first time, on 10 March 2009, China reportedly bagged a French offer to launch Eutelsat Communications in 2010, although it failed in putting into GEO a French-built Indonesian satellite in August 2009. China did launch satellites for external customers before, but this French offer opens up new international vistas for the Chinese space programme.

In this period, China launched a number of satellites of different varieties including communications, meteorological, Earth observation, etc. Two satellites were launched in November 2008.[763] In 2008 two meteorological satellites were launched– of the Feng Yun (FY) series. China had launched two-series of this class – five FY-2 series (FY-2A, B, C, D and E) of which three satellites (FY-2C, D and E) are operational. FY-2E was launched in December 2008 and will replace FY-2C.[764] In November 2008, China launched Chuangxin 1-02 and Shiyan Satellite 3 from Jiuquan by Long March 2D carriers for hydrological and meteorological data and data for disaster relief and experiments on new technologies respectively.[765] China also launched in November and December 2008 and in April 2009 Yao Gan-4, 5 and 6 surveillance satellites respectively, from Taiyuan by the Long March 4B launcher.[766] These all-weather military surveillance satellites, launched from 2006, have synthetic aperture radars.

In an effort to spruce-up, reorganise and reform the domestic communications industry, in April 2009, the China Aerospace Science and Technology Corporation acquired China Satellite Communications Corporation (which has nearly 966 million U.S. dollars in assets).[767] China's military satcoms are mainly through Zhongxing-20 satellites which support mobile communications. In addition, it was pointed out that China is also eyeing development of microsatellites for surveillance or data relay satellites etc to augment its military use of space.[768]

One of the main areas of focus in these launches is the quest for acquiring its own navigational system. As the Chinese interest in the European Galileo system waned, it intensified its own efforts in this field. By 2007, China launched four Beidou satellites as a part of an estimated 35 satellites in all to make up for the Compass global navigation system – with 30 to be in the medium Earth orbit and the rest in the geostationary orbit.[769] Ten of these satellites are expected to be launched by 2010 as part of the regional positioning system. The Beidou-series has ten services – five for civilian purposes and the other five for the military, with an accuracy of about 10 m. Xian Huaxun microelectronics company provides instruments for this series of satellites. The GPS Research Centre at Wuhan University is making efforts to integrate positioning signals from GPS, GLONASS and Compass systems.[770]

10.3. Launchers

One of the main space carriers of China is the Long March (LM) series of rockets. Seven derivatives were developed by China and launched from different locations. After the 1996 failure to launch a U.S. communications satellite, the LM series never failed in the 75 subsequent launches except on 31 August 2009 when the LM 3B failed to put the Indonesia Palapa D telecommunications satellite built by Thales Alenia Space of France into the designated orbit.[771]

Nevertheless, in December 2008, Great Wall Industry (the external commercial arm of the space industry) launched four LMs. In January–April 2009, eight launches were conducted successfully. Payloads included three communication satellites (one Simon Bolivar 1 for Venezuela), Earth observation spacecraft and Shenzhou 7.[772] Although one could argues that China still faces setbacks in developing a reliable new "heavier lift" launcher so far, certain experiments are being conducted in this direction, apart from the Kaitouzhe (KT) projects.[773]

10.4. Shenzhou 7 spacecraft

The highlight of this year's space programme in China was the successful launch of the manned spacecraft Shenzhou 7 in September 2008. National political considerations were apparent in the timing of the launch. According to the chief designer of the Chinese manned space programme Zhou Jianping, the Shenzhou 7 flight was shifted so that such an event could avoid "imping[ing] on the Olympics" that were held a month before.[774]

The stated goals of this mission are: "entry into orbit accurately, normal operation, complete success in the extravehicular activity, and returning safely and healthily."[775] This spacecraft was launched by a Long March 2F rocket, with a three-man mission. Among them, Taikonaut Zhai Zhigang performed a 20-minutes spacewalk. Other missions of the spacecraft included releasing a 40 kg microsatellite, manoeuvring to more than 100 miles away, data transmission experiments through the relay satellite Tianlian 1, etc.[776] These are expected to enhance the "covering power of China's middle- and low-orbit spacecraft in the future".[777]

After the successes of manned spacecraft launches, extra-vehicular space activity and aircraft rendezvous and docking techniques, China, according to Zhou Jianping, intends to launch "larger-scale and sustainable manned space flight activities". In future, Zhou suggested that China "can set up space laboratories and space stations, enabling people to fly in orbit for long periods. Spaceships can take people and cargo up to the space stations, [and] then return to the ground. In this way, we can achieve sustainable, relatively large-scale manned spaceflight activity". [778]

10.5. Lunar mission

So far nearly 70 Moon missions were conducted by different countries. China is the latest entrant into this field with its Chang'e programme. On 24 October 2007, Chang'e was launched into space. After 16 months, it hit the lunar surface in March 2009. China developed a new engine for this launch. Against the previous requirement of four-to-five after burns to reach distances of about 50,000 km, the propulsion system for this launch needed greater thrust of about nine-to-ten times so as to reach the target of about 380,000–400,000 km. To meet these requirements the Shanghai Institute of Space Propulsion developed a 490-newton orbital transfer engine.[779]

Soon after the first attempt to land on the Moon was successful, Ye Peijian, chief designer of Chang'e-1, identified a road map for further Moon missions in March 2009.[780] According to this plan, the Chang'e-2 mission is to be made by about 2011, Chang'e-3 mission of soft landing and probing Moon is likely to be realised in 2013, while Chang'e-4 is slated for launch in 2017.[781]

10.6. Space station

China also inched in this period towards having its own space station. Initiated in 2007, the space docking system of China is composed of an 8.5 ton Tiangong 1 module which is expected to perform, once launched in 2010–2011, long-term unattended operations in space.[782] One of the key techniques that China is currently engaged in is how to precisely control two high-speed spacecraft which meet and dock in space.

10.7. Impact on Asia

China is not the only country in Asia that has an active space programme. Japan and India are formidable players in this field that China has to consider. Besides, there are also other emerging space faring nations such as South Korea, Iran, North Korea and Israel. With such an array of new actors in the space field, China has to evolve policies either to cooperate, compete or marginalise some of these or their allies/friends such as the traditional actors (U.S./Europe) to retain its advantages and influence. China is in the process of implementing all these strategies in Asia and beyond in the recent period and is creating a niche for itself in the region. A brief review is made of the space-based events in the last one year in Asia to examine this aspect.

10.7.1. Japan

Undoubtedly, one of the more sophisticated space programmes in Asia rests with that of Japan. With the fast pace of developments in the Chinese space programme, Japan is a concerned country. Additionally, as the Chinese-led Six Party Talks on denuclearisation of the Korean Peninsula were not leading anywhere, North Korean nuclear and ballistic missile programmes are a major concern for Japan. Besides, North Korean launches in April 2009 posed direct challenges to Japan as the debris fell in its vicinity. Keeping these aspects in view, Japan adopted a "basic law on space" in May 2008. In August 2008, it set up a Space Development Strategy Headquarters in the cabinet. Japan decided on the new direction in its outer space policy in December 2008 for "wider use of satellites for national security purpose" and proposed to develop a midsize GX rocket, despite its rejection by the Strategic Headquarters for Space Development citing higher (triple) costs involved.[783] Japan's Aerospace Exploration Agency (JAXA) had unveiled a programme to land robots and human beings on the Moon by 2020 and 2030, respectively. In fact, Japan launched its lunar probe Kaguya a month before the Chinese attempt in October 2007. A year later, in October 2008, India also sent Chandrayaan 1.[784] The lunar missions of these three rising Asian countries indicate to the emerging scenarios in the region.

Japan has two launch vehicles – H 2A and M 5 for heavy and small payloads respectively. In January 2009, H 2A launched eight satellites. In addition, Japan intends to develop the H 2B launcher for commercial launches and for foreign customers.[785] Japan is also developing its Quasi-Zenith Satellite System with three satellites forming a regional constellation.

10.7.2. South Korea

With China and North Korea developing their space programme, industrialised South Korea would not like to be left alone in this field as the commercial angle of the space programme also becomes attractive. South Korea became the 13th country in the world in establishing a space centre in the southern island of Naro in June 2009 with Russian cooperation and in the words of the aerospace industry director Min Kyung-ju, his country has now taken "the first step on the path to making itself an aerospace powerhouse".[786] South Korea had launched, with foreign assistance, ten satellites, including Uribyol I, in 1992 and the indigenously built Arirang series from 1999.[787] South Korea proposed an ambitious space programme – including the Naro-1 launch on 30 July 2009, Naro-2 in April 2010, a Moon survey satellite launch in 2020 and a lunar landing explorer in 2025.[788] With about 419 million U.S. dollars, it developed Korea Space Launch Vehicle-1 (Naro-1) with Russian assistance. However, the 30 July 2009 launch (postponed to 19 August 2009 but finally suspended before launch), from the Naro Centre was unable to fulfil its aspirations.[789] Nevertheless, interestingly, the Korea Aerospace Research Institute, in its report in January 2009 expressed its willingness to conduct joint orbital experiments with Japan starting in 2012.[790]

10.7.3. North Korea

In April 2009, North Korea stated that it launched the communications satellite Kwangmyongsong 2 through the Unha 2 rocket from its Musudan-ri base.[791] Much controversy surrounded this launch as the U.S., Japan and South Korea expressed concerns on whether this is a ballistic missile launch or a civilian space launch. While Russia and China argued that this is meant for peaceful purposes, the U.S. and Japan argued that this launch violated the United Nations Security Council resolution 1718 of 2006.[792] In this regard, at least indirect support for North Korea is coming from its ally China in the space programme.

10.7.4. India

Currently, India is in the process of evolving a "Space Vision" until 2025 that charts out on effectively harnessing space for developmental activities. Priority areas include among others low cost access to space, development of new launchers and new generation of satellites.[793] As far as China is concerned, the Indian space programme is becoming a formidable one with at least the competitive dimension coming to the fore – in the short to medium term in terms of vying for commercial satellite launches worldwide – despite a MoU for space cooperation between the two countries that so far was largely been left ineffective.[794] Thus, of the 85 satellites launches worldwide in January–October 2008, India launched 12 satellites while China launched 10.

The period under study began with India launching 10 satellites in April 2008 with one rocket, indicating the dual-use capability of multiple independent launches. Indian successes in this field include the launch and recovery of the Space Capsule Recovery Experiment Module as well.[795] More importantly, ISRO launched its first spacecraft Chandrayaan 1 to the Moon in October 2008. Through the 11 instruments on-board, it had transmitted 70,000 images of the Earth's satellite. In addition, in August 2009, ISRO and NASA conducted probes on water ice. In August 2009 ISRO declared that it lost contact with the spacecraft, although it suggested that 100% of the technical jobs and 90–95% of the scientific experiments were completed.[796]

Besides, the Indian Gagan GPS system is expected to function before the Chinese system is in place. The Indian Regional Navigation Satellite Systems, approved in 2006 is to have about seven GEO satellites to form a regional navigation system by 2012 that is expected to provide accuracy of 20 m and signals transmitted in S-band frequency.[797]

In the commercial arena, Indian launches, although few in numbers, are significant in qualitative terms. Israel launched its TecSAR satellite with a synthetic aperture radar on board an Indian launch vehicle in January 2008.[798] In April 2008, ISRO launched the Cartosat 2A satellite for detailed mapping with less than 1 m resolution through the panchromatic camera. In April 2009, India launched an Israeli built Radar Imaging Satellite-2 (RISAT 2) with a synthetic aperture radar that is expected to enhance Indian surveillance capability over its neighbours, including China.[799]

10.7.5. Iran

Like South Korea, Iran is one of the latest entrants into the space field, but more successful as it is backed by the military infrastructure after the National Strategic Space Activities Programme was launched in 2006.[800] After the first launch in August 2008, Iran launched its lightweight telecommunications satellite Omid (Hope) with its launch vehicle Safir 2 from near Semnan city on 2 February 2009.[801] This satellite is equipped with two frequency bands and eight antennae and orbits the Earth 15 times a day. This is a significant launch for the Iranian quest to acquire strategic capabilities.[802] On this occasion, it was reported that Iran had launched a 12-year programme to send astronauts into space. [803]

10.8. Conclusions

Years of substantial work in space explorations had paid off well for China in 2008/2009. As outlined above, several significant space projects were launched and a number of initiatives were initiated that are expected to come to fruition in the short to medium terms. However, it could be argued that, as these are based on the programmes unveiled in 2000, incremental progress was achieved by China.

China's space launches have achieved notable successes in more than a decade, if we discount the failures related to communications satellite launches for Indonesia and Nigeria. Bidding for the French offers further elevates the Chinese prestige in space launches in the "mainstream" markets. However, one of the commercial constraints for Chinese launches is that these are launched from military installations.[804] Given the Chinese opaqueness, such launches may not see a major upsurge in the future.[805]

Overall, despite progress achieved, China (and India) is facing similar problems in the future prospects of the space programme. That is as commercial winds sweep their economies as a part of the globalisation, there is an emerging shortage of talented scientific personnel. China is trying to resolve this issue through the efforts of the party committees.[806] In September 2008, India's Defence Research & Development Organisation's Recruitment and Assessment Centre organised an international seminar inviting scientists and scholars from several countries to seek their advice on how to retain scientific talent in the organisation which had witnessed nearly 6–7% attrition annually.[807]

Another broad feature is that while Japan, South Korea and India have expressed cooperation in joint and multilateral fields in the space programme, China is carving out a lonely furrow or, alternatively, wishes to lead multilateral space activities. The Korea Aerospace Research Institute had expressed its opinion in a report in January 2009 that it would like to conduct joint orbital experiments with Japan starting in 2012. India had taken onboard satellites/systems from other countries such as the U.S. or from Europe. However, in this field, China's position appears to be *selective* in orientation. No doubt, there is also the Chinese sponsored Asia Pacific Space Cooperation Organisation, which attracted few members. It could then be argued that China's space activities are indirectly impacting on the other Asian space programmes, by way of alerting other countries such as Japan and India. As such the region reflects to a certain flux with the future holding for the emergence of different combinations.

Strategic/military aspects of the Chinese space programme have two inter-linked dimensions. The first one concerns the military applications of space asset and second one relates to China's recent spurt in the campaign against weaponisation of space after it conducted an anti-satellite (ASAT) test by DF-21/KT-1 missile to destroy its Feng Yun satellite in January 2007. Firstly, despite denials from the Chinese officials that its space programme (specifically Shenzhou 7) had no military applications,[808] it is clear from the extra vehicular activity and micro satellite projects that the space programme is of dual-use nature, not to mention the newest surveillance satellites that China had been launching, in addition to the military channels reserved in the Beidou navigation system.[809] Secondly, after the 2007 ASAT test, China is toying with a possible agreement at the global levels either in conjunction with Russia or preferably along with Russia and the U.S. on a treaty for peaceful uses of outer space.[810] In February 2008, for instance, China and Russia proposed a draft at the Geneva Conference on Disarmament (CD), drawing from a CD working paper of 2002, the 1967 Outer Space Treaty and the UN General Assembly 1993 resolution, aimed at peaceful uses of outer space.[811] In August 2009 Yang Jiechi, the foreign minister, addressing the CD, said:

> Outer space is now facing the looming danger of weaponisation ... Credible and effective multilateral measures must be taken to forestall the weaponisation and arms race in outer space .... Countries should neither develop missile defence systems that undermine global strategic stability nor deploy weapons in outer space.[812]

These renewed Chinese statements are coming out at a time when the Pentagon is preparing its Space Posture Review and China's space capabilities are being noted as in its annual reports submitted to the U.S. Congress. It is also possible that the Chinese efforts at making an arrangement with Russia (or possibly with the U.S.) is to marginalise Japan and India, specifically when both were attempting to make efforts in the military field (Japan had passed the basic law on space in May 2008 and India had set up the Integrated Space Cell in July 2008).

All in all, we see a China confident in its space programme, with a well laid out chart for the medium to long term projections and with the aim of becoming an undisputed leader in Asia and beyond in this field. Towards this goal, China had been concertedly modernising its space programme, partly with external help, and setting up space architectures to serve well its national interests.

---

[753] For recent studies on the subject see: Yuan, Jing-dong. "China's Ascendancy to a Space Power." China Brief 8.8 (2008); "The nature and extent of China's Space and Cyber Activities and their implications for U.S. Security." November 2008. U.S.–China Economic and Security Review Commission 2008 Report to Congress. Section 3. 15 Dec. 2009. http://www.uscc.gov: 156–181; Pollpeter, Kevin. Building for the Future: China's Progress in Space Technology during the Tenth 5–year Plan and the U.S. Response. Carlisle Barracks, PA: Strategic Studies Institute, 2008.

[754] Moss, Trefor. "The Asian Space Race." Jane's Defence Weekly 23 Oct. 2008.

[755] Interestingly, according to a North Korean study, 0.1 gram of helium-3 can contain the same amount of energy obtained from 4.3 tons of coal. This report suggested that Indian Chandrayaan 1 project is aimed at ultimately developing "Moon mine". See "For the Peaceful Use of Space." 9 Feb. 2009. Korean Central News Agency (KCNA) accessed from U.S. Foreign Broadcast Information Services available at: http://wnc.fedworld.gov; at NewsEdge Document Number: 200902091477.1_fbb906f03a190f4b File Number 985 Accession Number 275900500 [Hereafter FBIS].

[756] According to Han Liyan of School of Economics and Management of Beihang University, 80% of the 1000 types of new materials China developed came from space technology. Again, nearly 2000 space technology items were transferred to the civilian economy. See "Space Mission a Boost for Civil Industries." 26 Sept. 2008. Xinhua. Another report suggested making available space food, chocolates, ear plugs etc for the civilian market. See "China Turns Space Technology Into Civilian Products." 26 Sept. 2008. Xinhua. See also "Successful Manned Space Mission Triggers Booming Aerospace Economy." 21 Oct. 2005. Xinhua. FBIS.

[757] It is interesting to note the financial allocations to civilian space activities globally. In 2007, the U.S. spent 18.82 billion U.S. dollars, the European Union 3.57 billion U.S. dollars, China 2.5 billion U.S. dollars, Japan 1.9 billion U.S. dollars, France 1.8 billion U.S. dollars, Russia 1.34 billion U.S. dollars and India 0.84 billion U.S. dollars. See "Russia Sixth in Terms of Financing Civilian Space Programs." Agentstvo Voyennykh Novostey. 7 June 2008. FBIS.

[758] See Hiramatsu, Shigeo. "North Korea's Nuclear Missiles Are a 'Political Weapon'." Sankei Shimbun. 20 Apr. 2009. FBIS.

[759] Zhang identified output value generated specifically in the broadcast and communication satellites production, a new production chain of "Big Dipper" command system, etc. See Xuanliang, Li. "China's Space Program Said Unaffected by Financial Crisis." 5 Apr. 2009. Xinhua Domestic Service. FBIS.

[760] See Westlake, Michael. "China on Course in Space." Aerospace America (April 2009): 8–9 and 17.

[761] "China Expected to Launch 15 to 16 Satellites This Year." People's Daily 10 Mar. 2009. FBIS.

[762] While Thales Alenia Space is expected to garner 9.2% of the global share, European Satellite Nav about 4.1%, Loral Space Systems 5.3% and EADS Astrium 7.2% in the same period. See Edwards, John S. "Finding a Niche: Satellite Manufacturers See New Opportunities in Evolving Market." Aviation Week & Space Technology 26 Jan. 2009: 141–43.

[763] "China's Space Industry Takes Off." 5 Nov. 2008 Xinhua. FBIS.

[764] "FengYun-2E (FY-2E) Delivered in Orbit." 25 May 2009. China National Space Administration. 15 Dec. 2009. http://www.cnsa.gov.cn/n615709/n620682/n639462/168784.html.
[765] This was the 112th launch of the LM series. See "China Puts Two Satellites Into Orbit." 6 Nov. 2008. China National Space Administration. 15 Dec. 2009. http://www.cnsa.gov.cn/n615709/n620682/n639462/167771.html.
[766] "China Launches Yaogan V remote-sensing Satellite." 15 Dec. 2008. China National Space Administration. 15 Dec. 2009. http://www.cnsa.gov.cn/n615709/n620682/n639462/167774.html.
[767] Grevatt, Jon. "CASC Acquires China Satcom as Telecommunications Reform Gathers Momentum." Jane's Defence Industry 14 Apr. 2009.
[768] See Keymer, Eleanor. "Space-saving Devices: Militaries Look to Make the Most of Commercial Satellites." International Defence Review 7 Oct. 2008.
[769] See Caceres, Marco. "Navigation Satellites Fuel Payload Growth." Aerospace America (June 2009): 14–16.
[770] Butterworth-Hayes, Philip. "Satellite Navigation Newcomers: Cooperation or Competition?" Aerospace America (March 2009): 32–37.
[771] Indonesia spent nearly 500 million U.S. dollars on this satellite. See Clark, Stephen. "Chinese Rocket Launches Satellite to Wrong Orbit." 31 Aug. 2009. Spaceflightnow.com. 15 Dec. 2009.
[772] See Caceres, Marco. "GEO Comsats Up, Launch Programs Down." Aerospace America (April 2009): 20–22.
[773] For instance, it was reported that in December 2008, China conducted a hybrid rocket test to test an engine from Jiuquan range. It was reported that the rocket measured 3.417 m long and 0.22 m in diameter. See "China Conducts First Successful Test of Hybrid Rocket." 6 Dec. 2008. China National Space Administration. 15 Dec. 2009. http://www.cnsa.gov.cn/n615709/n620682/n639462/167773.html.
[774] One of the reasons for delay in this launch was the preparation for an appropriate space suit. See for the interview, Chunguang, Yu and Zhao Bo. "What Will the Flight of 'Shen VII Bring Us? – Chief Designer of China's Manned Space Flight Project Zhou Jianping explains 'Shen VII." Liberation Army Daily 8 Oct. 2008. On the issues related to the space suit see Gang, Liu and Zhao Bo. "Make Life Armor for Space Walk – Interview With Chen Shanguang, Commander-In-Chief and Chief Designer of Astronaut System." Liberation Army Daily 6 Oct. 2008. FBIS.
[775] "A Stirring Song of Victory in Space Flight – Warm Congratulations on the Complete Success of the Shenzhou 7 Manned Space Flight of Our Country." People's Daily Editorial 29 Sept. 2009. FBIS.
[776] Saichun, Zoo and Zhao Bo. "Shen Seven [Shenzhou Seven] Will Achieve Technological Breakthroughs in Four Respects." Liberation Army Daily 5 Nov. 2008; Jie, Tan and Chai Yongzhong. "Integrated Observation and Control Takes First Step Forward – An Interview with Observation and Control System Chief Designer Qian Weiping." Liberation Army Daily 5 Nov. 2008. FBIS.
[777] Chunguang, Yu and Zhao Bo. "Zhou Jianping Talks About Shenzhou VII." Liberation Army Daily 29 Sept. 2008. FBIS.
[778] See for the interview, Chunguang, Yu and Zhao Bo. "What Will the Flight of 'Shen VII Bring Us? – Chief Designer of China's Manned Space Flight Project Zhou Jianping explains 'Shen VII." Liberation Army Daily 8 Oct. 2008. FBIS.
[779] The average age of the members of the team which developed this engine is said to be in their thirties. See Fangjia, Hao, Ming, Ji and Gao Lu. "Speaking From the Heart: Chang'e-1 Satellite Engine Production Team – Every Success Signals an Even Bigger Challenge." 8 Nov. 2008. Xinhua Domestic Service. FBIS.
[780] "China's Lunar Probe Chang'e-1 Impacts Moon." 1 Mar. 2009. Xinhua. FBIS.
[781] "China To Land Probe on Moon at Latest in 2013, Chief Designer Says." 2 Mar. 2009. Xinhua. FBIS.
[782] "Unmanned Space Module To Be Launched in 2010, Await Space Docking." 28 Feb. 2009. Xinhua. FBIS; "China Strives for Completion of Manned Space Station by 2020." People's Daily 8 Dec. 2008. FBIS.

[783] See "Japan Urged To Stop Its Moves To Turn Itself Into Military Power." KCNA 8 Dec 2008. FBIS.
[784] Moss, Trefor. "The Asian Space Race." Jane's Defence Weekly 23 Oct. 2008.
[785] See Caceres, Marco. "GEO Comsats Up, Launch Programs Down." Aerospace America (April 2009): 20–22. H-2 is operated by Mitsubishi Heavy Industries.
[786] "Aerospace Industry." The Korea Herald Editorial 13 June 2009. FBIS.
[787] "Space Dream – Korea Takes First Step Toward Exploring Outer World." The Korea Times 12 June 2009. FBIS.
[788] Westlake, Michael. "Joining the Space Race, Carefully." Aerospace America (July–August 2009): 8–10.
[789] "Software Caused S. Korean Rocket Launch Problems." 20 Aug. 2009. AFP. 15 Dec. 2009. http://news.id.msn.com/regional/article.aspx?cp-documentid=3539634.
[790] "S. Korea Pushing For Joint Space Experiment With Japan." Yonhap 18 Jan. 2009. FBIS.
[791] The first Kwangmyongsong 1 was launched in August 1998. However, the U.S. suggested that this was a Taepodong 1 launch. See "DPRK Launches Rocket." 5 Apr. 2009. Xinhua. FBIS.
[792] "UNSC Presidential Statement Condemns DPRK's Rocket Launch." Yonhap 14 Apr. 2009. FBIS; Sue-young, Kim "China, Russia Lukewarm Over Sanctioning N. Korea". The Korea Times 29 Mar. 2009. FBIS.
[793] "Enhance Space Vision: Manmohan." The Hindu 26 Aug. 2009.
[794] A few writings which contend for an emerging arms race between the two are Day, Dwayne A. "The New Path to Space: India and China Enter the Game." 13 Oct. 2008. The Space Review. 15 Dec. 2009. http://www.thespacereview.com/article/1231/1; Singh, Madhur. "India Gains on China in Asia's Space Race." 21 Oct. 2008. Time. 15 Dec. 2009. http://www.time.com/time/world/article/0.8599,1852608.00.html.
[795] "India Confident of Manned Space Programme: ISRO." The Hindu 13 June 2009.
[796] See Subramanian, T. S. "ISRO Loses Radio Contact With Chandrayaan-1." The Hindu 30 Aug. 2009.
[797] See Caceres, Marco. "Navigation Satellites Fuel Payload Growth." Aerospace America (June 2009): 14–16.
[798] Reportedly no country possesses such a satellite. See Ben-David, Alon. "Israel Boosts Imagery Capability with SAR Satellite Launch." Jane's Defence Weekly 30 Jan. 2008.
[799] See Bedi, Rahul. "India Launches Israeli-built Spy Satellite." Jane's Defence Weekly 23 Apr. 2009.
[800] "Senior Space Science Expert Interview by Iran: 'Omid Satellite Was Designed, Put Into Operation in Two, Half Years." Iran Online 16 Apr. 2009. FBIS.
[801] Some suggest to the Chinese help in the Iranian rocket technology. See Westlake, Michael. "China on Course in Space." Aerospace America (April 2009): 8–9 and 17.
[802] Richardson, Doug. "Iran Satellite Launch Suggests Advances in Indigenous Missile Technology." Jane's Missiles & Rockets 5 Mar. 2009.
[803] "Iran Kicks off 12-Year Project To Send Astronaut Into Space." 12 Feb. 2009. Xinhua. FBIS. See also for a North Korean perspective critiquing U.S. position on Iran's launch, Hak-nam, Ri. "Progressive Mankind's Struggles To Build an Independent New World Cannot be Stopped." Rodong Sinmun 2 Mar. 2009. FBIS.
[804] See the interview with Chairman and CEO of Arianespace, Jean Yves Le Gall at Aerospace America (March 2009) pp. 12–15 on the difficulties in revising ITAR [U.S. law on International Traffic in Arms Regulations].
[805] Jones, Morris. "My Decade With Shenzhou." 3 Aug. 2009. Spacedaily.com. 15 Dec. 2009. http://www.spacedaily.com/reports/My_Decade_with_Shenzhou_999.html.
[806] Pengfei, Liang, Huaze, Li and Li Xiaomei. "Building 'National Team' in Space Field – Stories About Research Institute Under General Armament Department Using Scientific Development Concept to Guide Innovations in Its Scientific Research Work." Liberation Army Daily 2 Mar. 2009. FBIS.

[807] Bedi, Rahul. "Indian DRDO Strives to Stem Loss of Military Scientists." Jane's Defence Weekly 5 Sept. 2008.
[808] See the arguments of Cui Jijun director of the Jiuquan Satellite Launch Center: "Experts Dismiss Concerns Over China's Manned Space Program." 22 Sept. 2008. Xinhua. FBIS.
[809] See "China's Space Army Program vs. Taiwanese Satellite Targets." Kanwa News 20 Dec. 2008.
[810] See Westlake, Michael. "China on Course in Space." Aerospace America (April 2009): 8–9 and 17.
[811] Petrov, Vladimir. "China and Russia Outline Opposition to Space Weaponry." Jane's Defence Weekly 27 Feb.2008.
[812] Yang cited in "China Warns of 'Arms Race in Outer Space'." 12 Aug. 2009. AP. 15 Dec. 2009. http://www.google.com/hostednews/ap/article/ALeqM5gqYd7ilaVIa0UGe7wE5BsHn5CzrwD9A1A1N01.

PART 3

FACTS AND FIGURES

1. Chronology: July 2008–June 2009

Christophe Venet

1.1. Access to space

Europe	Other countries
LAUNCH LOG	
July 08	
07 Ariane 5ECA – ProtoStar 1 and Arabsat 4AR (C)	16 Zenit 3SL – EchoStar 11 (C) 22 Kosmos 3M – SAR Lupe 5 (I) 26 Soyuz 2.1b – Kosmos 2441 (I)
August 08	
14 Ariane 5ECA – Superbird 7 and AMC 21 (C)	03 Falcon 1 – Celestis 07, Trailblazer, PreSat and NanoSail D (failure due to collision between stages 1 and 2) (D) 17 Safir – Omid (failure of second stage) (D) 18 Proton M – Inmarsat 4F3 (C) 29 Dnepr 1 – Rapideye 1–5 (R)
September 08	
	06 Long March 2C – Huangjing 1A and Huangjing 1B (R) 06 Delta 2 7420 10C – GeoEye 1 (R) 10 Soyuz U – Progress ISS 30P (ISS) 19 Proton M – Nimiq 4 (C) 24 Zenit 3SL – Galaxy 19 (C) 25 Proton K – Kosmos 2442, Kosmos 2443 and Kosmos 2444 (N) 25 Long March 2F – Shenzhou 7 (MF) and Banxing (D) 28 Falcon 1 – Ratsat (D)
October 08	
	01 Dnepr 1 – THEOS (R) 12 Soyuz FG – Soyuz ISS 17S (ISS)

	19 Pegasus XL – IBEX (S)
	22 PSLC – Chandrayaan 1 and MIP (S)
	25 Long March 4B – Shijian 6–3A and 6–3B (S)
	25 Delta 2 7420 10C – Cosmo-Skymed 3 (R)
	25 Long March 3B – VeneSat 1 (C)
November 08	
	05 Long March 2D – Shyian 3 (R) and Chuangxin 1–02 (C)
	05 Proton M – Astra 1M (C)
	14 Soyuz U – Kosmos 2445 (I)
	14 Shuttle Endeavour – STS 126, MPLM 5 (ISS) and PSSC-Testbed (D)
	26 Soyuz U – Progress ISS 31P (ISS)
December 08	
20 Ariane 5ECA – Hot Bird 9 and Eutelsat W2M (C)	01 Long March 2D – Yaogan 4 (R)
	02 Molnyia M – Kosmos 2446 (EW)
	10 Proton M – Ciel 2 (C)
	15 Long March 4C – Yaogan 5 (R)
	23 Long March 3A – Fengyun 2E (M)
	25 Proton K – Kosmos 2447, Kosmos 2448 and Kosmos 2449 (N)
January 09	
	18 Delta 4 Heavy – NROL 26 (I)
	23 H2A – GOSAT and SpriteSat (R), PRISM, SDS 1, SOHLA 1, Kagayaki, KKS 1 and STARS 1 (D)
	30 Tsiklon 3 – Koronas Foton (S)
February 09	
12 Ariane 5ECA – Hot Bird 10 and NSS 9 (C), Spirale A and Spirale B (D)*	02 Safir – Omid (D)
	06 Delta 2 7320 10C – NOAA 19 (M)
	10 Soyuz U – Progress ISS 32P (ISS)
	11 Proton M – Express AM 44 and Express MD 1 (C)
	24 Taurus XL – OCO (failure: payload fairing did not separate) (S)
	26 Zenit 3SLB – Telstar 11 N (C)
	28 Proton K – Raduga 1 (C)*

March 09	
	07 Delta 2 7925 10L – Kepler (S)
	15 Shuttle Discovery – STS 119 and ISS 15A (ISS)
	17 Rokot KM – GOCE (S)
	24 Delta 2 7925 – Navstar 2 RM7 (N)
	26 Soyuz FG – Soyuz ISS 19 (ISS)
April 09	
	03 Proton M – Eutelsat W2A (C)
	04 Atlas 5 421 – WGS F2 (C)*
	05 Taepodong 2 – Kwangmyongsong (probable failure) (C)
	14 Long March 3C – Beidou 2 (N)
	20 PSLV C12 – RISAT 2 (R) and ANUSAT (D)
	20 Zenit 3SL – Sicral 1B (C)*
	22 Long March 2C – Yaogan 6 (R)
	29 Soyuz U – Kosmos 2450 (I)
May 09	
14 Ariane 5ECA – Herschel and Planck (S)	05 Delta 2 7929 10C – STSS-ATRR (D)*
	07 Soyuz U – Progress ISS 33P (ISS)
	11 Shuttle Atlantis – Hubble servicing (other)
	16 Proton M – Protostar 2 (C)
	19 Minotaur 1 – TacSat 3*, PharmaSat 1, HawkSat 1, PolySat CP 6 and AeroCube 3 (D)
	21 Soyuz 2.1a – Meridian 2 (C)*
	27 Soyuz FG – Soyuz ISS 19S (ISS)
June 09	
	18 Atlas 5 401 – LRO and LCROSS (S)
	21 Zenit 3SLB – Measat 3a (C)
	27 Delta 4M – GOES O (M)
	30 Proton M – Sirius FM5 (C)
TECHNOLOGY	
23 October 08 First successful test of Vega's Zefiro 9-A solid-fuel rocket motor	*29 September 08* First successful launch of SpaceX's Falcon 1
	20 November 08 First successful test of the Orion launch abort motors by NASA and ATK

	12 February 09 Mitsubishi Heavy Industries unveiled the design of the new H 2B rocket *April 2009* Roscosmos established the preliminary design of the next generation Rus-M launcher
BUSINESS	
25 March 09 ESA and CNES signed a 435 million euros contract on the use of the GSC for ESA activities *15 June 09* ESA and Arianespace signed a Frame Contract for the procurement of launch services	*31 October 08* NASA awarded 206 million U.S. dollars contract to United Space Alliance for next-generation services for its space exploration programme *23 December 08* NASA awarded ISS resupply contracts to Orbital Sciences (1,9 billion U.S. dollars) and SpaceX (1,6 billion U.S. dollars) *27 February 09* NASA awarded the contract for the Constellation Spacesuit for the Moon to Oceaneering International *April 09* Roscosmos selected a team composed of TsSKB Progress, Energya and Makeev to develop the future Rus-M launcher

C: Communications – D: Development – I: Intelligence – ISS: International Space Station – M: Meteorological – MF: Manned Flight – N: Navigation – R: Remote Sensing – S: Scientific – EW: Early Warning System – *: military

1.2. Space science and exploration

Europe	Other countries
EARTH SCIENCES	
February 09 ESA pre-selected three missions for the next Earth explorer mission (BIOMASS, PREMIER, CoReH2O)	*23 January 09* Launch of GOSAT (Greenhouse Gases Observing Satellites) (JAXA) Monitors the distribution of the density of carbon dioxide
17 March 09 Launch of GOCE (Global Ocean Circulation Explorer) (ESA) Studies the Earth's gravity field	*24 February 09* Failed launch of OCO (Orbiting Carbon Observatory) (NASA)
05 June 09 ESA extended Envisat mission	
ASTRONOMY	
10 February 09 ESA extended the Cluster mission until 31 December 2009 Studies the Sun-Earth connection	*19 October 08* Launch of IBEX (Interstellar Boundary Explorer) (NASA) Studies the global interaction between the solar wind and the interstellar medium
	30 January 09 Launch of Koronas Foton (Roscosmos) Investigates solar-terrestrial connections physics
14 May 09 Launch of Herschel and Planck (ESA) Investigate the origins of the Universe	*11–24 May 09* Hubble servicing mission (NASA)
26 June 09 End of ESA/NASA Ulysses mission (solar orbiter)	
EXPLORATION	
05 September 08 ESA's Rosetta spacecraft flew by asteroid Steins	*22 October 08* Launch of Chandraayan 1 and MIP (Moon Impact Probe) (ISRO) High resolution remote sensing of the lunar surface
10 February 09 ESA extended the missions Mars Express and Venus Express until 31 December 2009	*07 March 09* Launch of Kepler (NASA) Explores the structure and diversity of planetary systems
June 09 Inauguration of a life support pilot plant in Barcelona	*18 June 09* Launch of LRO (Lunar Reconnaissance Orbiter) and LCROSS (Lunar CRater Observation and Sensing Satellite) (NASA) Studies the Moon's surface, radiation levels and polar regions
30 June 08 Start of the extended mission for Cassini (ESA/NASA/ASI) Studies the Saturnian moons Titan and Enceladus	

MANNED SPACEFLIGHT	
29 September 08 ATV Jules Verne completed its 6-months successful mission with a controlled destructive re-entry *20 May 09* ESA selected six new astronauts *27 May–1 December 09* Frank de Winne mission to the ISS *11 October 09* ESA Astronaut Frank de Winne became the first European ISS Commander	*25–28 September 08* Chinese Shenzhou 7 mission featuring the first Chinese spacewalk *14–30 November 08* STS-126 mission (NASA) Delivery of the Multi-Purpose Logistic Module *15–28 March 09* STS-119 mission (NASA) Delivery of the Space Station Truss and the Photo Voltaic Module *31 March–14 July 09* Second phase of the Mars 500 experiment

1.3. Applications

Europe	Other countries
EARTH OBSERVATION	
August 08 29 Launch of the 5 Rapideye satellites (Rapideye AG, Germany) *September 08* Launch of the first GMES pre-operational services *October 08* 25 Launch of Cosmo-Skymed 3 (Italy)	*September 08* *06 Launch of Huangjing 1A and Huangjing 1B (China)* *06 Launch of GeoEye 1 (GeoEye, USA)* *October 08* 01 Launch of THEOS (Thailand) *November 08* 05 Launch of Shyian 3 (China) *December 08* 01 Launch of Yaogan 4 (China) 15 Launch of Yaogan 5 (China) 23 Launch of Fengyun 2E (M) (China) *January 09* 23 Launch of GOSAT and SpriteSat (Japan) *February 09* 06 Launch of NOAA 19 (M) (USA) *April 09* 20 Launch of RISAT 2 (India) 22 Launch of Yaogan 6 (China) *June 09* 27 Launch of GOES O (M) (USA)
INTELLIGENCE AND EARLY WARNING	
July 08 22 Launch of SAR Lupe 5 (Germany)	*July 08* 26 Launch of Kosmos 2441 (Russia) *November 08* 14 Launch of Kosmos 2445 (Russia) *December 08* 02 Launch of Kosmos 2446 (EW) (Russia) *January 09* *18* Launch of NROL 26 (USA)

	April 09 29 Launch of Kosmos 2450 (Russia)
NAVIGATION	
July 08 01 Launch of the procurement phase for the Full Operation Capability Galileo satellites	*September 08* 25 Launch of Kosmos 2442, Kosmos 2443 and Kosmos 2444 (Russia) *December 08* 25 Launch of Kosmos 2447, Kosmos 2448 and Kosmos 2449 (Russia) *March 09* 24 Launch of Navstar 2 RM7 (USA) *April 09* 14 Launch of Beidou 2 (China)
TELECOMMUNICATIONS/BROADCASTING	
	July 08 07 Launch of ProtoStar 1 (ProtoStar, USA) and Arabsat 4AR (Arabsat, Saudi Arabia) 16 Launch of EchoStar 11 (EchoStar, USA) *August 08* 14 Launch of Superbird 7 (SCC, Japan) and AMC 21 (SES Americom, USA) 18 Launch of Inmarsat 4F3 (Inmarsat, International) *September 08* 19 Launch of Nimiq 4 (Telesat Canada, Canada) 24 Launch of Galaxy 19 (Intelsat, USA) *October 08* 25 Launch of VeneSat 1 (Venezuela)
November 08 05 Launch of Astra 1M (SES Astra, Luxembourg)	*November 08* 05 Launch of Chuangxin 1–02 (China) *December 08* 10 Launch of Ciel 2 (Ciel satellite, Canada) 20 Hot Bird 9 and Eutelsat W2M (Eutelsat, International)
February 09 12 Launch of NSS 9 (SES New Skies, Netherlands)	*February 09* 11 Launch of Express AM 44 and Express MD 1 (RSCC, Russia)

	26 Launch of Telstar 11 N (Loral Skynet, USA) 28 Launch of Raduga 1* (Russia) 12 Launch of Hot Bird 10 (Eutelsat, International)
April 09 20 Launch of Sicral 1B* (Italy)	*April 09* 03 Launch of Eutelsat W2A (Eutelsat, International) 04 Launch of WGS F2* (USA) 05 Failed launch of Kwangmyongsong (North Korea) *May 09* 16 Launch of Protostar 2 (ProtoStar, USA) 21 Launch of Meridian 2* (Russia) *June 09* 21 Launch of Measat 3a (Measat Satellite Systems, Malaysia) 30 Launch of Sirius FM5 (Sirius Satellite Radio, USA)
TECHNOLOGY DEVELOPMENT	
February 09 12 Launch of Spirale A* and Spirale B* (DGA, France)	*August 08* 03 Failed launch of Celestis 07 (Space Services Inc., USA/Malaysia), Trailblazer (MDA, USA), PreSat and NanoSail D (NASA, USA) *September 08* 25 Launch of Banxing (China) *November 08* 14 Launch of PSSC–Testbed (USAF, USA) *January 09* 23 Launch of PRISM (University of Tokyo, Japan), SDS 1 (JAXA, Japan), SOHLA 1 (SOHLA, Japan), Kagayaki (Sorun Corporation, Japan), KKS 1 (Tokyo Metropolitan College of Industrial Technology, Japan) and STARS 1 (Kagawa University, Japan) *April 09* 20 Launch of ANUSAT (Anna University, India) *May 09* 05 Launch of STSS-ATRR* (MDA, USA)

	19 Launch of TacSat 3* (Air Force Research Laboratory, USA), PharmaSat 1 (NASA Ames, USA), HawkSat 1 (Hawk Institute for Space Sciences, USA), PolySat CP 6 (CalPoly State University, USA) and AeroCube 3 (Aerospace Corp., USA)
BUSINESS	
November 08 ESA signed the Small GEO Platform and the Small GEO Mission contracts with OHB-System and Hispasat respectively *15 June 2009* Contracts for the first constellation of Galileo satellites awarded to EADS Astrium and OHB-System	*16 December 08* NASA/NOAA selected Lockheed Martin for GOES-R contract worth 1 billion U.S. dollars

1.4. Policy and international cooperation

<table>
<tr><th colspan="2">GENERAL POLICY</th></tr>
<tr><td>03 June 08 Adoption of the French Space Operation Act

17 December 08 Official release of the Draft Code of Conduct for Outer Space Activities by the Council of the EU</td><td>06 March 09 South African National Space Policy officially launched

May 09 Australia announced the creation of a Space Policy Unit and the launch of a new Australian Space Research Programme

02 June 09 Japan released its Basic Space Plan</td></tr>
<tr><th colspan="2">GENERAL COOPERATION</th></tr>
<tr><td>07 July 08 Launch of the Network of European Regions Using Space Technologies (NEREUS)

13–14 October 08 10th European Interparliamentary Space Conference (EISC) in Prague

May 09 Framework agreement between CNES and Luxembourg</td><td>June 08 Declaration of intent signed by South Africa, Nigeria and Algeria to launch the African Resource Management and Environmental Constellation (ARMC) project

30 June–01 July 08 CNSA/NASA meeting to discuss U.S.–China space cooperation

November 08 Cooperation agreement between the Indonesian and the Ukrainian space agencies

16 December 08 The Asia-Pacific Space Cooperation Organisation (APSCO) started its operations

June 09 Russia signed agreements on space with Egypt and Nigeria

30 June 09 10th meeting of the Russian-Chinese Subcommittee on Space Cooperation</td></tr>
<tr><td colspan="2">September 08 French-Indian agreement on space cooperation

05–07 November 08 Launch of the United Nations Spatial Data Infrastructure (UNSDI) at the 9th meeting of the United Nations Geographic Information Working Group (UNGIWG)

19–20 November 08 5th plenary session of the Group on Earth Observation (GEO)

08 December 08 The International Mobile Satellite Organization (IMSO) signed a MoU with the International Association of Marine Aids to Navigation and Lighthouse Authorities (IALA), the International Chamber of Shipping (ICS) and the International Organization for Standardisation (ISO)

08–12 December 08 3rd meeting of the International Committee on GNSS (ICG)

22 December 08 Launch of three new space cooperation programmes between France and Brazil

27 May 09 IMSO signed a LRIT (Long Range Identification and Tracking) services agreement with the European Maritime Safety Agency (EMSA)

16 June 09 ASI and the Israeli Space Agency (ISA) signed a cooperation agreement

17 June 09 ASI and Roscosmos signed a letter of intent for cooperation</td></tr>
</table>

<table>
<tr><th colspan="2">SPACE SCIENCE</th></tr>
<tr><td colspan="2">July 08 International Moon Research Agreement signed by several countries (USA, France, Germany, UK, Japan, India, Canada, Italy, South Korea)

December 08 Agreement on space science between the DLR and the China Manned Space Engineering Office (CMSEO)

18 February 09 NASA and ESA decided to cooperate on a Jupiter mission and plan for another mission to visit Saturn's moons Titan and Enceladus

17 June 09 CNES and Roscosmos signed a cooperation agreement for the Phobos-Grunt mission

30 June 09 NASA and ESA established a Mars Exploration Joint Initiative</td></tr>
<tr><th colspan="2">APPLICATIONS</th></tr>
<tr><td>9 July 08 Adoption of the Regulation on the further implementation of the European satellite navigation programmes by the European Parliament

12 November 08 Adoption of the Communication on GMES "We care for a safer planet" by the European Commission

26 November 08 Adoption of the Council conclusions on GMES "Towards a GMES programme"

28 January 09 ESA and the EC signed an amendment to the EC/ESA agreement on GMES

14 May 09 The European Commission granted licenses to operate MSS in Europe to Inmarsat and Solaris Mobile

20 May 09 Proposal for a Regulation on GMES and its initial operations by the European Commission

June 09 Report on the Galileo programme by the European Court of Auditors

26 June 09 Adoption of the Report on the implementation of the GNSS programmes and on future challenges by the European Commission</td><td>July 08 Russia and India launched cooperation to develop next-generation navigation satellites

05 February 09 Russia adopted the Federal law on navigation activities

30 April 09 GAO Report on GPS

20 May 09 China and Brazil signed an agreement to provide EO data to African countries through a joint space programme</td></tr>
<tr><td colspan="2">02 July 08 EUMETSAT signed a cooperation agreement with the China Meteorological Administration (CMA)

02 July 08 EUMETSAT extended its data exchange agreement with NOAA

10 December 08 EUMETSAT endorsed the concept of cooperation with NOAA on a Joint Polar System</td></tr>
</table>

2. Country profiles

AUSTRIA

Population[813]	8.2 million
GDP[814]	276 billion euros
Responsibility[815]	The Austrian Space Programme is funded by the Federal Ministry for Transport, Innovation and Technology (BMVIT) and managed by the Agency for Aeronautics and Space (ALR) of the Austrian Research Promotion Agency (FFG).
Activities[816]	In addition to ESA programmes, Austria has a national programme, the Austrian Space Applications Programme (ASAP), with an additional budget line for GMES activities in 2008.
Budget	In 2008, 56.4 million euros (ESA 32.8 million, national 20.1 million, EUMETSAT 3.4 million).
Staff	ALR – 13
Direct employment in the space manufacturing industry[817]	301

BELGIUM

Population[818]	10.4 million
GDP[819]	344 billion euros
Responsibility[820]	The Belgian Federal Science Policy Office manages Belgian space activities and the Belgian participation in national and international programmes through its Department for space research and applications.
Activities[821]	In addition to ESA programmes (mainly telecommunications, Proba, Launchers, Prodex, ISS), there are bilateral cooperation projects with the U.S. on STEREO, France on COROT, SPOT and Pleiades and Argentina on the SAOCOM programme.
Budget	In 2008, 192.8 million euros (ESA 138.4 million, national 46.6 million, EUMETSAT 4.2 million).
Staff	Department for space research and applications – About 20
Direct employment in the space manufacturing industry[822]	1284

CZECH REPUBLIC

CZECH SPACE OFFICE

Population[823]	10.2 million
GDP[824]	130 billion euros
Responsibility[825]	The Ministry of Education, Youth and Sports supervises space activities and the cooperation with ESA. The Czech Space Office (CSO), a non-profit association, coordinates space activities.
Activities[826]	In addition to the ESA PECS programme, the Czech space activities focus on astronomy, magnetospheric, ionospheric and atmospheric research, microgravity research experiments, scientific instruments and micro-satellites.
Budget[827]	In 2008, 7.4 million euros (ESA programme PECS 2,4 million, EUMETSAT 0.94 million, other 4.1 million).
Staff	CSO – 17

DENMARK

DTU Space
Institut for Rumforskning og -teknologi

Population[828]	5.5 million
GDP[829]	225 billion euros
Responsibility[830]	The Ministry of Science, Technology and Innovation is responsible for the national space policy and space activities.
	The National Space Institute (DTU) at the Technical University of Denmark undertakes most space-related activities.
Activities	In addition to ESA programmes, bilateral cooperation is undertaken with the U.S. (HEAO-3), Sweden (Viking), Russia (Spectrum X-G) and France (Granat).
Budget	In 2008, 31.4 million euros (ESA 23.9 million, national 4.6 million, EUMETSAT 2.9 million).
Staff	DTU – About 110
Direct employment in the space manufacturing industry[831]	167

FINLAND

Population[832]	5.2 million
GDP[833]	181 billion euros
Responsibility[834]	The Ministry of Trade and Industry, the Funding Agency for Technology and Innovation (Tekes) and the Academy of Finland are funding space activities in Finland. The Finnish Space Committee consists of representatives of all stakeholders and coordinates all the activities. Tekes is the executive body for space activities and, together with the Academy of Finland for basic research, manages the Finnish participation within ESA programmes and other international projects.
Activities	In addition to ESA programmes, Finland has bilateral activities with the USA (TWINS, Mars Science Laboratory, Phoenix, ISS), France (Pleiades) Germany (TanDEM-X and TerraSAR-X) and Japan (ISS), as well as a national space technology programme.
Budget	In 2008, 55 million euros (ESA 16.4 million, national 36.4 million, EUMETSAT 1.4 million).
Staff	Within Tekes – Space Technology and Environmental Monitoring – 9
Direct employment in the space manufacturing industry[835]	153

FRANCE

Population[836]	64 million
GDP[837]	1902 billion euros
Responsibility	The Centre National d'Etudes Spatiales[838] (CNES) is responsible for the French space activities. It is under the shared responsibility of the Ministry of Education and Research and of the Ministry of Defense. The Office National d'Etudes et de Recherches Aérospatiales (ONERA)[839] is also responsible for space-related research.
Activities	In addition to ESA programmes, national civil and military programmes are undertaken (Pleiades, Syracuse, Helios, Essaim, e-Gorce, Athena-Fidus) as well as bilateral cooperation with the USA (Calipso, Jason 2 and 3) and India (Megha Tropiques, Saral, Oceansat 3, Altika-Argos).
Budget	In 2008, CNES had a budget of 1.731 billion euros (including ESA 685 million). ONERA had a budget of 202 million euros with space representing 13% of its activities. A contribution of around 24.6 million euros for EUMETSAT should be added.
Staff	CNES – 2376
Direct employment in the space manufacturing industry[840]	11,641

GERMANY

Population[841]	82.3 millions
GDP[842]	2458 billion euros
Responsibility[843]	The German Space Agency within the German Aerospace Center (DLR) is responsible for German space activities. It is under the responsibility of the Ministry of Economics and Technology.
Activities	In addition to ESA programmes, Germany has national civil and commercial programmes in the field of Earth observation (RapidEye, TerraSAR-X, TanDEM, EnMAP), smallsat platforms (AstroBus), manned space flight (ISS, microgravity experiments), launch services (Eurockot, OHB-Cosmos), associated ground systems, space technologies (such as intersatellite links). Germany is involved in bilateral cooperation with the U.S. (GRACE, Dawn, Sofia) and its military programmes include remote sensing satellites (SAR-Lupe radar satellites) and satcoms (Satcom BW).
Budget	In 2008, 895 million euros (ESA 533.4 million, national 328 million, EUMETSAT 33.6 million).
Staff	DLR for space activities – about 1600
Direct employment in the space manufacturing industry[844]	4962

GREECE

Population[845]	10.7 million
GDP[846]	231 billion euros
Responsibility[847]	The General Secretariat for Research and Technology (GSRT) of the Ministry of Development is responsible for Greek space activities.
Activities[848]	The Greek space activities cover mainly the fields of space physics, ionospheric physics, remote sensing and telecommunications.
Budget	In 2008, about 15 million euros (ESA 11.4 million, national 1.3 million, EUMETSAT 1.44 million).
Staff	For space <5

Magyar Űrkutatási Iroda

HUNGARY

Population[849]	9.9 millions
GDP[850]	101 billion euros
Responsibility[851]	The Hungarian Space Office, under the responsibility of the Ministry of Environment and Water, manages the Hungarian space activities.
Activities[852]	Participation in microgravity, Earth observation, life and material sciences and GSTP programmes of ESA.
Budget[853]	In 2008, 3.65 million euros (ESA PECS 2 million, national 0.65 million, EUMETSAT about 1 million).
Staff[854]	Total space activities – 250

ENTERPRISE IRELAND

IRELAND

Population[855]	4.2 million
GDP[856]	184 billion euros
Responsibility[857]	Enterprise Ireland, in association with the Office of Science and Technology of the Department of Enterprise, Trade and Employment, manages and coordinates space activities in Ireland.
Activities	Irish space activities are in the fields of software systems and services, precision mechanical components, advanced materials, electronics/microelectronics and telecommunications systems and service engineering.
Budget	In 2008, 16.4 million euros (ESA 13.3 million, national, national 1.4 million, EUMETSAT 1.68 million)
Staff	For space <5
Direct employment in the space manufacturing industry[858]	42

ITALY

Population[859]	58.3 millions
GDP[860]	1529 billion euros
Responsibility[861]	The Italian Space Agency (ASI), under the Ministry of University and Research, manages the Italian space activities.
Activities[862]	Italian civil space activities include both ESA programmes and a national civil programme according to the Activity Plan 2008–2010. Italy's national activities include small scientific missions (AGILE, PRISMA, MIOSAT), dual-use Earth observation satellites (Cosmo-Skymed) and military satcoms (Sicral), as well as commercial telecommunications and radar satellites. Italy has developed bilateral cooperation with France (Vega, Orfeo, Athena Fidus) and Argentina (SIASGE).
Budget	In 2008, 752.1 million (ESA 343 million, national 389.3 million, EUMETSAT 19.85 million).
Staff	ASI – About 200
Direct employment in the space manufacturing industry[863]	3985

LUXEMBOURG

Population[864]	491,755
GDP[865]	37.4 billion euros
Responsibility[866]	Luxinnovation, the National Agency for Innovation and Research, under the responsibility of the Ministry of Culture, Higher Education and Research, coordinates space activities in Luxembourg.
Activities	Luxembourg focuses mainly on telecommunications with a major player in the field, SES Astra.
Budget	In 2008, 12,1 million euros (ESA 11.1 million, national 0.7 million, EUMETSAT 0.33 million).
Staff	Luxinnovation space <5
Direct employment in the space manufacturing industry[867]	27

NETHERLANDS

Population[868]	16.7 millions
GDP[869]	580 billion euros
Responsibility	The Netherlands Agency for Aerospace Programmes (NIVR),[870] under the Ministry of Economic Affairs, is responsible for the industrial space activities, while the Institute for Space Research (SRON)[871] manages research activities (To be replaced by the Netherlands Space Office from 1 September 2009).
Activities	The space research fields include astrophysics, astronomy, microgravity and Earth observation.
Budget	In 2008, 146.4 million euros (ESA 98 million, national 41.4 million, EUMETSAT 7 million).
Staff	NSO – 25; SRON – 142
Direct employment in the space manufacturing industry[872]	460

NORWAY

Population[873]	4.6 millions
GDP[874]	288 billion euros
Responsibility[875]	The Norwegian Space Centre (NSC), under the Ministry of Trade and Industry, manages Norwegian space activities.
Activities	In addition to ESA programmes (in particular Earth observation, telecommunications, and launchers), Norway has national support programmes and commercial activities (Telenor). Moreover, Norway operates the Andøya rocket range and the Svalbard ground station. Norway has also a bilateral agreement with Canada on the use of Radarsat 2 data.
Budget	In 2008, 60 million euros (ESA 43.9 million, national 13 million, EUMETSAT 3.1 million).
Staff	NSC – 31
Direct employment in the space manufacturing industry[876]	254

POLAND

Population[877]	38.4 millions
GDP[878]	326 billion euros
Responsibility	Polish space activities are under the joint responsibility of the Ministry for Scientific Research and Information Technology and the Ministry of Economic Affairs and Labour. Within the Academy of Sciences, the Space Research Centre[879] coordinates space activities and hosts the Polish space office (PSO).[880]
Activities	Polish space activities are mainly in the fields of space science, navigation and remote sensing applications. Poland is participating in several ESA missions such as Herschel and Planck, Rosetta, ExoMars, Beppi Columbo.
Budget	In 2008, around 3 million euros (ESA PECS 1.2 million, national 2 million).
Staff	n.a.

PORTUGAL

Population[881]	10.6 millions
GDP[882]	154 billion euros
Responsibility	Portuguese space activities are coordinated by a Portuguese Space Office within the Foundation for Science and Technology (FCT)[883] depending from the Ministry of Science, Technology and Higher Education (GRICES).
Activities	Mainly participation in ESA programmes (telecommunications systems, technology developments, Earth observation, exploration, tracking station of Santa Maria).
Budget	In 2008, 21.4 million euros (ESA 16.6 million, national 2.8 million, EUMETSAT 2 million).
Staff	n.a.
Direct employment in the space manufacturing industry[884]	109

ROMANIA

Population[885]	22.2 millions
GDP[886]	132 billion euros
Responsibility[887]	The Romanian Space Agency (ROSA), under the responsibility of the Ministry of Education and Research, is managing the Romanian space activities.
Activities	ESA PECS and national activities defined by the National Plan for R&D and Innovation are covering the fields of space science (space physics and astronomy), space systems (construction of nanosatellites and microgravity experiments) and space applications (telemedicine, Earth observation and navigation space-based services).
Budget	In 2008, about 14 million euros for the programme "Space and Security" within the National Plan for R&D and Innovation 2007–2013, and 2 million euros for ESA PECS activities.
Staff	ROSA – about 40

SPAIN

Population[888]	43.8 millions
GDP[889]	1061 billion euros
Responsibility[890]	The Centre for the Development of Industrial Technology (CDTI), under the Ministry of Industry, Commerce and Tourism, is funding and coordinating Spanish space activities.
Activities	In addition to ESA programmes, Spain has a national space programme including governmental and commercial programmes, especially in civil and military telecommunications (Hispasat, Spainsat) and Earth observation (SEOSAT/INGENIO, SEOSAR/PAZ, INTAuSAT 1) and is involved in bilateral cooperation projects, like with France in military observation systems (Helios). Moreover, Spain manages national, ESA and NASA ground facilities.
Budget	In 2008, 225 million euros (ESA 152.8 million, national 60.8 million, EUMETSAT 11.4 million).
Staff	CDTI – around 150
Direct employment in the space manufacturing industry[891]	2095

SWEDEN

RYMDSTYRELSEN
Swedish National Space Board

Population[892]	9.0 millions
GDP[893]	330 billion euros
Responsibility[894]	The Swedish National Space Board (SNSB), under the Ministry of Industry, Employment and Communication, is responsible for space activities in Sweden. Basic research is funded via the Ministry of Education and Research.
Activities	In addition to ESA programmes, Sweden has national programmes (subsystems, satellites and sounding rockets), and bilateral cooperation mainly with France (Proteus, Pleiades, Spot, Vulcain) and Germany (Texus, Rexus, Bexus). Both countries are partners of the technology demonstration project Prisma. Sweden also operates the test range of Kiruna and a satellite ground station at Esrange.
Budget	In 2008, 92.8 million euros (ESA 54.6 million, national 34.2 million, EUMETSAT 4 million).
Staff	17
Direct employment in the space manufacturing industry[895]	641

SWITZERLAND

Schweizerische Eidgenossenschaft
Confédération suisse
Confederazione Svizzera
Confederaziun svizra

Population[896]	7.6 millions
GDP[897]	322 billion euros
Responsibility[898]	The Space affairs division (or Swiss Space Office) of the State Secretariat for Education and Research of the Federal Department of Home Affairs is responsible for Swiss space activities and cooperates closely with the Swiss Department of Foreign Affairs on that topic. The Federal Commission for space affairs (CFAS) is preparing a new Swiss space policy. The Interdepartmental coordination committee for space (IKAR) is responsible for the coordination of the activities.
Activities	Most of the Swiss activities are undertaken within ESA programmes (space science, human spaceflight, launchers, Earth observation, Prodex, navigation).
Budget	In 2008, 99.3 million euros (ESA 87.1 million, national 7.4 million, EUMETSAT 4.8 million).
Staff	n.a.
Direct employment in the space manufacturing industry[899]	580

UNITED KINGDOM

BNSC

Population[900]	61.1 millions	
GDP[901]	1947 billion euros	
Responsibility[902]	The British National Space Centre (BNSC) coordinates UK civil space policy and programmes. It is a partnership of 7 UK government departments and 2 research councils and is hosted by the Department for Innovation, Universities and Skills (DIUS).	
Activities	In addition to ESA programmes, the main fields of activities are space science, Earth observation systems (Topsat), military and commercial communications systems and microsatellites.	
Budget	In 2008, 341.4 million euros (ESA 264.9 million, national 50.4 million, EUMETSAT 26.1 million).	
Staff	BNSC headquarters – 45	
Direct employment in the space manufacturing industry[903]		3437

European Space Agency

Responsibility[904]

The European Space Agency is an inter-governmental organisation with the mission to provide and promote, for exclusively peaceful purposes, the exploitation of space science, research and technology, as well as space applications.
ESA achieves this through:

- Space activities and programmes
- A long-term space policy defined with the EC
- A specific industrial policy
- Coordinating European with national space programmes

Activities

ESA activities are divided into mandatory programmes (mainly scientific programmes) and optional programmes including:

- Telecommunications
- Launchers
- Human Spaceflight, Microgravity and Exploration
- Navigation

Budget

In 2008, 3028.3 million euros (including 2421.73 million from Member States' contributions).

2008 Contributions from ESA Member States and Canada

	Contribution (in thousand euros)	Percentage
France	556.4	23.0%
Germany	533.4	22.0%
Italy	343	14.2%
United Kingdom	264.9	10.9%
Spain	152.8	6.3%
Belgium	138.4	5.7%
Netherlands	98	4.0%
Switzerland	87.1	3.6%
Sweden	54.6	2.3%
Norway	43.9	1.8%
Austria	32.8	1.4%
Denmark	23.9	1.0%
Canada	22.3	0.9%
Portugal	16.6	0.7%
Finland	16.4	0.7%
Ireland	13.3	0.5%
Greece	11.4	0.5%
Luxembourg	11.1	0.5%
Czech Republic	1.43	0.1%
Total	2421.73	100.0%

Staff

Over 2000

European Commission

Responsibilities[905]	The European Commission responsibilities include: • Defining the priorities and requirements for space based systems at the service of the EU's main objectives and policies and citizens' needs • Aggregating the political will and user demand in support of these • Ensuring the availability and continuity of services supporting EU policies • Ensuring integration of space-based systems with related ground and in-situ systems to promote the development of user-driven application services supporting EU policies • Creating an optimum regulatory environment to facilitate innovation • Promoting coordination of the European position in international cooperation
Activities	Space-related activities in the Commission take place in different Directorates. The Directorate GMES, Security and Defence of the Enterprise and Industry Directorate General is responsible for the coordination of the European Commission's space policy and the GMES activities. The Directorate General for Research is responsible for space-related research activities. The Directorate General for Energy and Transport is responsible for most Galileo-related activities.
Budget	The Framework Programme 7 has a budget dedicated to the Theme "Space" of 1430 million euros[906] over the period 2007–2013. This budget covers mainly GMES-related activities.
	The budgetary resources for EGNOS and Galileo were set at 3.4 million euros for the period from 1 January 2007 to 31 December 2013. The Galileo programme also benefits from additional funding from the Trans-European Transports Networks (TEN-T) programme.

EUMETSAT

Responsibility[907]	The European Organisation for the Exploitation of Meteorological Satellites (EUMETSAT), founded in 1986, is an inter-governmental organisation which operates a fleet of meteorological satellites and their related ground systems.
Activities	EUMETSAT's main purpose is to continuously deliver weather and climate-related satellite data, images and products. This information is supplied to the National Meteorological Services of the organisation's 21 Member and 9 Cooperating States in Europe, as well as other users world-wide. Through EUMETCast (EUMETSAT's Broadcast System for Environmental Data), EUMETSAT disseminates data and products to a wide user community. In addition, EUMETSAT provides training to help users exploit satellite data.
Budget	The Member States' contributions are based on a scale which is proportional to the gross national income of the individual Member States. In addition, each of the Cooperating States contributes 50% of the full membership fee.

2008 Contributions of EUMETSAT's Member States (in %)

	Percentage	Contribution (in million euros)
Germany	21.4%	33.6
United Kingdom	16.6%	26.14
France	15.7%	24.65
Italy	12.6%	19.85
Spain	7.3%	11.43
Netherlands	4.4%	6.94
Switzerland	3.1%	4.8
Belgium	2.7%	4.23
Sweden	2.6%	4.06
Austria	2.2%	3.46
Norway	2.0%	3.1
Turkey	1.9%	3.02
Denmark	1.8%	2.86
Greece	1.4%	2.27
Finland	1.4%	2.22
Portugal	1.2%	1.98
Ireland	1.0%	1.68
Slovak Republic	0.2%	0.41
Slovenia	0.2%	0.36
Croatia	0.2%	0.35
Luxembourg	0.2%	0.33
Total	100.0%	157.76

Staff[908]	223 as of 31 Dec. 2008

CANADA

Population[909]	33.5 millions
GDP[910]	1602 billion Canadian dollars (1137 million euros)
Responsibility[911]	The Canadian Space Agency (CSA) manages Canadian space activities. It operates like a government department and reports to the Parliament through the Ministry of Industry.
Activities	The CSA activities include: • Four key programmes: Earth observation (Radarsat), Space Science and Exploration (MOST, ePOP on Cassiope, NEOSSat), Satellite Communications (Anik), and Space Awareness and Learning. • Manned spaceflight (Canada has an astronaut corps) • Cooperation on a variety of projects with NASA (ISS, CloudSat, Dextre robotic arm, Phoenix, JWST, APXS) • Participation in ESA programmes as a Cooperating State Canada also develops – and cooperates with the U.S. on – military space capabilities.
Budget[912]	In 2008/2009, CSA had a budget of 382 million Canadian dollars, i.e. about 248 million euros (including 22.3 million euros for ESA)
Staff	CSA – 575
Employment in the space sector[913]	6742

CHINA

Population[914]	1338.6 millions
GDP[915]	30.067 billion yuan/renminbi (3124 billion euros)
Responsibility[916]	The CNSA (Chinese National Space Administration) coordinates the civilian space programmes and the cooperation with foreign space agencies. Until March 2008, CNSA was under the responsibility of the COSTIND (Commission on Science, Technology and Industry for National Defence).
	The GAD (General Armament Department), under the auspices of the Central Military Commission (CMC), runs the military space programmes, the launch infrastructure and manned spaceflight activities. The Chinese Academy of Science (CAS) is responsible for space research and the elaboration of the National Programmes.
Activities	The Chinese activities include: • Satellites for science and technology demonstration (Shi Jian), Earth observation (Zi Yuan), navigation (Beidou), meteorological (Feng Yun), telecommunications (DFH) as well as recoverable capsules. • Lunar orbiter (Chang'e 1) • Commercial telecommunications satellites (sold for instance to Nigeria, Venezuela and Pakistan) • Ground facilities for satellites • Launch services, launchers (Long March) and launch facilities • Manned spaceflight • Bilateral cooperation with Brazil on CBERS satellites, and with ESA on the scientific mission Double Star and on the Dragon programme in applications. China is hosting the Asia-Pacific Space Cooperation organisation (APSCO) Headquarters.
Staff	n.a.
Budget	Estimated between 1.5 and 2.5 billion U.S. dollars (between 1 and 2 billion euros).

INDIA

Population[917]	1166 millions
GDP[918]	52,655 billion rupees (858 billion euros)
Responsibility[919]	The Space Commission defines Indian space policy and the Department of Space is responsible for India's space activities. The Indian Space Research Organisation (ISRO) implements the space programmes.
Activities	The Indian activities include: • Remote sensing satellites (IRS) and multi-purpose satellites (INSAT) with telecommunications and meteorological functions • Dual-use satellites (CartoSat) • Launch vehicles (PSLV and GSLV, RLV-TSTO reusable launcher programme) and services • Sounding Rockets • Associated ground systems • Cooperation mainly with Bulgaria, ESA and NASA on Chandrayaan-1, with CNES on Megha-Tropiques and Oceansat-3, with ASI on Oceasat-2, with Russia on Coronas-Foton, with Israel on TAUVEX and with Canada on UVIT • Small satellites (IMS family) • Commercial activities, through the commercial arm of ISRO, Antrix (Contracts for W2 A and Hylas) • Military satellites (Launch of the SAR satellite)
Staff	n.a.
Budget	In 2009/2010, about 45.9 billion rupees (about 656 million euros).

JAPAN

Population[920]	127 millions
GDP[921]	507,614 billion yen (3496 billion euros)
Responsibility[922]	The Japanese Aerospace Exploration Agency (JAXA), under the responsibility of the Ministry of Education, Culture, Sports, Science and Technology (MEXT), manages Japanese space activities.
Activities	The Japanese activities include: • Scientific missions (Kaguya, Solar-B, Planet-C, Astro, Hayabusa, Astro-H) • ISS and manned spaceflight (Kibo module, HTV) • Earth observation (GOSAT, GPM, GCOM) • Telecommunications (ETS-VIII, Winds, QZSS) • Navigation (QZSS) • Launch services (H2 A and M5) • New launchers (Galaxy Express) • Cooperation with ESA on Bepi-Colombo, EarthCare and ALOS and with NASA on AQUA, TRMM, Geotail, GPM
Budget	In 2008, 2 billion euros including 1.1 billion euros for JAXA
Staff	JAXA – about 1650

RUSSIA

ROSCOSMOS

Population[923]	140 millions
GDP[924]	41,668 billion rubles (1189 billion euros)
Responsibility[925]	Russia's Federal Space Agency, Roscosmos, under the direct responsibility of the Government, manages the Russian civil space activities, while the Military Space Forces (VKS), under the Ministry of Defence, manage the military space programmes.
Activities	Russian civil activities defined in the Federal Space Programme for 2006–2015 include: • Satellites for science (Kompass), remote sensing (Resurs), communications • Launchers (Soyuz, Rockot, Cosmos, Start, Volna, Shtil, and Cyclone, Zenit, as well as Dnepr with Ukraine, and Angara and Rus M in development) • Cooperation with international partners in commercial ventures to commercialise launch services (ILS, Starsem, Rockot, Cosmos, Sea Launch) • Manned spaceflight (ISS) in cooperation with NASA and ESA including commercial activities (space tourists) • Associated ground facilities Military activities include dual-use navigation systems (Glonass), surveillance, early-warning, ELINT, and communications missions.
Staff	n.a.
Budget	In 2008, around 1 billion euros, and 305 billion rubles (about 7.3 billion euros) for the Federal Space Programme's budget for 2006–2015.

UKRAINE

Population[926]	45.7 millions
GDP[927]	950 billion hryvnia (127 billion euros)
Responsibility[928]	The National Space Agency of Ukraine (NSAU) is responsible for space activities in Ukraine, as well as about 30 design offices, research institutes and enterprises that represent most of the Ukrainian space sector. The National Academy of Sciences of Ukraine supervises space research in its institutes and has joint institutions with NSAU.

Activities	Ukraine has a National Space Programme that includes:
	• Scientific research (astrophysics and astronomy, ionosphere and magnetosphere research, microgravity and life sciences)
	• Launch vehicles and launch services
	  ◦ Operational launch vehicles (Zenit, Cyclone, Dnepr)
	  ◦ Launch vehicles under development (Cyclone 4, Dnepr M, Mayak)
	• Remote sensing satellites (Okean and Sich series), telecommunications satellites (Lybid), automatic multi-purpose space platforms,
	• Rocket and spacecraft engines, as well as advanced materials and technologies
	• Ground facilities (National Space Facilities Control and Test Centre in Yevpatoriya)
	• Cooperation with Brazil for the utilisation of the Alcantara launch base

Budget	In 2008, 44.7 million euros
Staff	NSAU HQ – 115

USA

Population[929]	307.2 millions
GDP[930]	14,264 billion U.S. dollars (10,127 billion euros)
Responsibility[931]	The National Aeronautics and Space Administration (NASA) is responsible for most U.S. civil space programmes. The National Oceanic and Atmospheric Agency (NOAA), under the Department of Commerce, manages meteorological and oceanographic programmes.
	The Department of Defense (DoD), and in particular the Air Force, manages most of the military space programmes. The National Reconnaissance Office (NRO) is responsible for the intelligence programmes.
Activities	American civil space activities include: • Manned space flight (ISS) • Science (including Hubble, Stereo, THEMIS, Dawn, GLAST, Kepler, SDO) • Exploration (New Horizons, Phoenix with other international partners, LCROSS) • Earth observation (CloudSat, OCO, AIM, Aqua, Terra, Aura, Calipso, CloudSta, Jason 2 with France) and meteorological satellites (GOES) • Commercial telecommunications satellites • Launchers and launch services (Shuttle, Pegasus, Minotaur, Falcon, Ares, Taurus) • Associated ground systems Military systems includes launchers (Atlas, Delta), as well as dual-use navigation (GPS), meteorological (DMSP), surveillance, intelligence, communications, early-warning, and technology demonstration systems.
Staff	n.a.
Budget	In 2008, about 66 billion U.S. dollars (about 47 billion euros) including 17.3 billion U.S. dollars for NASA (12.2 billion euros).

[813] CIA World Factbook. 2 Nov. 2009. https://www.cia.gov/library/publications/the-world-factbook/geos/au.html.
[814] "Gross Domestic Product at market prices, 2008." 25 Nov. 2009. Eurostat. 10 Dec. 2009. http://epp.Eurostat.ec.Eurospa.eu/portal/page/portal/statistics/search_database.
[815] FFG website. 10 Dec. 2009. www.ffg.at/content.php?cid=232.
[816] FFG website. 10 Dec. 2009. www.ffg.at/content.php?cid=232.
[817] ASD Eurospace. "Facts and Figures. The European Space Industry in 2008." 13th edition, June 2009. Paris: ASD Eurospace.
[818] CIA World Factbook. 2 Nov. 2009. https://www.cia.gov/library/publications/the-world-factbook/geos/be.html.
[819] "Gross Domestic Product at market prices, 2008." 25 Nov. 2009. Eurostat. 10 Dec. 2009. http://epp.Eurostat.ec.Eurospa.eu/portal/page/portal/statistics/search_database.
[820] Belgian Federal Science Policy Office website. 10 Dec. 2009. www.belspo.be/belspo/res/rech/spatres_en.stm.
[821] ESD Partners. European Space Directory 2009. 24th Edition. Paris: ESD Partners Publication, 2009: 79.
[822] ASD Eurospace. "Facts and Figures. The European Space Industry in 2008." 13th edition, June 2009. Paris: ASD Eurospace.
[823] CIA World Factbook. 2 Nov. 2009. https://www.cia.gov/library/publications/the-world-factbook/geos/e.html.
[824] "Gross Domestic Product at market prices, 2008." 25 Nov. 2009. Eurostat. 10 Dec. 2009. http://epp.Eurostat.ec.Eurospa.eu/portal/page/portal/statistics/search_database.
[825] Czech Space Office website.. www.czechspace.cz.
[826] Ibid.
[827] Ibid.
[828] CIA World Factbook. 2 Nov. 2009. https://www.cia.gov/library/publications/the-world-factbook/geos/da.html.
[829] "Gross Domestic Product at market prices, 2008." 25 Nov. 2009. Eurostat. 10 Dec. 2009. http://epp.Eurostat.ec.Eurospa.eu/portal/page/portal/statistics/search_database.
[830] National Space Institute website. 10 Dec. 2009. http://www.space.dtu.dk/English.aspx.
[831] ASD Eurospace. "Facts and Figures. The European Space Industry in 2008." 13th edition, June 2009. Paris: ASD Eurospace.
[832] CIA World Factbook. 2 Nov. 2009. https://www.cia.gov/library/publications/the-world-factbook/geos/fi.html.
[833] "Gross Domestic Product at market prices, 2008." 25 Nov. 2009. Eurostat. 10 Dec. 2009. http://epp.Eurostat.ec.Eurospa.eu/portal/page/portal/statistics/search_database.
[834] ESD Partners. European Space Directory 2009. 24th Edition. Paris: ESD Partners Publication, 2009: 83.
[835] ASD Eurospace. "Facts and Figures. The European Space Industry in 2008." 13th edition, June 2009. Paris: ASD Eurospace.
[836] CIA World Factbook. 2 Nov. 2009. https://www.cia.gov/library/publications/the-world-factbook/geos/fr.html.
[837] "Gross Domestic Product at market prices, 2008." 25 Nov. 2009. Eurostat. 10 Dec. 2009. http://epp.Eurostat.ec.Eurospa.eu/portal/page/portal/statistics/search_database.
[838] CNES website. 11 Dec. 2009. www.cnes.fr.
[839] ONERA website. 11 Dec. 2009. www.onera.fr.
[840] ASD Eurospace. "Facts and Figures. The European Space Industry in 2008." 13th edition, June 2009. Paris: ASD Eurospace.
[841] CIA World Factbook. 2 Nov. 2009. https://www.cia.gov/library/publications/the-world-factbook/geos/gm.html.
[842] "Gross Domestic Product at market prices, 2008." 25 Nov. 2009. Eurostat. 10 Dec. 2009. http://epp.Eurostat.ec.Eurospa.eu/portal/page/portal/statistics/search_database.

[843] DLR website. 11 Dec. 2009. www.dlr.de.
[844] ASD Eurospace. "Facts and Figures. The European Space Industry in 2008." 13th edition, June 2009. Paris: ASD Eurospace.
[845] CIA World Factbook. 2 Nov. 2009. https://www.cia.gov/library/publications/the-world-factbook/geos/gr.html.
[846] "Gross Domestic Product at market prices, 2008." 25 Nov. 2009. Eurostat. 10 Dec. 2009. http://epp.Eurostat.ec.Eurospa.eu/portal/page/portal/statistics/search_database.
[847] GSRT website. 11 Dec. 2009. www.gsrt.gr.
[848] Ibid.
[849] CIA World Factbook. 2 Nov. 2009. https://www.cia.gov/library/publications/the-world-factbook/geos/hu.html.
[850] "Gross Domestic Product at market prices, 2008." 25 Nov. 2009. Eurostat. 10 Dec. 2009. http://epp.Eurostat.ec.Eurospa.eu/portal/page/portal/statistics/search_database.
[851] Hungarian Space Office website. 11 Dec. 2009. www.hso.hu.
[852] ESD Partners. European Space Directory 2009. 24th Edition. Paris: ESD Partners Publication, 2009: 89.
[853] Ibid.
[854] Hungarian Space Office website. 11 Dec. 2009. www.hso.hu.
[855] CIA World Factbook. 2 Nov. 2009 https://www.cia.gov/library/publications/the-world-factbook/geos/ei.html.
[856] "Gross Domestic Product at market prices, 2008." 25 Nov. 2009. Eurostat. 10 Dec. 2009. http://epp.Eurostat.ec.Eurospa.eu/portal/page/portal/statistics/search_database.
[857] Entreprise Ireland. 11 Dec. 2009. www.enterprise-ireland.com.
[858] ASD Eurospace. "Facts and Figures. The European Space Industry in 2008." 13th edition, June 2009. Paris: ASD Eurospace.
[859] CIA World Factbook. 2 Nov. 2009. https://www.cia.gov/library/publications/the-world-factbook/geos/it.html.
[860] "Gross Domestic Product at market prices, 2008." 25 Nov. 2009. Eurostat. 10 Dec. 2009. http://epp.Eurostat.ec.Eurospa.eu/portal/page/portal/statistics/search_database.
[861] ASI website. 11 Dec. 2009. www.asi.it.
[862] "Italy Space Activities Year 2008." ASI. 11 Dec. 2009. http://www.asi.it/files/COPUOS%20-%20Italian%20Space%20Activities%20Report%202008.pdf.
[863] ASD Eurospace. "Facts and Figures. The European Space Industry in 2008." 13th edition, June 2009. Paris: ASD Eurospace.
[864] CIA World Factbook. 2 Nov. 2009. https://www.cia.gov/library/publications/the-world-factbook/geos/lu.html.
[865] "Gross Domestic Product at market prices, 2008." 25 Nov. 2009. Eurostat. 10 Dec. 2009. http://epp.Eurostat.ec.Eurospa.eu/portal/page/portal/statistics/search_database.
[866] Luxinnovation website. 11 Dec. 2009. www.innovation.public.lu.
[867] ASD Eurospace. "Facts and Figures. The European Space Industry in 2008." 13th edition, June 2009. Paris: ASD Eurospace.
[868] CIA World Factbook. 2 Nov. 2009. https://www.cia.gov/library/publications/the-world-factbook/geos/nl.html.
[869] "Gross Domestic Product at market prices, 2008." 25 Nov. 2009. Eurostat. 14 Dec. 2009. http://epp.Eurostat.ec.Eurospa.eu/portal/page/portal/statistics/search_database.
[870] NVIR website. 1 July 2009. www.nivr.nl.
[871] SRON website. 1 July 2009. www.sron.nl.
[872] ASD Eurospace. "Facts and Figures. The European Space Industry in 2008." 13th edition, June 2009. Paris: ASD Eurospace.
[873] CIA World Factbook. 2 Nov. 2009. https://www.cia.gov/library/publications/the-world-factbook/geos/no.html.

[874] "Gross Domestic Product at market prices, 2008." 25 Nov. 2009. Eurostat. 14 Dec. 2009. http://epp.Eurostat.ec.Eurospa.eu/portal/page/portal/statistics/search_database.
[875] NSC website. 14 Dec. 2009. http://www.spacecentre.no/english/.
[876] ASD Eurospace. "Facts and Figures. The European Space Industry in 2008." 13th edition, June 2009. Paris: ASD Eurospace.
[877] CIA World Factbook. 2 Nov. 2009. https://www.cia.gov/library/publications/the-world-factbook/geos/pl.html.
[878] "Gross Domestic Product at market prices, 2008." 25 Nov. 2009. Eurostat. 14 Dec. 2009. http://epp.Eurostat.ec.Eurospa.eu/portal/page/portal/statistics/search_database.
[879] Space Research Centre website. 14 Dec. 2009. e.cbk.waw.pl/cbk.
[880] Polish Space Office website. 14 Dec. 2009. www.kosmos.gov.pl.
[881] CIA World Factbook. 2 Nov. 2009. https://www.cia.gov/library/publications/the-world-factbook/geos/pl.html.
[882] "Gross Domestic Product at market prices, 2008." 25 Nov. 2009. Eurostat. 14 Dec. 2009. http://epp.Eurostat.ec.Eurospa.eu/portal/page/portal/statistics/search_database.
[883] FCT website. 14 Dec. 2009. http://alfa.fct.mctes.pt/index.phtml.en.
[884] ASD Eurospace. "Facts and Figures. The European Space Industry in 2008." 13th edition, June 2009. Paris: ASD Eurospace.
[885] CIA World Factbook. 2 Nov. 2009. https://www.cia.gov/library/publications/the-world-factbook/geos/ro.html.
[886] "Gross Domestic Product at market prices, 2008." 25 Nov. 2009. Eurostat. 14 Dec. 2009. http://epp.Eurostat.ec.Eurospa.eu/portal/page/portal/statistics/search_database.
[887] ROSA website. 16 Dec. 2009. www.rosa.ro.
[888] CIA World Factbook. 2 Nov. 2009. https://www.cia.gov/library/publications/the-world-factbook/geos/sp.html.
[889] "Gross Domestic Product at market prices, 2008." 25 Nov. 2009. Eurostat. 14 Dec. 2009. http://epp.Eurostat.ec.Eurospa.eu/portal/page/portal/statistics/search_database.
[890] CDTI website. 15 Dec. 2009. www.cdti.es/index.asp?MP=15&MS=192&MN=3.
[891] ASD Eurospace. "Facts and Figures. The European Space Industry in 2008." 13th edition, June 2009. Paris: ASD Eurospace.
[892] CIA World Factbook. 2 Nov. 2009. https://www.cia.gov/library/publications/the-world-factbook/geos/ro.html.
[893] "Gross Domestic Product at market prices, 2008." 25 Nov. 2009. Eurostat. 14 Dec. 2009. http://epp.Eurostat.ec.Eurospa.eu/portal/page/portal/statistics/search_database.
[894] Swedish National Space Board website. 16 Dec. 2009. www.snsb.se.
[895] ASD Eurospace. "Facts and Figures. The European Space Industry in 2008." 13th edition, June 2009. Paris: ASD Eurospace.
[896] CIA World Factbook. 2 Nov. 2009. https://www.cia.gov/library/publications/the-world-factbook/geos/ro.html.
[897] "Gross Domestic Product at market prices, 2008." 25 Nov. 2009. Eurostat. 14 Dec. 2009. http://epp.Eurostat.ec.Eurospa.eu/portal/page/portal/statistics/search_database.
[898] Space Affairs Division website. 16 Dec. 2009. www.sbf.admin.ch/htm/themen/weltraum_fr.html.
[899] ASD Eurospace. "Facts and Figures. The European Space Industry in 2008." 13th edition, June 2009. Paris: ASD Eurospace.
[900] CIA World Factbook. 2 Nov. 2009. https://www.cia.gov/library/publications/the-world-factbook/geos/ro.html.
[901] "Gross Domestic Product at market prices, 2008." 25 Nov. 2009. Eurostat. 14 Dec. 2009. http://epp.Eurostat.ec.Eurospa.eu/portal/page/portal/statistics/search_database.
[902] BNSC website. 16 Dec. 2009. http://www.bnsc.gov.uk/.
[903] ASD Eurospace. "Facts and Figures. The European Space Industry in 2008." 13th edition, June 2009. Paris: ASD Eurospace.
[904] ESA website. 16 Dec. 2009. www.esa.int.

[905] European Space Policy website. 17 Dec. 2009. http://ec.europa.eu/enterprise/policies/space/.
[906] FP7 website. 17 Dec. 2009. http://cordis.europa.eu/fp7/cooperation/space_en.html.
[907] EUMETSAT website. 17 Dec. 2009. http://www.eumetsat.int/Home/index.htm.
[908] EUMETSAT. "Annual Report 2008." 17 Dec. 2009. http://www.eumetsat.int/Home/Main/AboutEUMETSAT/Publications/AnnualReport/groups/cps/documents/document/pdf_ar_2008.pdf.
[909] CIA World Factbook. 2 Nov. 2009. https://www.cia.gov/library/publications/the-world-factbook/geos/ca.html.
[910] IMF. Gross Domestic Product at current price 2008. 17 Dec. 2009. http://www.imf.org/external/pubs/ft/weo/2009/01/weodata/index.aspx.
[911] CSA website. 17 Dec. 2009. www.space.gc.ca.
[912] CSA. 2008–2009 Departmental Performance Report. 17 Dec. 2009. http://www.asc-csa.gc.ca/pdf/pr-2009.pdf.
[913] Ibid.
[914] CIA World Factbook. 2 Nov. 2009. https://www.cia.gov/library/publications/the-world-factbook/geos/ch.html.
[915] IMF. Gross Domestic Product at current price 2008. 17 Dec. 2009. http://www.imf.org/external/pubs/ft/weo/2009/01/weodata/index.aspx.
[916] CNSA website. 17 Dec. 2009. www.cnsa.gov.cn.
[917] CIA World Factbook. 2 Nov. 2009. https://www.cia.gov/library/publications/the-world-factbook/geos/in.html.
[918] IMF. Gross Domestic Product at current price 2008. 17 Dec. 2009. http://www.imf.org/external/pubs/ft/weo/2009/01/weodata/index.aspx.
[919] ISRO website. 18 Dec. 2009. www.isro.org.
[920] CIA World Factbook. 2 Nov. 2009. https://www.cia.gov/library/publications/the-world-factbook/geos/ja.html.
[921] IMF. Gross Domestic Product at current price 2008. 17 Dec. 2009. http://www.imf.org/external/pubs/ft/weo/2009/01/weodata/index.aspx.
[922] JAXA website. 18 Dec. 2009. www.jaxa.jp/index_e.html.
[923] CIA World Factbook. 2 Nov. 2009. https://www.cia.gov/library/publications/the-world-factbook/geos/rs.html.
[924] IMF. Gross Domestic Product at current price 2008. 17 Dec. 2009. http://www.imf.org/external/pubs/ft/weo/2009/01/weodata/index.aspx.
[925] Roscosmos website. 18 Dec. 2009. www.roscosmos.ru.
[926] CIA World Factbook. 2 Nov. 2009. https://www.cia.gov/library/publications/the-world-factbook/geos/up.html.
[927] IMF. Gross Domestic Product at current price 2008. 17 Dec. 2009. http://www.imf.org/external/pubs/ft/weo/2009/01/weodata/index.aspx.
[928] NSAU website. 18 Dec. 2009. www.nkau.gov.ua.
[929] CIA World Factbook. 2 Nov. 2009. https://www.cia.gov/library/publications/the-world-factbook/geos/us.html.
[930] IMF. Gross Domestic Product at current price 2008. 17 Dec. 2009. http://www.imf.org/external/pubs/ft/weo/2009/01/weodata/index.aspx.
[931] NASA website. 18 Dec. 2009. www.nasa.gov.

3. Bibliography of space policy publications. July 2008–June 2009

Blandina Baranes

3.1. Monographs

Bates, Jason, ed. Artificial Satellites in Communication. Rockville: Access Intelligence, 2009.
Bell, Jim. Moon 3-D: the Lunar Surface Comes to Life. New York: Sterling Publishing, 2009.
Bentley, Matthew. Spaceplanes: from Airport to Spaceport. Berlin: Springer, 2008.
Besser Bruno P. Österreichs Weg in den Weltraum. Paris: Beauchesne, 2009.
Bormann, Natalie and Sheehan, Michael. Securing Outer Space. London, New York: Routledge Chapmann & Hall, 2009.
Burgess, Coli and Hall, Rex. The First Soviet Cosmonaut Team: Their Lives and Legacies. Berlin: Springer Praxis Books, 2008.
Buzz, Aldrin and Ken, Abraham. Magnificent Desolation: The Long Journey Home from the Moon. New York: Harmony Books, 2009.
Chaikin, Andrew. Man on the Moon: The Voyages of the Apollo Astronauts. New York: Penguin, 2009.
——. Voices from the Moon: Apollo Astronauts Describe Their Lunar Experiences. New York: Viking Studio, 2009.
Chesley, Alfred T., ed. Next Generation of Human Space Flight Systems. Hauppauge NY: Nova Science Publishers Inc., 2009.
Codignola, Luca and Schrogl, Kai-Uwe, eds. Humans in Outer Space – Interdisciplinary Odysseys. Wien: SpringerWienNewYork, 2008.
Cooke, Antony. Make Time for Stars: Fitting Astronomy Into Your Busy Life. New York: Springer, 2009.
Czysz, Paul A. and Claudio Bruno. Future Spacecraft Propulsion Systems: Enabling Technologies for Space Exploration. New York: Springer Praxis Books, 2008.
Dickson, Paul. A Dictionary of the Space Age (New Series in NASA History). Baltimore: Johns Hopkins University Press, 2009.
Dorado Gutiérrez, José María ed. Spain and the European Space Effort. Paris: Beauchesne, 2008.
Eickhoff, Jens. Simulating Spacecraft Systems. Berlin: Springer, 2009.
Elbert, Bruce. Introduction to Satellite Communication. 3rd Edition. Boston: Artech House, 2008.
Evans, Ben. Escaping the Bonds of Earth: The Fifties and the Sixties. New York: Springer Praxis Books/Space Exploration, 2009.
Finarelli, Margaret G., Alexander Joseph K. Space Science and the International Traffic in Arms Regulations: Summary of a Workshop. Washington D.C.: The National Academies Press, 2008.
Fisher, Steven C. and Rahman, Shamim A. Remembering the Giants: Apollo Rocket Propulsion Development. Washington D.C.: NASA History Division, 2009.
Gallantine, Jay. Ambassadors from Earth: Pioneering Explorations with Unmanned Spacecraft. Nebraska: University of Nebraska, 2009.
Gerl, Florian Albin. Outer Space – The Emerging Market: Prospects and Implications on the World Economy. Analysis of the Market Potential. München: Grin Verlag, 2008.
Griffin, Michael Douglas. Leadership in Space: Selected Speeches of NASA Administrator Michael Griffin, May 2005–October 2008. Washington D.C.: National Aeronautics and Space Administration, 2008.

Harris, Philip R. Space Enterprise: Living and Working Offworld in the 21st Century. New York: Springer Praxis Publishing, 2009.

Hertenberger, Gerhard. Aufbruch in den Weltraum: Geheime Raumfahrtprogramme, dramatische Pannen und faszinierende Erlebnisse russischer Kosmonauten. Wien: Seifert Verlag, 2009.

Hetherington, Edith W. and Hetherington, Norriss S. Astronomy and Culture. Santa Barbara, Calif.: Greenwood Press/ABC-CLIO, 2009.

Hitt, David, Garriott, Owen K. and Kerwin, Joe P. Homesteading Space: The Skylab Story. Nebraska: University of Nebraska Press, 2008.

Hotakainen, Markus. Mars: From Myth and Mystery to Recent Discoveries. Berlin: Springer, 2008.

Icon Group International. The 2009 Report on Space and Satellite Insurance Services: World Market Segmentation by City. San Diego, CA.: Icon Group International, 2009.

Jaumann, Ralf, et al. Der Mond – Entstehung, Erforschung, Raumfahrt. Köln: Fackeltraeger-Verlag, 2009.

Johnson-Freese, Joan. Heavenly Ambitions: America's Quest to Dominate Space. Pennsylvenia: University of Pennsylvania Press, 2009.

Launius Roger D. and Johnston, Andrew K. Smithsonian Atlas of Space Exploration. New York: Collins, 2009.

Lele, Ajey, and Sin, Gunjan, eds. Space Security and Global Cooperation. New Delhi: Academic Foundation in association with the Institute for Defence Studies and Analyses, 2009.

Ley, Wilfried, Wittmann, Klaus, and Hallmann, Willi, eds. Handbook of Space Technology. Chichester: Wiley-Blackwell, 2009.

Macauley, Molly K. and Alexander, Joseph K. United States Civil Space Policy. Summary of a Workshop. Washington D.C.: National Academy Press, 2008.

Mari, Christopher, ed. The Next Space Age. New York: H. W. Wilson, 2009.

McBrewster, John, Miller Frederic P. and Vandome Agnes F., eds. Satellite: Spaceflight, Natural satellite, Timeline of artificial satellites and space probes, Communications satellite, Sputnik 2, Anti-satellite weapon.... Global navigation satellite system. Beau Bassin: Alphascript Publishing, 2009.

——. Space Shuttle: Spacecraft, Human spaceflight, Space Shuttle Solid Rocket Booster, Space Shuttle orbiter, Spaceplane, International Space Station, Maxime... Shuttle Columbia, Space Shuttle Challenger. Beau Bassin: Alphascript Publishing, 2009.

McDonald, Allan J. and James R. Hansen. Truth, Lies, and O-Rings: Inside the Space Shuttle Challenger Disaster. Gainesville: University Press of Florida, 2009.

Mix, Lucas John. Life in Space: Astrobiology for Everyone. Cambridge, Mass.; London: Harvard University Press, 2009.

Moltz, James Clay. The Politics of Space Security: Strategic Restraint and the Pursuit of National Interest. Stanford, Calif.: Stanford University Press, 2009.

Musgrave, Gary E., Larsen Axel and Sgobba, Tommaso, eds. Safety Design for Space Systems. Oxford [i.a.]: Butterworth-Heinemann, 2009.

Neufeld, Michael. Wernher von Braun: Visionär des Weltraums Ingenieur des Krieges. München: Siedler, 2009.

Nolan, Kevin. Mars. A Cosmic Stepping Stone. Uncovering Humanity's Cosmic Context. New York: Springer Praxis Books, 2008.

Olla, Phillip. Space Technologies for the Benefit of Human Society and Earth. New York: Springer, 2009.

Osborn, Stephanie. Burnout: the mystery of Space Shuttle STS-281. Kingsport Tenn.: Twilight Times Books, 2009.

Pilch, Frances and Damon, Coletta, eds. Space and Defense Policy. London; New York: Routledge Chapman & Hall, 2009.

Pop, Virgiliu. Who owns the Moon? Extraterrestrial Aspects of Land and Mineral Resources Ownership. New York: Springer, 2009.

Puttkamer, Jesko Freiherr von. Abenteuer Apollo 11: von der Mondlandung zur Erkundung des Mars. 4th Edition. München: Herbig, 2009.

Reichel, Eugen. Typenkompass Bemannte Raumfahrzeuge seit 1960. Stuttgart: Motorbuch Verlag, 2008.

Sabathier, Vincent, Faith Ryan G. and Bander, Ashley. Mid- and Long-term Prospects for Human Spaceflight: Mitigating the Gaps. Washington DC.: CSIS, 2009.
Salla, Michael. E. Exposing U.S. Government Policies on Extraterrestrial Life: The Challenge of Exopolitics. Kealakekua, Hawaii: Exopolitics Institute, 2008.
Sandau, Rainer, Rösner Hans-Peter and Valenzuela, Arnoldo, eds. Small Satellites for Earth Observation. Selected Proceedings of the 6th International Symposium of the International Academy of Astronautics. Berlin: Springer, 2008.
Schrogl, Kai-Uwe, Mathieu, Charlotte and Lukaszczyk, Agnieszka, eds. Threats, Risks and Sustainability – Answers by Space. Wien: SpringerWienNewYork, 2009.
Seedhouse, Eric. Lunar Outpost. The Challenges of Establishing a Human Settlement on the Moon. New York: Springer Praxis Books, 2008.
——. Martian Outpost: The Challenges of Establishing a Human Settlement on Mars. Berlin: Sringer Praxis Books, 2009.
Seppinen, Ilkka and Pellinen, Risto. The History of Finnish Space Activities. Paris: Beauchesne, 2009.
Sgobba, Tommaso, Larsen, Axel and Musgrave, Gary, eds. Safety Design for Space Systems. Burlington, MA; Oxford: Butterworth Heinemann, 2009.
Shayler David J. Space Rescue: Ensuring the Safety of Manned Spacecraft. New York: Springer Praxis Books, 2009.
Sheehan, Michael and Bormann, Natalie. Securing Outer Space: International Relations Theory and the Politics of Space. London: Routledge, 2008.
Smith, Andrew. Moonwalker: Wie der Mond das Leben der Apollo-Astronauten veränderte. Frankfurt: Fischer (S.), 2009.
Space Foundation. The Space Report 2009. Washington DC.: Space Foundation, 2009.
Swinered, Graham. How Spacecraft Fly. New York: Springer Praxis Books, 2008.
Turner, Martin, J.L. Rocket and Spacecraft Propulsion: Principles, Practice and New Developments. Berlin: Springer, 2008.
Ulivi, Paolo and Harland, David M. Robotic Exploration of the Solar System: Part II: Hiatus and Renewal, 1993–1996. New York: Springer Praxis Books, 2008.
United Nations Institute for Disarmament Research, ed. Security in Space: The Next Generation, Conference Report, 31 March–1 Apr. 2008. New York: United Nations, 2008.
Van Pelt, Michel. Space Tethers and Space Elevators. NewYork: Springer, 2009.
von Croy, Alexis. Der Mond und die Abenteuer der Apollo-Astronauten. München: Herbig, 2009.
von Rauchhaupt, Ulf. Der neunte Kontinent: Die wissenschaftliche Eroberung des Mars. Frankfurt: Fischer (S.), 2009.
Wijker, Jaap J. Spacecraft Structures. Berlin: Springer, 2008.
Yun, Zhao. Space Commercialization and the Development of Space Law from a Chinese Legal Perspective. Hauppauge, NY: Nova Science Publishers Inc., 2009.
Zellmeyer, Stephan. A Place in Space. The History of Swiss Participation in European Space Programmes 1960–1987. Paris: Beauchesne, 2008.
Zimmerman, James. Approaches to Future Space. Washington D.C.: The National Academies Press, 2009.
Zubrin, Robert. How to Live on Mars: A Trusty Guidebook to Surviving and Thriving on the Red Planet. New York: Three Rivers Press, 2008.

3.2. Articles

Aoki, Setsuko. "Introduction to the Japanese Basic Space Law of 2008." ZLW – Zeitschrift für Luft- und Weltraumrecht (German Journal of Air and Space Law) 57 (2008): 585–589.
Arnould, Jacques. "Arts in Space: French Experiences and Elements of Prospective." Acta Astronautica 64 (2009): 714–717.
Baum, Seth D. "Cost-benefit Analysis of Space Exploration: Some Ethical Considerations." Space Policy 25 (2009): 75–80.

Becerra Ortiz, Jairo A. "A Survey of Colombia's New Outer Space Policy: Reforms in Colombian Law." Acta Astronautica 63 (2008): 560–563.

Benkö, Marietta and Schrogl, Kai-Uwe. "The UN Committee on the Peaceful Uses of Outer Space: Adoption of the Resolution on Enhancing Registration Practice and of the UNCOPUOS Space Debris Mitigation Guidelines. ZLW – Zeitschrift für Luft- und Weltraumrecht (German Journal of Air and Space Law) 57 (2008): 335–353.

Bianchi, Stefano, et al. "VEGA, the European Small Launcher: Development Status, Future Perspectives and Applications." Acta Astronautica 63 (2008): 416–427.

Bommakanti, Kartik. "Satellite integration and multiple independently retargetable reentry vehicles technology: Indian-United States civilian space cooperation." Astropolitics 7 (2009): 7–31.

Broniatowski, David A. and Weigel, Annalisa. "The Political Sustainability of Space Exploration." Space Policy 24 (2008): 148–157.

Broniatowski, David A., et al. "A Framework for Evaluating International Cooperation in Space Exploration." Space Policy 24 (2008): 181–189.

Buckley, Nicole, Johnson-Green, Perry and Lefebvre. "Worms to Astronauts: Canadian Space Agency Approach to Life Sciences in Support of Exploration." Acta Astronautica 63 (2008): 1148–1151.

Bugos, Glenn E. and Boyd, John W. "Accelerating Entrepreneurial Space: The Case for an NACA-style Organization." Space Policy 24 (2008): 140–147.

Couston, Mireille. "La loi française sur les optérations spatiales." ZLW – Zeitschrift für Luft- und Weltraumrecht (German Journal of Air and Space Law) 58 (2009): 253–282.

David, James. "What Should Nations Reveal about their Spying from Space? An Examination of the US Experience." Space Policy 25 (2009): 117–127.

de Bruijn Ferdi J., Ashford, Edward W. and Larson, Wiley J. "SpaceTech – Postgraduate Space Education." Acta Astronautica 63 (2008): 486–492.

Dempsey, Paul Stephen. "The Evolution of US Space Policy." Annals of Air and Space Law XXXIII (2008).

Diwakar, P.G., et al. "Empowering the Rural Poor Through EO Products and Services – An Impact Assessment." Acta Astronautica 63 (2008): 551–559.

Fisk, Len A. "The Impact of Space on Society: Past, Present and Future." Space Policy 24 (2008): 175–180.

Froehlich, Annette. "Analyse de la proposition sur la communautarisation du Centre Spatial Guyanais (CSG)." ZLW – Zeitschrift für Luft- und Weltraumrecht (German Journal of Air and Space Law) 57 (2008): 558–570.

——. "ESA-EU-Vereinbarung zur Umsetzung des gemeinsamen GMES-Projekts" ZLW – Zeitschrift für Luft- und Weltraumrecht (German Journal of Air and Space Law) 57 (2008): 354–366.

Fuss, Sabine, Szolgayova, Jana and Obersteiner, Michael. "A Real Options Approach to Satellite Mission Planning." Space Policy 24 (2008): 199–207.

Gaubert, Alain and Lebeau, André. "Reforming European Space Governance." Space Policy 25 (2009): 37–44.

Gbem, Anastasia A. "Space Developments in African Countries." Annals of Air and Space Law XXXIV (2009).

Harris, Philip R. "Overcoming Obstacles to Private Enterprise in Space." Space Policy 24 (2008): 124–127.

——. "Space Theme Parks: Promoting Space to the Public." Space Policy 25 (2009): 88–89.

Henry, Philippe, et al. "Europe's Major Challenge for the 21st Century: Access to Space." Space Policy 25 (2009): 99–108.

Hermida, Julian. "Law Reform and National Space Law: A Participatory Approach to Space Law Making in Developing Countries." Annals of Air and Space Law: XXXIV (2009).

Hitchens, Theresa and Chen, David. "Forging a Sino-US "grand bargain" in Space." Space Policy 24 (2008): 128–131.

Hobe, Stephan, et al. "Ten Years of Cooperation Between ESA and EU: Current Issues." ZLW – Zeitschrift für Luft- und Weltraumrecht (German Journal of Air and Space Law) 58 (2009): 49–73.

Jakhu, Ram and Buzdugan, Maria. "Development of Natural Resources of the Moon and Other Celestial Bodies: Economic and Legal Aspects." Astropolitics 6 (2008): 201–250.
Jakhu, Ram and Singh, Karan. "Space Security and Competition for Radio Frequencies and Geostationary Slots." ZLW – Zeitschrift für Luft- und Weltraumrecht (German Journal of Air and Space Law) 58 (2009): 74–93.
Jakhu, Ram. "The Case for Enhanced India-Canada Space Cooperation." Space Policy 25 (2009): 9–19.
Jayaraman, V., Parihar J.S. and Srivastava, S.K. "Rejuvenation of Agriculture in India. Cost Benefits in Using EO Products." Acta Astronautica 63 (2008): 493–502.
Joyce, Sarah, Ferguson, Chantal and Weinstein, Philip. "Public Support for Mars Missions: The Importance of Informing the Next Generation." Acta Astronautica 64 (2009): 718–723.
Jurist, John M. "Commercial Suborbital Sounding Rocket Market: A Role for Reusable Launch Vehicles." Astropolitics 7 (2009): 32–49.
Kaiser, Stefan A. "Chinese Anti-Satellite Weapons: New Power Geometry and New Legal Policy." Astropolitics 6 (2008): 313–323.
Kim, Doo Hwan. "Space Law in Korea: Existing Regulations and Future Tasks." ZLW – Zeitschrift für Luft- und Weltraumrecht (German Journal of Air and Space Law) 57 (2008): 571–584.
Launius, Roger D. "Managing the Unmanageable: Apollo, Space Age Management and American Social Problems." Space Policy 24 (2008): 158–165.
——. "United States Space Cooperation and Competition: Historical Reflections." Astropolitics 7 (2009): 89–100.
Maccone, Claudio. "SETI, Extrasolar Planets Search and Interstellar Flight: When are They Going to Merge?" Acta Astronautica 64 (2009): 724–734.
MacLeish, Marlene Y., et al. "Education for the Journey to the Moon, Mars, and Beyond." Acta Astronautica 63 (2008): 1158–1167.
MacPhail, Donald. "Increasing the Use of Earth Observations in Developing Countries." Space Policy 25 (2009): 6–8.
Makarov, Yuri and Payson, Dmitry. "Russian Space Programmes and Industry: Defining the New Institutions for New Conditions." Space Policy 25 (2009): 90–98.
Masutti Anna. "Proposals for the Regulation of Unmanned Air Vehicle Use in Common Airspace." Air and Space Law 34 (2009): 1–12.
Mejía-Kaiser, Martha. "Informal Regulations and Practices in the Field of Space Debris Mitigation." Air and Space Law 34 (2009): 21–34.
Montluc, Bertrand de. "The New International Political and Strategic Context for Space Policies." Space Policy 25 (2009): 20–28.
Mugellesi Dow, Roberta, et al. "Overview of the Knowledge Management System in ESA/ESOC." Acta Astronautica 63 (2008): 448–457.
National Security Space Office. "Documentation: Space-based Solar Power as an Opportunity for Strategic Security." Astropolitics 7 (2009): 83–88.
Neufeld, Michael J. "von Braun and the Lunar-Orbit Rendezvous Decision: Finding a Way to Go to the Moon." Acta Astronautica 63 (2008): 540–550.
Noble, Michael J. "Export Controls and United States Space Power." Astropolitics 6 (2008): 251–312.
Novykov, O., et al. "Continous Space Education System and Its Role in Increasing Efficiency of Engineering Staff Training for Ukraine Space Rocket Industry." Acta Astronautica 64 (2009): 33–35.
Nyman-Metcalf, Katrin. "Space for the Benefit of Mankind." Annals of Air and Space Law XXXIV (2009).
Oei, Ing and Mirra, Carlo. "Europe's Human Research Experiments Integration on the International Space Station." Acta Astronautica 63 (2008): 1126–1136.
Onoda, Masami. "Satellite Observation of Greenhouse Gases: Monitoring the Climate Change Regime." Space Policy 24 (2008): 190–198.
Oosterlinck, Rene. "The Evolution of the Temporary Presence Exception to Patent Infringement in Relation to Space Applications and the French Space Act." Annals of Air and Space Law XXXIV (2009).

Palmerini Giovanni B., Sabatini, Marco and Perrotta, Giorgio. "En Route to the Moon Using GNSS Signals." Acta Astronautica 64 (2009): 467–483.

Payson, Dmitry. "Institutional Design for the National Implementation of Space Activities in the Russian Federation." ZLW – Zeitschrift für Luft- und Weltraumrecht (German Journal of Air and Space Law) 57 (2008): 367–370.

Peter, Nicolas and Logsdon, John. "Reinvigorating Transatlantic Space Relations: The Joint ESPI-SPI Memorandum." Space Policy 24 (2008): 119–123.

Peter, Nicolas and Stoffl, Katharina. "Global Space Exploration 2025: Europe's Perspectives for Partnerships." Space Policy 25 (2009): 29–36.

Petroni, Giorgio, et al. "Discovering the Basic Strategic Orientation of Big Space Agencies." Space Policy 25 (2009): 45–62.

Podhajsky, Sandra and Negueruela, Cristina de. "From Dreams to Reality: 20 Years of ISU Team Projects." Acta Astronautica 63 (2008): 1152–1157.

Pop, Virgiliu. "Space and Religion in Russia: Cosmonaut Worship to Orthodox Revival." Astropolitics 7 (2009): 150–163.

Rapp Lucien. "When France puts its Own Stamp on the Space Law Landscape: Comments on Act No. 2008–518 of 3 June 2008 Relative to Space Operations." Air and Space Law 34 (2009): 87–103.

Reiman, Saara. "Is Space an Environment?" Space Policy 25 (2009): 81–87.

Reiter, Thomas and Mirra, Carlo. "The ASTROLAB Mission: A Milestone in European ISS Operations." Acta Astronautica 63 (2008): 198–203.

Robinson, George L. "The Search for Biogenesis and the Lurch towards Space Law Secularism." Annals of Air and Space Law XXXIV (2009).

Rodrigues, Ivette and Carvalho, Himilcon. "Space Education and Outreach Symposium (E1). Structures for Space Education (2)" Acta Astronautica 63 (2008): 475–485.

Rose, Lars. "Review and Assessment of Select US Space Security Technology Proposals." Space Policy 24 (2008): 208–223.

Ross, Martin, et al. "Limits on the Space Launch Market Related to Stratospheric Ozone Depletion." Astropolitics 7 (2009): 50–82.

Sachdeva, Gulshan S. "Space Mines: Dialectics of Legality." Astropolitics 7 (2009): 135–149.

Sadeh, Eligar. "Space Policy Challenges Facing the Barack Obama Administration." Space Policy 25 (2009): 109–116.

Sandrone, Silvio and Wagner, Alain. "Henry the Navigator and the Moon." Acta Astronautica 64 (2009): 484–493.

Scott, Henry T. "Mitigating the Danger of Space Debris by Enforcing and Developing Already Existing Space Law." Annals of Air and Space Law XXXIV (2009).

Shapiro, Robert. "A New Rationale for Returning to the Moon? Protecting Civilization with a Sanctuary." Space Policy 25 (2009): 1–5.

Siddiqi, Asif. "*Sputnik* 50 Years Later: New Evidence on Its Origins." Acta Astronautica 63 (2008): 529–539.

Slijper, Frank. "The EU should Freeze its Military Ambitions in Space." Space Policy 25 (2009): 70–74.

Sridhara Murthi, K.R., Madhusudan, H.N. "Strategic Considerations in Indian Space Programme – Towards Maximising Socio-Economic Benefits." Acta Astronautica 63 (2008): 503–508.

Stone, Dennis, et al. "NASA's Approach to Commercial Cargo and Crew Transportation." Acta Astronautica 63 (2008): 192–197.

Suresh, Byrana Nagappa. "History of Indian Launchers." Acta Astronautica 63 (2008): 428–434.

Suzuki, Kazuto. "A Brand New Space Policy or Just Papering Over a Political Glitch? Japan's New Space Law in the Making." Space Policy 24 (2008): 171–174.

Tronchetti, Fabio. "The Non-appropriation Principle as a Structural Norm of International Law: A New Way of Interpreting Article II of the Outer Space Treaty." Air and Space Law 33 (2008): 277–305.

von der Dunk, Frans G. "A Sleeping Beauty Awakens: The 1968 Rescue Agreement after Forty Years." Journal of Space Law 34 (2008): 411–434.

——. "A European "Equivalent" to United States Export Controls: European Law on the Control of International Trade in Dual-Use Space Technologies." Astropolitics 7 (2009): 101–134.

——. "EGNOS: Testing the Waters for Galileo – Certifying Europe's First Satellite Navigation Services." Journaal LuchtRecht 7 (2008): 75–84.

——. "Europe and the 'Resolution Revolution': 'European' Legal Approaches to Privacy and their Relevance for Space Remote Sensing Activities." Annals of Air and Space Law XXXIV (2009).

——. "Impementing the United Nations Outer Space Treaties – The Case of the Netherlands." Nationales Weltraumrecht/National Space Law (2008): 81–104.

Zevako, V.S. "Modern Space Technology for Ukrainian Industry." Acta Astronautica 64 (2009): 81–83.

Zhao, Haifeng. "The *Status Quo* and the Future of Chinese Space Legislation." ZLW – Zeitschrift für Luft- und Weltraumrecht (German Journal of Air and Space Law) 58 (2009): 94–122.

Zhou, Yi. "Perspectives on Sino-US Cooperation in Civil Space Programs." Space Policy 24 (2008): 132–139.

List of figures and tables

Figures

Part 1: The Year in Space 2008/2009

Part 2: Views and Insights

Tables

Part 1: The Year in Space 2007/2008

Part 2: Views and Insights

About the authors

Josef Aschbacher is the Head of the GMES Space Office of the European Space Agency, which is located at ESA ESRIN in Frascati, Italy. Before, from 2001 to 2006 he was Programme Coordinator in the Directorate of Earth Observation Programmes at ESA's HQ in Paris. He was at the Space Applications Institute at the EC Joint Research Centre in Ispra, Italy, from 1993 to 2000, where he was the Scientific Assistant to the Director in his last position. Through his work at the EC and ESA he earned a profound knowledge of the Earth Observation landscape in Europe and has actively contributed to the advancement of European space programmes, most notably the Global Monitoring for Environment and Security (GMES) initiative. Josef Aschbacher has a Ph.D. and Masters degree in Natural Sciences from the University of Innsbruck, Austria. He has 25 years of working experience in the domain of satellite Earth observation and has published more than 100 articles, many in peer-reviewed journals. He has been a lecturer at University Innsbruck, Austria, the Asian Institute of Technology in Bangkok, Thailand, the Politecnico Milano, Italy, and the University of Wuerzburg, Germany.

Blandina Baranes joined the European Space Policy Institute (ESPI) in Vienna in February 2005. She is the librarian and documentalist of ESPI. Before she was the chief librarian of the Jewish Studies Department at the University of Vienna. During the past years she has been working as documentalist and librarian for different institutions such as, The Austrian Broadcasting Corporation, DER SPIEGEL and others. She conducted her studies and research in Austria and Israel and graduated with an Master from Vienna University, Faculty of Philosophy, Department of Social and Cultural Anthropology.

Frans G. von der Dunk is the Harvey and Susan Perlman Alumni/Othmer Professor of Space Law at the University of Nebraska-Lincoln's L.M. Programme on Space and Telecommunication Law. He also is Director of Black Holes BV, Consultancy in space law and policy, based in Leiden. Previously, he was Co-Director, then Director of Space Law Research at the International Institute of Air and Space Law at Leiden University since 1990. He has written well over 120 articles and published papers, has given more than 100 presentations at international meetings and was visiting professor at some 25 foreign universities across the world. As of 2006, he is the Series Editor of 'Studies in Space Law', published by Brill. Finally, he has given a range of interviews to the international media on issues of space law and policy. Prof. von der Dunk has served as adviser to the Dutch Government, several foreign Governments, the European Commission, the European Space Agency (ESA) and other international organisations. He is Director Public Relations of the International Institute of Space Law (IISL), Member of the Board of the European Centre for Space Law (ECSL), and Member for the Netherlands in the International Law Association's (ILA) Committee on Space Law. He is also Member of the International Editorial Board of "Space Policy".

Florian Hafner is Research Associate at the Department of European, International and Comparative Law at the Law Faculty (intlaw.univie.ac.at) of the University of Vienna since January 2009. He studied law and business administration in Vienna and Sydney. In 2008 he finished his law degree at the University of Vienna in 2006 he finished his degree in Business Education at the Vienna University of Economics and Business Administration. He has been working for different Austrian law firms and tax consultant offices. Currently he is working at the National Point of Contact (NPOC), Austria of the European Center for Space Law (ECSL) and writing his doctoral thesis on a new Austrian space law.

Ram S. Jakhu is Associate Professor at the Institute of Air and Space Law, Faculty of Law, McGill University, Montreal, Canada. He teaches and conducts research in international space law, law of space applications, law of space commercialisation, government regulation of space activities, law of telecommunications, Canadian communications law, and public international law. He is also Director of Research for the Space Security Index, the managing editor of Space Regulations Library Series, a member of the Editorial Board of the Annals of Air and Space Law, a member of the Board of Director of the International Institute of Space Law and a Fellow and the Chairman of the Legal and Regulatory Committee of the International Association for the Advancement of Space Safety. He has served as Director of the Centre for the Study of Regulated Industries at McGill University (1999–2004) and as Professor and the first Director of the MSS Programme of the International Space University in Strasbourg, France (1995–1998). Professor Jakhu holds a BA, an LLB and an LLM from Panjab University and an LLM and a DCL from McGill University.

Srikanth Kondapalli is Professor in Chinese Studies at Jawaharlal Nehru University. He is also an Honorary Fellow at Institute of Chinese Studies, Delhi and Research Associate at Centre for Chinese Studies, University of Stellenbosch, South Africa. He served at Institute for Defence Studies & Analyses, New Delhi for nearly 12 years. He is educated in Chinese studies in India and China with a specialisation in Modern Chinese History in Master of Arts at Osmania University at Hyderabad, Master of Philosophy and Doctor of Philosophy in 1989 and

1995 respectively in Chinese Studies Division of School of International Studies, Jawaharlal Nehru University, New Delhi. He learnt Chinese language at Beijing Language & Culture University and was a post-Doctoral Visiting Fellow at People's University, Beijing from 1996 to 1998. He was a Visiting Professor at National Chengchi University, Taipei in 2004, a Visiting Fellow at China Institute of Contemporary International Relations, Beijing in May 2007 and an Honorary Professor at Shandong University, Jinan. He has published two books, two monographs, co-edited two volumes and several articles in edited books and in national and international journals and newspapers. He is a guest faculty member at India's Foreign Service Institute, National Defence College, Army War College, College of Naval Warfare, College of Air War, Indo-Tibetan Border Police Academy and Border Security Force Academy.

Sascha Lange is a research associate and political consultant at the German Institute for International and Security Affairs (SWP) in Berlin since 2005. He is currently working at the research unit European and Atlantic Security and conducting research on global military capabilities, Aero-, space and armaments industries and chemical and biological weapons. This includes a project for the „Zentrum für Transformation der Bundeswehr" (The Bundeswehr Transformation Centre). The topic of the research project is the German equivalent to NCW, „Vernetzte Operationsführung – NetOpFü". Selected publications in English include following papers: Europe's Reaction to the "Augustine Report", The End for the Airbus A400M?, Missile Defense in and for Europe?, SAR-Lupe Satellites launched, Partial Capability Loss through MEADS. In previous positions, he served as an intern at the Institute for Peace Research and Security Policy University Hamburg (1999) and at the SWP, before getting a scholarship from the Forum Ebenhausen to conduct research within the SWP. He holds a biology degree from Bremen University (Germany) and wrote his thesis on the "Detection and analysis of biological weapons".

Irmgard Marboe is Associate Professor of International Law at the Department of European, International and Comparative Law at the Law Faculty of the University of Vienna. She is the head of the National Point of Contact for Space Law in Austria and member of ECSL and IISL. She regularly participates in the Legal Subcommittee of UNCOPUOS and has been elected to be the chairperson of the Working Group under the agenda item on "National legislation relevant to the peaceful exploration and use of outer space" between 2009 and 2011. Her research in the area of space law has so far focused on telecommunication and space, national space legislation, European space policy and authorisation of space activities. Her current teaching activities at the University of Vienna include courses and seminars on general international law, international economic law, international courts and tribunals, and space law.

Bertrand de Montluc is presently Associate Professor at "Sciences Po" (I.E.P.) Paris school of higher education specialised in economics and political sciences (Europe-Asia Undergraduate cycle, lecture on Science, Technology & Society). He is research fellow at the CERI, University of Paris Center for International Relations Studies (international governance of nano-biotechnologies). He also acts as president of the strategy and international affairs committee of the non-profit professional organisation "3AF" (Aeronautical and Astronautics Association of France). In his previous functions he was Space and Technology senior analyst at the French Foreign Affairs Ministry's Policy planning staff (C.A.P.) and advisor for the director of Strategic affairs (D.A.S.) of the French Ministry of Defense. Before that, for four years he held the position of space & technology counsellor at the Brussels French Diplomatic Delegation to the E.U. (1999–2003). From 1981 to 1999 he worked at the French Space Agency C.N.E.S. as head of the international relations department, then as Deputy Director for strategy and European cooperation (ESA and bilateral relations). Dr. de Montluc graduated from the Sciences Po and has a doctorate in law (PHD in international law). He has been honoured as "chevalier" of the National Order of Merit. He is the author of more than 100 publications in the field of space policy, international affairs, defense and security.

Scott Pace is the Director of the Space Policy Institute and a Professor of Practice in International Affairs at George Washington University's Elliott School of International Affairs. From 2005 to 2008, he served as the Associate Administrator for Programme Analysis and Evaluation at NASA. Prior to NASA, Dr. Pace was the Assistant Director for Space and Aeronautics in the White House Office of Science and Technology Policy (OSTP). From 1993 to 2000, Dr. Pace worked for the RAND Corporation's Science and Technology Policy Institute (STPI). From 1990 to 1993, Dr. Pace served as the Deputy Director and Acting Director of the Office of Space Commerce, in the Office of the Deputy Secretary of the Department of Commerce. He received a Bachelor of Science degree in Physics from Harvey Mudd College in 1980; Masters degrees in Aeronautics & Astronautics and Technology & Policy from the Massachusetts Institute of Technology in 1982; and a Doctorate in Policy Analysis from the RAND Graduate School in 1989. Dr. Pace received the NASA Outstanding Leadership Medal in 2008, the U.S. Department of State's Group Superior Honor Award, *GPS Interagency Team*, in 2005, and the NASA Group Achievement Award, *Columbia Accident Rapid Reaction Team*, in 2004. Dr. Pace is a currently a member of the Board of Trustees, Universities Space Research Association and a Corresponding Member of the International Academy of Astronautics.

Walter Peeters is Dean and Vice-President for Academic Affairs at the International Space University (ISU) at Strasbourg, France. After initial management positions in construction and the petrochemical industry, he joined the European Space Agency (ESA) in 1983 in a number of project control and management functions, among others in the HERMES project in Toulouse, France. From 1990, he was involved in astronaut activities as Head of the Coordination Office of the European Astronaut Center in Cologne, with strong involvement in the EUROMIR missions. He joined ISU in 2000, after serving as Visiting Professor (Non-profit Marketing) at the University of Louvain, Belgium and was nominated as Dean of ISU in 2005. He has published more than 50 articles on incentive contracting, project management, space commercialisation and organisation in the space sector and is author of the book *Space Marketing* (Kluwer, 2000). Recent consultancy assignments include activities in the field of space economy (OECD, EC), space policy (Luxemburg, Estonia) and space tourism (Singapore spaceport, Gallactic Suite, Excalibur). He holds bachelor's degrees in engineering and applied economics (Catholic University of Louvain), a Master of Business Administration degree (Louvain, Cornell University) and a Ph.D. in Industrial Organisation (TU Delft, the Netherlands). He has been recently appointed as Director of the International Institute of Space Commerce (IISC), a think-tank based on the Isle of Man as a joint project between ISU and the Isle of Man government.

Wolfgang Rathgeber studied Electrical Engineering at the universities of Erlangen (Germany), Campinas (Brazil) and Valencia (Spain). He has been with the German Aerospace Centre (DLR) since 1999. Until 2003, he was a doctorate student at the Institute of Radio Frequency Engineering and Radar Systems in Oberpfaffenhofen (Germany) with a thesis on high resolution signal processing for Synthetic Aperture Radar (SAR). From 2003 until 2005, he was scientific staff member at the European Academy of Technology Assessment Bad Neuenahr-Ahrweiler (Germany). Since 2005 he has been seconded to the European Space Policy Institute (ESPI) in Vienna (Austria), where he is in charge of security topics.

Kai-Uwe Schrogl is the Director of the European Space Policy Institute (ESPI) in Vienna, Austria since 1 September 2007. Prior to this, he was the Head of the Corporate Development and External Relations Department in the German Aerospace Center (DLR). Previously he also worked with the German Ministry for Post and Telecommunications and the German Space Agency (DARA). He has been a delegate to numerous international forums and recently served as the chairman of various European and global committees (ESA International Relations Committee and two UNCOPUOS plenary working groups). He presented, respectively testified, at hearings of the European Parliament and the U.S. House of Representatives. Kai-Uwe Schrogl has written or co-edited 12 books and more than 130 articles, reports and papers in the fields of space policy and law as well as telecommunications policy. He is editor in chief of the "Yearbook on Space Policy" and the book series "Studies in Space Policy" both published by ESPI at SpringerWienNewYork. In addition he sits on editorial boards of various international journals in the field of space policy and law (Acta Astronautica, Space Policy, Zeitschrift für Luft- und Weltraumrecht, Studies in Space Law/Nijhoff). Kai-Uwe Schrogl is a Member of the Board of Directors of the International Institute of Space Law, Member of the International Academy of Astronautics (recently chairing its Commission on policy, economics and regulations) and the Russian Academy for Cosmonautics. He holds a doctorate degree in political science and lectures international relations at Tübingen University, Germany (as an Honorary Professor). He has been a regular lecturer at, for example, the International Space University (where he serves as Adjunct Faculty) and the European Centre for Space Law's Summer Courses.

Jürgen Turek is Deputy Director of the Center for Applied Policy Research (C•A•P) at the Ludwig-Maximilians-University in Munich. He there also heads from 1998 up today the Research Group on the Global Future. In addition, from 2001 to 2005, he was Book Editor for the book review section of the German journal "Internationale Politik", which was published monthly by Deutsche Gesellschaft für Auswärtige Politik (DGAP; German Council on Foreign Relations) in Berlin in this period. Previously and besides his studies, from 1981 to 1987, he was a Documentary Editor at the Bundespresseamt der deutschen Bundesregierung (BPA Press and Information Office of the German Federal Government), from 1988 to 1991, a Research Fellow at the Department of Political Science at the University of Mainz and, from 1991 to 1992, a Press Officer at the Cologne Fair Ltd. in Cologne. Between 1993 and 1995, Jürgen Turek was involved in various projects of the C•A•P as a Research Fellow. Before his appointment to his current positions there, he was the Deputy Director of Bertelsmann Research Group on Policy Questions from 1995 to 1998. He has published three books (Co-Author with Werner Weidenfeld) on the current state of competition policy in Europe and social. In addition, he published numerous articles on European research and technology policy (e.g. since 1993 in the Yearbook of European Integration (Jahrbuch der Europäischen Integration change as well as a Study Letter on the Research and Technology Policy in Europe for the Fachhochschule Fernstudienverbund der Länder, East Germany) and important issues of the future have been published in books, journals and newspapers, including the "Frankfurter Allgemeine Zeitung", "Handelsblatt", "Das Parlament" and "Internationale Politik" Jürgen Turek holds an M.A. from the Johannes Gutenberg University of Mainz in political science, journalism and sociology.

Christophe Venet is a Ph.D. candidate at the Institute for Political Science at Tübingen University, Germany and an Associate Fellow of the European Space Policy Institute (ESPI) since January 2010. In 2009, he was Research Assistant at the Institute. In this capacity, he contributed to the Report "Space Policies, Issues and Trends in 2008/2009". He also worked on the issue of space commerce and space entrepreneurship, preparing several presentations on these issues. He was invited to become a peer-reviewer for the journal "Acta Astronautica" in August 2009. He graduated from the Institut d'Etudes Politiques de Strasbourg, France, and studied international relations at the Moscow State Institute of International Relations (MGIMO), Russia. He also holds a Masters degree in Peace Studies and International Politics from Tübingen University. His dissertation deals with EU policy in the field of space security, focusing on the actorness of Europe and on interests and norms underlying the policy processes within the European Space Policy.

Contributors to the Yearbook on Space Policy 2008/2009 during the author's conference on 17 September 2009 at ESPI in Vienna. From left: Kai-Uwe Schrogl, Frans G. von der Dunk, Sascha Lange, Scott Pace, Florian Hafner, Srikanth Kondapalli, Christophe Venet, Blandina Baranes, Josef Aschbacher, Wolfgang Rathgeber, Irmgard Marboe, Walter Peeters, Bertrand de Montluc.

Index

F

G

H

I

J

K

L

M

N

O

P

R

S

T

U

V

MIX
Papier aus verantwortungsvollen Quellen
Paper from responsible sources
FSC® C105338

If you have any concerns about our products,
you can contact us on
ProductSafety@springernature.com

In case Publisher is established outside the EU,
the EU authorized representative is:
Springer Nature Customer Service Center GmbH
Europaplatz 3, 69115 Heidelberg, Germany

Printed by Libri Plureos GmbH
in Hamburg, Germany